The Birds of CITES

and How to Identify Them

Johannes Erritzoe

illustrated by
Helga Boullet Erritzoe and the Author

Foreword by
H.R.H. Prince Philip
President of the World Wide Fund for Nature

Sponsored by
Aage V Jensen Charity Foundation, Denmark

(L)
The Lutterworth Press
Cambridge

The Lutterworth Press
P.O. Box 60
Cambridge
CB2 2NT

British Library Cataloguing-in-Publication Data
A catalogue record is available for this book from the British Library.

First published in Great Britain by The Lutterworth Press 1993

ISBN 0 7188 28917 - Cased Trade edition
0 7188 28915 - Ring Bound
0 7188 28941 - Cased Non-net edition
0 7188 2895X - Leather bound edition

Printed in Belgium by
Proost

Cover illustration by Helga B. Erritzoe

Front cover:
Toco Toucan (Ramfastos toco)
Long-whiskered Owlet (Xenoglaux Loweryi - first described 1977)

Back cover:
Toucan Barbet (Semnornis ramphastinus)
Guianan Cock of the Rock (Rupicola rupicola)

CONTENTS

WWF® World Wide Fund For Nature

Office of the President

CH-1196 Gland Switzerland
Telephone: (022) 64 91 11
Telex: 419 618 wwf ch
Telefax: (022) 64 54 68

In the last three hundred years alone, at least 350 mammal and bird species have become extinct. There are many more obvious threats to the survival of wild species of plants and animals, but among the most dangerous are the poaching and smuggling of wild animals and plants and their derivatives. The trade in wild animals and their derivatives has been going on for centuries. For example, the elephant population of North Africa was virtually exterminated by the demand for ivory from imperial Rome, but it has now reached crisis proportions.

In recent times, the demand for ivory and rhino horn has encouraged poaching and smuggling, and this has had a devastating effect on elephant populations in other parts of Africa and in Asia. Rhinos in both Africa and Asia are presently on the brink of extinction. However, the trade goes far beyond those species. Reptile skins, furs, ornamental birds, tropical fish, tortoiseshell, orchids, medicinal plants, apes and monkeys, and most recently, tigers, whose bones are in demand for traditional Chinese medicine, are all traded internationally, legally or illegally.

As tourist activity increased in the late 1960s, the trade in wild animals and plants grew to an alarming extent and began to affect previously flourishing populations and even the remotest regions. The major conservation organisations, such as the International Flora and Fauna Preservation Society (IFFPS), the International Council for Bird Preservation (ICBP) and the World Wide Fund for Nature (WWF) soon realised that something drastic needed to be done to control and regulate the international trade in wildlife.

In 1973 the International Union for the Conservation of Nature and Natural Resources (IUCN-The World Conservation Union) convened a conference of interested organisations in Washington DC, which was sponsored by the United States government. The outcome was that the Convention on the International Trade in Endangered Species (CITES) was founded on 3 March 1973 and signed by twenty-one states. CITES came into force on 1 July 1975 and today more than a hundred and twenty states are parties to the Convention.

CITES lists all threatened and endangered species and regulates the conditions under which they can be exported and imported. Needless to say, the success of the whole system depends on the willingness of the member states to enforce the CITES regulations and to take severe action against poachers and illegal traders. However, before that can happen, the police, wardens and customs authorities in both exporting and importing states must be able to recognise the species listed in the Convention.

WWF® World Wide Fund For Nature

Office of the President

CH-1196 Gland Switzerland
Telephone: (022) 64 91 11
Telex: 419 618 wwf ch
Telefax: (022) 64 54 68

Previous efforts have been made in some countries to provide recognition material for these officials, but this book is the first attempt to provide a means to positively identify some 1,400 bird species that are at varying degrees of risk of global extinction and appear in the CITES lists. I am confident that the information in this book will provide invaluable help for the officials directly involved, and other conservation professionals, in the task of implementing and enforcing the protective legislation. I am sure that it will also be of great interest to bird-lovers of all nations. I hope it will help them to appreciate the importance of CITES and encourage them to become actively involved in this critical sector in the battle for the conservation of nature.

Philip

CONVENTION ON INTERNATIONAL TRADE IN ENDANGERED SPECIES
OF WILD FAUNA AND FLORA

SECRETARIAT
6, rue du Maupas
Case postale 78
CH-1000 Lausanne 9, Switzerland

PREFACE

of the book

THE BIRDS OF CITES

and how to identify them

by Johannes Erritzoe

The Convention on International Trade in Endangered Species (CITES), signed in 1973 and entered into force in 1975, now involves more than 115 member countries. The agencies charged with applying the controls are hampered by the difficulty of identifying the protected species. CITES has published a basic manual in black and white for identification purposes, covering all animals, but has been held back by lack of funds so that only a few less than half the number of bird species have been covered to date. The lack of colour, especially with regard to birds is a serious handicap.

Now a Danish artist and scientific conservationist, Johannes Erritzoe and his wife, Helga B. Erritzoe have prepared this manual identifying all the birds in the CITES appendices. Beautiful and exact full-colour paintings are paired with concise descriptions of each species and subspecies. Range and names in other languages are given, source references being identified in all cases. Technical language is avoided as far as possible since Customs Officers are not always familiar with it.

A glossary is accompanied by a table of translations of more than 650 English words used in the book into three other major world languages. As an introduction you will not only find a plate with 54 colours identified by their English names, but also a plate showing the topography of a bird, explained in English terms.

The book has been produced in two forms - hardcover for ornithological bibliophiles; and ring-bound for Customs Officers and other management authorities, taxidermists, aviculturists, importers and exporters, hunters, scientists, etc. Ring-binding will allow the work to be updated every two years when the meeting of the Conference of the Parties agrees changes to its lists; changes made at the 8th meeting of the Conference of the Parties (Japan) have been incorporated into the manuscript.

The CITES Secretariat has had the pleasure of studying all the colour plates and some samples of the text. We find that the coloured illustrations are beautifully painted and the updated descriptions are very well done and easy to understand. It will become a useful implementation tool for all involved in the rules of this organization, and we whole-heartedly recommend "The BIRDS of CITES".

CITES Secretariat
Lausanne, Switzerland
October 1992

CITES SECRETARIAT UNEP PNUE SECRETARIA PNUMA CITES

Jean-Patrick Le Duc
Infraction Prevention and
Information Officer

ACKNOWLEDGEMENTS

It goes without saying that a book of this nature requires the assistance of a great number of people. *The Birds of CITES* would never have materialized if it had not been for the enormous help I have received ever since I commenced work on the book 5 years ago.

Having this opportunity I am anxious to express my profound gratitude to all those who helped me *professionally, scientifically* and not forgetting, *financially.*

Professionally, morally and scientifically:

In the initial stage I received valuable help from Mr Christopher K Swann, Managing Director of the well-known English bookseller, Wheldon & Wesley. I owe thanks to Mr Swann for his guidance and introduction to many an interesting and useful contact.

Thanks to Mr Tim Inskipp of the Wildlife Trade Monitoring Unit for supplying me with references to a wealth of material, which otherwise would have been very difficult, if not impossible, for me to unearth because of its obscurity.

Thanks to Mr Ben Charles Van Damme from Institut Royal des Sciences Naturelles de Belgique for sending me important papers and information.

Thanks to Anders Pape Moeller and Thorkil Duch, my old Danish friends for their encouragement during the difficult periods producing the book, and not forgetting all the stimulating talks we have had together.

My dear brother-in-law, Dr Peder Moesgaard, has given me good advice particularly in the early days, when I not only needed help in gaining command of the English language but also on how I should plan my work. Quite honestly I do not think I would have overcome the initial difficulties if I had not had his support.

I am deeply indebted to the many libraries and their staff who helped me in my research of books and papers. I particularly want to thank Mrs Marion Jacobsen of the German library in Haderslev, Denmark and Mrs Birgit Olesen of the library of Christianfeld, also Denmark.

Special thanks to the staff of The British Museum (Natural History), Tring, particularly Mr Graham Cowles and Mr Michael Walters for permission to study the vast bird collection belonging to this world famous museum.

Talking about museums, I cannot help also extending warm thanks to Dr Jon Fjelsaa of the Zoological Museum of Copenhagen, not only for allowing me to make all my studies of their bird-skins, but also for many a piece of advice and very positive criticism.

Apart from the privilege and honour of having the President of the World Wide Fund for Nature, HRH Prince Philip, Duke of Edinburgh to write the Foreword, I feel a great desire to express my sincere thanks to other distinguished members within WWF. Mr Simon Lyster, WWF-UK was the first within WWF to show great interest for my project, and his support has given me much needed comfort. General Secretary Lene Witte and Tommy Dybbro, both from WWF-Denmark, have also given me much encouragement and an indispensable helping hand. Mrs Witte arranged for my book to be introduced to HRH Prince Philip, and Mr Dybbro gave me other valuable introductions and many a professional piece of advice, whenever these were needed.

Within the organization of CITES my project has been followed and observed with the greatest interest and, dare I say, respect. Initially I had extremely useful co-operation from Dr Peter Dollinger of CITES Secretariat in Lausanne, for which I am, and have always been, very grateful.

May I record my gratitude to Mr David H W Morgan of the UK CITES Scientific Authority for Animals. He read the draft of the manuscript and gave me many suggestions for its improvement.

Another great encouragement to my work on the project was the Introduction to the book given in October 1992 by Mr Jean-Patrick Le Duc, the Infraction Prevention and Information Officer of CITES, Lausanne. I thank Mr Le Duc for the confidence he showed in allowing me to produce the book under the auspices of CITES. I am looking forward to a continuous close co-operation with him and his friendly staff.

The assistance and information I have received over the years from the National Forest and Nature Agency, Ministry of Environment in Denmark, have all been of the greatest help to me. I wish to use this opportunity to express my genuine thanks to all those who gave assistance to me and my book. Special thanks to Mrs Karen Westerbye-Juhl, Director General, and Mr Veit Koester, Head of Division, whom both guided me, whether I needed international help and contacts or explanations of difficult text and legislation.

Thanks to Mr Bent Sunesen, a private and very good friend of mine. Bent has read every word of the text, has an eye for every little detail and his knowledge of English grammar should make the book easier to read and understand. On top of this his love and interest for birds have proven to be of great help in the overall presentation.

Thanks to Lutterworth for their great help in producing this book. It has been a pleasure and inspiration for me to work with this nearly 200 year old English publishing company. A special thanks to Mr Adrian Brink, Managing Director, and Mr Colin Lester, Editorial Director, who have both put great personal interest and devotion into the project, which I believe everyone will see reflected in the final result.

Personally I wish to pay tribute to my dear wife, Helga, for her companionship and great tolerance during the 5 years I have been working on this book. The artwork of Helga requires no foreword from me, but of course I wish to express my gratitude to her for helping me in illustrating the book.

I cannot write these lines without sending my heartfelt thanks to the late Erik Petersen, Denmark. Many years ago he was responsible for my development as a naturalist.

Financially:

CONSUL GENERAL THORKILD ERRITZOE, Denmark/UK.

First and foremost I wish to express my deepest thanks to my dear brother, Consul General Thorkild Erritzoe, member of the presidium of WWF-Denmark. Without his help morally as well as materially ever since I started, I would never have been able to finish this project. Thorkild not only arranged the funding, but he also introduced me to, and actually pushed me into, a number of important international connections, which I, to be honest, would not have dared on my own. I am proud to see him amongst "The Erritzoe-brothers", and no words are adequate to express the depth of my gratitude and affection.

AAGE V JENSEN CHARITY FOUNDATION, Denmark

As you have probably already seen on the title page, the sponsor of this book is Aage V Jensen Charity Foundation, Denmark. My brother and I wish to thank the Chairman of the Board of Aage V Jensen Charity Foundation, Mr Leif Skov, and the other board-members for supporting *The Birds of CITES* financially. Without the generosity of this well-known foundation, the book just would not have been published.

Personally I am proud and delighted that my book has been supported through the inheritance of a personality like Aage V Jensen, who amongst bird-lovers was, and still is, known as one of the most distinguished.

ALLINGHAM & HANSEN A/S, Denmark

In our attempts to find the most outstanding ring-binders, when quality and design were given the highest priority, my brother and I had to choose the well-known Danish manufacturer, Allingham & Hansen A/S.

Throughout our cooperation with the owners and management of Allingham & Hansen, Messrs Jan Oyvind Hansen and Andreas Ejlersen showed our project a growing interest and affection.

Having shared and understood the economic perspective, which for a long period had nearly caused the abandonment of publication of the book, these gentlemen generously presented free of charge 3,000 ring-binders of which 2,000 were given to CITES, Geneva for free-distribution amongst the member states who, according to their assessment, have the greatest need and the weakest economy.

All people and organisations working in observance of the Washington Convention will, I am sure, wish to join me in expressing my greatest gratitude to:

Consul General Thorkild Erritzoe,
Aage V Jensen Charity Foundation,
and
Allingham & Hansen A/S.

Christiansfeld, Denmark
August 1993

Johannes Erritzoe

INTRODUCTION

Scope of the book

This book is intended to be of use to all those within the purview of the rules laid down by the Convention on International Trade in Endangered Species of wild fauna and flora, also known as CITES (e. g. aviculturists, importers, taxidermists, scientific institutions, conservationists, tourists, and the controlling authorities).

The aim of this work is to enable readers to identify the birds of CITES, only when they are available for protracted close-range inspection. Therefore there is no description of field characters, habits, breeding population, behaviour, food or voice. Readers requiring such information can find it in the books and papers mentioned in the Reference List.

The CITES Appendix Lists

There are now (1993) 1478 bird species and subspecies listed by CITES. List I of the CITES Appendices includes all species threatened with extinction which are, or may be, affected by trade. Trade in specimens of these species must be subject to particularly strict regulation in order that their survival should not be endangered further, and must only be authorised in exceptional circumstances.

List II includes: (a) all species which, although not necessarily threatened with extinction now, may become so unless trade in specimens of such species is subject to strict regulation; and (b) other species which must be subject to regulation, in order that trade in specimens of certain species referred to in (a) may be brought under effective control, i.e. 'look-alike' species.

List III includes all species which any party has identified as being subject to regulation within its jurisdiction for the purpose of preventing or restricting exploitation, and as needing the cooperation of other parties in the control of trade.

(Please note that within the European Community there are different provisions which are subject to revision. Please ask your Management Authority for information. An asterisk (*) placed against the scientific name of a family or an order indicates that one or more species or subspecies are included in List I and are therefore excluded from List II. After List III the name of the State which has included the bird is mentioned. Only birds from that State require the prior grant and presentation of an export permit. From other countries the prior presentation of a certificate of origin is sufficient.)

Which birds are included?

In this book, 406 birds have been described and illustrated in colour, including all the birds in Lists I, II and III, except for those members of the 'look-alike' groups which are not in List I. These birds, which are included in List II, comprise the birds of prey (269 species), the cranes (9 species), the parrots (283 species), the owls (145 species), the hummingbirds (326 species) and the birds of paradise (40 species).

All 'look-alike' birds are mentioned only by their scientific and English names. Almost all genera which are not included in List I are shown by a black and white illustration of one representative only, to give a general impression of the bird. For readers interested in this genera, more extensive references are given.

Classification

Until 1992 CITES had adopted Morony, J. J., W. J. Bock, & J. Farrand 1975: *Reference List of the Birds of the World*, AMNH, New York, with corrections and additions 1978. Regarding subspecies, the English names and ranges, Howard, R., & A. Moore, 1984: *A Complete Checklist of the Birds of the World*, Macmillan, London, had been consulted. In March 1992 the conference of the parties to the convention decided to adopt Sibley, C., & B. L. Monroe, Jr. 1991: *Distribution and Taxonomy of Birds of the World*, Yale University Press, New Haven, as the standard reference to the genus and species names of birds listed in the Appendices.

In order that the book should be up to date, I decided not only to include the 37 new bird species listed at the 1992 conference, but also to revise the whole text according to Sibley & Monroe's new checklist. Owing to the large number of changes, the order of the numbered items within the lists of related families (the 'look-alike' species) has had to be rearranged in some cases.

Where the subspecies named by Howard & Moore (1984) are not accepted by Sibley & Monroe, and where the differences are more than minor ones, I have mentioned such subspecies under the heading of Geographical Variations. As a rule, all subspecies mentioned in Sibley & Monroe's book are described in order to ensure reliable identification.

The 'superspecies' concept used by Sibley & Monroe is not mentioned in this book for reasons of simplicity. (A superspecies is a group of species derived from a common ancestor, each occuring in different geographical areas and considered too distinct in form and structure to be regarded as a single species).

For birds which earlier in this century had another scientific name, not accepted by most authorities today, yet which still often figures in accessible books, the older name is also mentioned.

Distribution

The range of each species and subspecies is indicated in rough outline, generally in accordance with Sibley & Monroe and Howard & Moore. Introduced birds are rarely mentioned. Readers who want more exact information may consult: Peters, J. L. et al. 1931-1986: *Checklist of Birds of the World.* Vol. I-XV. Harvard University Press, Cambridge, Mass., and Long, J. L. 1981: *Introduced Birds of the World*, David & Charles, Newton.

Migration

Only in one case (a crane) is it indicated that the bird is sedentary, because it is essential to know its origin to determine this subspecies with any certainty.

Description

Length is measured in centimetres (1 inch = 2.54cm) i.e. the total length of the bird stretched out from the tip of the bill to the tip of the tail. In cases where the difference in length between the sexes is significant, the two are given separately. In a few cases wing length is also given when necessary to identify a bird or subspecies. This measurement is taken between the bend of the wing (carpal joint) and the tip of the longest flight-feather (primary), and is always given in millimetres (1 inch = 25.4mm). The wing is straightened and flattened to give the maximum length.

If plumage differs between the sexes, males are described first. When plumage differs between summer and winter, the adult in breeding plumage is described first, and to save space it is not stated explicitly that this is the breeding dress. Plumage outside the breeding season is always referred to as 'non breeding'.

In order to save space only adults and young birds (juveniles) are described. Eggs, nestlings or downy young birds are seldom of interest for the readership of this book and are mostly undescribed. Large birds often have several plumages, called 'immature', between juvenile and the adult stage. They are mostly a mixture of the two plumages and therefore in most cases are not mentioned.

Plumages are described in the usual sequence: head and neck, upperparts, wings, underparts and tail. In some of the briefer descriptions, this sequence has been modified.

Colours

Strictly defined and standardized colour names are not used, the aim being to use common colour names. Smithe, F. B. 1975: *Naturalist's Color Guide*, AMNH, New York, has been used to some extent. Where two colours are combined the last-named colour is the most dominant (e.g. 'red-brown' has a weaker tinge of red than brown; 'reddish-brown' means that there is still a weaker tinge of red). Please see the Colour Plate (page 1) with all the colours and their English names used in this book.

Characters from allied forms

The text describing each species has been made concise so that it can be placed opposite the plate depicting the species. Therefore there is no special section for describing other species with which a bird can be confused, although the distinguishing characteristics that set the species in question apart from others is always stated.

Illustrations

In the cases of sexual difference, both male and female are depicted. A separate illustration also occurs if a race is distinct from the nominate, (the one first described with the two last scientific names identical). Markedly distinct juveniles are also pictured. In the colour plates, males are marked A, females B, young birds C, and colour-phases D, E and F. The scale of each plate applies to all birds on that plate, except for a few cases, where the difference is pointed out. Most of the birds have been painted from museum skins.

Populations

If a bird is so threatened that it is known there are fewer than 200 left in the wild or in captivity, this is mentioned after the main text.

References

A list of references is given at the end of the book. The titles are arranged alphabetically under the author's name, and code-numbered. At the end of each species entry the code-numbers of the books and papers used are given.

Future alterations

Because of the changing nature of CITES the publisher will print substitute pages including every amendment, as well as colour plates of the new birds. Please contact The Lutterworth Press, PO Box 60, Cambridge, CB1 2NT (Fax: (0)223 66951) for more information.

QUICK GUIDE TO ALL BIRD FAMILIES

(not to scale)*

Struthioniformes
1 Ostriches
Struthionidae
No. of species: 1
Range: Africa
Plate No: 1

Rheiformes
2 Rheas
Rheidae
No. of species: 2
Range: Brazil, Bolivia, Peru
Plate No: 1

Casuariiformes
3 Cassowaries
Casuariidae
No. of species: 3
Range: Australia, New Guinea

4 Emus
Dromaiidae
No. of species: 1
Range: Australia

Apterygiformes
5 Kiwis
Apterygidae
No. of species: 3
Range: New Zealand

Tinamiformes
6 Tinamous
Tinamidae
No. of species: 45-46
Range: Mexico, S. America
Plate No: 2

Sphenisciformes
7 Penguins
Spheniscidae
No. of species: 16-18
Range: Coast of Antarctica, S. America, S.Africa
Plate No: 2

Gaviiformes
8 Divers
Gaviidae
No. of species: 4-5
Range: N. Eurasia, N. America

Podicipediformes
9 Grebes
Podicipedidae
No. of species: 19-22
Range: Worldwide except Antarctica
Plate No: 3

Procellariiformes
10 Albatrosses
Diomedeidae
No. of species: 13-14
Range: Southern Oceans
Plate No: 3

11 Petrels, Shearwaters, Fulmars
Procellariidae
No. of species: 53-75
Range: Worldwide

12 Storm Petrels
Hydrobatidae
No. of species: 20-21
Range: Worldwide

*Sibley & Monroe, Morony, Bock & Farrand and Howard & Moore have been consulted. The arrangement of orders and families follows, with a few exceptions, Howard & Moore (1984).

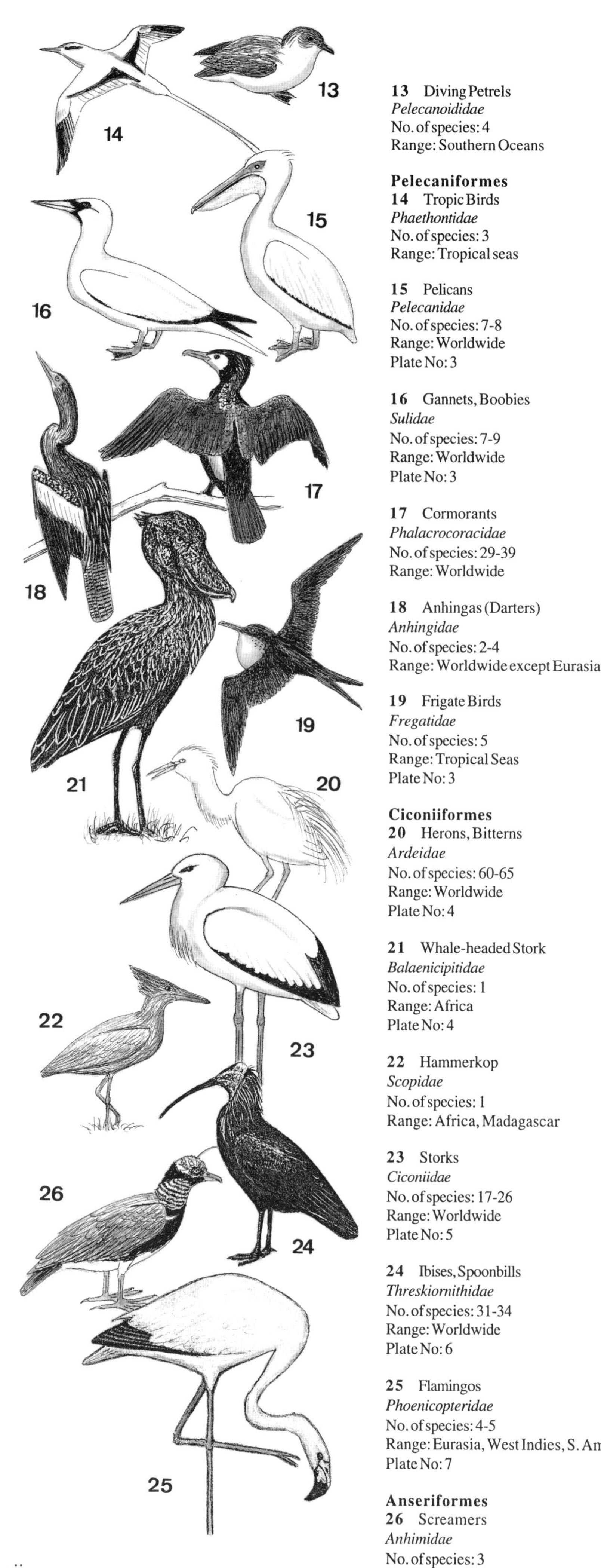

13 Diving Petrels
Pelecanoididae
No. of species: 4
Range: Southern Oceans

Pelecaniformes
14 Tropic Birds
Phaethontidae
No. of species: 3
Range: Tropical seas

15 Pelicans
Pelecanidae
No. of species: 7-8
Range: Worldwide
Plate No: 3

16 Gannets, Boobies
Sulidae
No. of species: 7-9
Range: Worldwide
Plate No: 3

17 Cormorants
Phalacrocoracidae
No. of species: 29-39
Range: Worldwide

18 Anhingas (Darters)
Anhingidae
No. of species: 2-4
Range: Worldwide except Eurasia

19 Frigate Birds
Fregatidae
No. of species: 5
Range: Tropical Seas
Plate No: 3

Ciconiiformes
20 Herons, Bitterns
Ardeidae
No. of species: 60-65
Range: Worldwide
Plate No: 4

21 Whale-headed Stork
Balaenicipitidae
No. of species: 1
Range: Africa
Plate No: 4

22 Hammerkop
Scopidae
No. of species: 1
Range: Africa, Madagascar

23 Storks
Ciconiidae
No. of species: 17-26
Range: Worldwide
Plate No: 5

24 Ibises, Spoonbills
Threskiornithidae
No. of species: 31-34
Range: Worldwide
Plate No: 6

25 Flamingos
Phoenicopteridae
No. of species: 4-5
Range: Eurasia, West Indies, S. America
Plate No: 7

Anseriformes
26 Screamers
Anhimidae
No. of species: 3
Range: S. America

27 Ducks, Geese, Swans
Anatidae
No. of species: 145-158
Range: Worldwide except Antarctica
Plate No. 8, 9, 10, 11, 12, 13, 81

Falconiformes
28 New World Vultures
Cathartidae
No. of species: 7
Range: N. & S. America
Plate No: 14

29 Osprey
Pandionidae
No. of species: 1
Range: Worldwide
Plate No: 19

30 Hawks, Eagles
Accipitridae
No. of species: 211-240
Range: Worldwide except Antarctica
Plate No: 14, 15, 18, 19, 20, 21

31 Secretary Bird
Sagittariidae
No. of species: 1
Range: Africa
Plate No: 21

32 Falcons, Caracaras
Falconidae
No. of species: 60-63
Range: Worldwide except Antarctica
Plate No: 16, 17, 21

Galliformes
33 Magapodes
Megapodiidae
No. of species: 9-19
Range: Australia, New Guinea, Borneo, Philippines
Plate No: 22

34 Curassows, Guans, Chachalacas
Cracidae
No. of species: 42-50
Range: Texas, S. America
Plate No: 22, 23, 80

35 Pheasants, Grouse, Guineafowl, Turkeys
Phasianidae
No. of species: 183-213
Range: Nearly worldwide
Plate No: 24, 25, 26, 27, 28, 29, 30, 31

36 Hoatzin
Opisthocomidae
No. of species: 1
Range: Northern S. America

Gruiformes
37 Mesites
Mesitornithidae
No. of species: 3
Range: Madagascar

38 Button Quails
Turnicidae
No. of species: 14-17
Range: Worldwide except N. & S. America
Plate No: 35

39 Plains Wanderer
Pedionomidae
No. of species: 1
Range: Australia
Plate No: 35

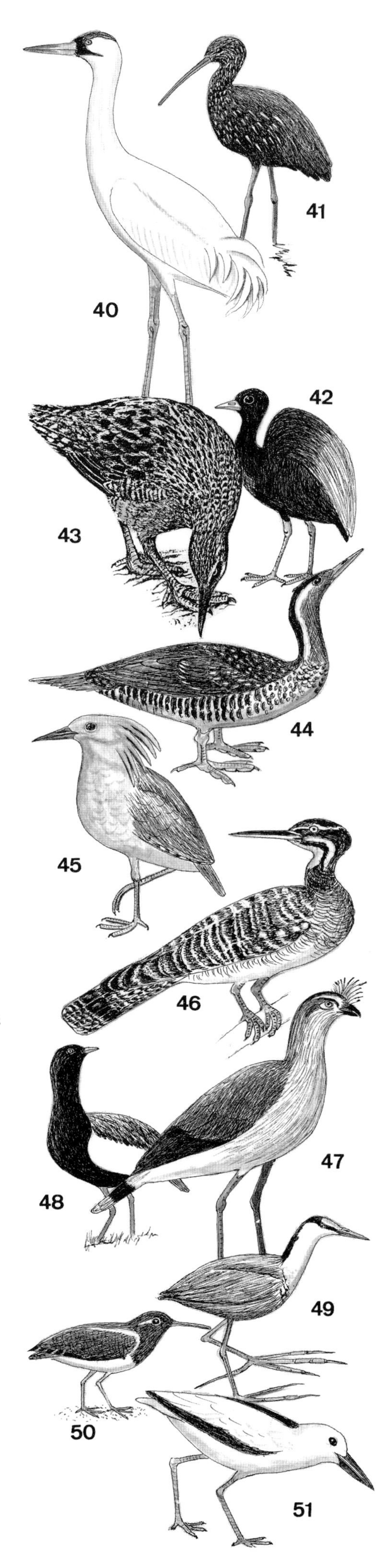

40 Cranes
Gruidae
No. of species: 14-15
Range: N. America & Old World exc. Malayan Archipelago, Polynesia & New Zealand
Plate No. 32, 33, 34

41 Limpkin
Aramidae
No. of species: 1
Range: Southern N. America & S. America

42 Trumpeters
Psophiidae
No. of species: 3
Range: S. America

43 Rails, Coots
Rallidae
No. of species: 123-142
Range: Worldwide
Plate No: 35

44 Sungrebes (Finfoots)
Heliornithidae
No. of species: 3
Range: Mexico, S. America, Africa, India, Malaysia

45 Kagu
Rhynochetidae
No. of species: 1
Range: New Caledonia
Plate No: 35

46 Sun Bittern
Eurypygidae
No. of species: 1
Range: Mexico to S. America

47 Seriemas
Cariamidae
No. of species: 2
Range: S. America

48 Bustards
Otididae
No. of species: 21-25
Range: S. Europe, Africa, S. Asia, Australia
Plate No: 36, 37, 38

Charadriiformes
49 Jacanas
Jacanidae
No. of species: 7-8
Range: Mexico, S. America, Africa, S. Asia Philippines, Malaysia, Australia

50 Painted Snipes
Rostratulidae
No. of species: 2
Range: Africa, S. Asia to Java, Philippines, Australia, S. America

51 Crab Plover
Dromadidae
No. of species: 1
Range: Africa, India

52 Oystercatchers
Haematopodidae
No. of species: 6-11
Range: Nearly worldwide

53 Ibis Bill
Ibidorhynchidae
No. of species: 1
Range: C. Asia, Himalayas, India

54 Avocets, Stilts
Recurvirostridae
No. of species: 7-13
Range: Worldwide

55 Stone-Curlews (Thick-knees)
Burhinidae
No. of species: 9
Range: Worldwide exc. N. America
Plate No: 39

56 Coursers, Pratincoles
Glareolidae
No. of species: 16-17
Range: Eurasia, Africa, Australia

57 Plovers
Charadriidae
No. of species: 62-67
Range: Worldwide

58 Sandpipers, Snipes, Phalarope
Scolopacidae
No. of species: 81-91
Range: Worldwide
Plate No: 39

59 Seed Snipes
Thinocoridae
No. of species: 4
Range: S. America

60 Sheathbills
Chionididae
No. of species: 2
Range: S. Atlantic & W. Indian Ocean and southernmost S. America

61 Skuas, Jaegers
Stercorariidae
No. of species: 5-8
Range: Worldwide

62 Gulls, Terns
Laridae
No. of species: 86-103
Range: Worldwide
Plate No: 39

63 Skimmers
Rynchopidae
No. of species: 3
Range: Coast of N. & S. America, Africa, India, Burma, Kampuchea

64 Auks
Alcidae
No. of species: 21-23
Range: N. Pacific, N. Atlantic, Arctic oceans

Columbiformes
65 Sandgrouse
Pteroclididae
No. of species: 16
Range: S. W. Europe, Africa, S. Asia

66 Doves, Pigeons
Columbidae
No. of species: 280-310
Range: Worldwide exc. Antarctic
Plate No: 40, 41, 42, 43

Psittaciformes
67 Lories, Cockatoos, Parrots
Psittacidae
No. of species: 328-358
Range: Tropical & subtropical parts of N. Hemisphere and most parts of S. Hemisphere exc. Antarctica
Plate No: 44, 45, 46, 47, 48, 49, 50, 51, 52, 79, 81

Cuculiformes
68 Turacos
Musophagidae
No. of species: 18-23
Range: Africa
Plate No: 53

69 Cuckoos, Coucals
Cuculidae
No. of species: 127-130
Range: Worldwide

Strigiformes
70 Barn Owls
Tytonidae
No. of species: 10-17
Range: Nearly worldwide
Plate No: 54

71 Owls
Strigidae
No. of species: 125-159
Range: Worldwide exc. Antarctica
Plate No: 54, 55, 56

Caprimulgiformes
72 Oilbird
Steatornithidae
No. of species: 1
Range: Trinidad, Venezuela to Peru

73 Frogmouths
Podargidae
No. of species: 12-14
Range: India, Burma, Thailand, Malaysia Papuan Region, Australia, Solomons, Philippines

74 Potoos
Nyctibiidae
No. of species: 5-7
Range: Mexico, Honduras, Panama, Trinidad, S. America

75 Owlet-Nightjars
Aegothelidae
No. of species: 7-8
Range: Australia, New Guinea, New Caledonia, Moluccas

76 Nightjars
Caprimulgidae
No. of species: 70-76
Range: Nearly worldwide

Apodiformes
77 Swifts
Apodidae
No. of species: 72-99
Range: Nearly worldwide

78 Tree Swifts
Hemiprocnidae
No. of species: 3-4
Range: India, Burma, Thailand, Malaysia, Papuan Region, Philippines

Trochiliformes
79 Hummingbirds
Trochilidae
No. of species: 315-338
Range: N. C. & S. America
Plate No: 57, 58, 59

Coliiformes
80 Mousebirds
Coliidae
No. of species: 6
Range: Africa

Trogoniformes
81 Trogons
Trogonidae
No. of species: 35-39
Range: Mexico, C. & S. America, West Indies, Africa, India, S.E. Asia, Malaysia, Philippines
Plate No: 59

Coraciiformes
82 Kingfishers
Alcedinidae
No. of species: 87-94
Range: Worldwide

83 Todies
Todidae
No. of species: 5
Range: Greater Antilles

84 Motmots
Momotidae
No. of species: 8-9
Range: Mexico, C. & S. America, Cozumel, Trinidad, Tobago

85 Bee-Eaters
Meropidae
No. of species: 21-26
Range: Temperate and tropical parts of Old World and Australia

86 Rollers, Ground-Rollers
Coraciidae
No. of species: 16-17
Range: Africa, Madagascar, Eurasia, East Indies, Australia, Philippines, Solomons

87 Courols
Leptosomatidae
No. of species: 1
Range: Madagascar, Comores Islands

88 Hoopoes
Upupidae
No. of species: 1-2
Range: Europe, Asia, Africa, Madagascar

89 Wood Hoopoes
Phoeniculidae
No. of species: 5-8
Range: Africa

90 Hornbills
Bucerotidae
No. of species: 44-56
Range: Africa, S. Asia, Malaysia, Philippines, Solomons
Plate No: 60, 83, 84, 85

Piciformes
91 Jacamars
Galbulidae
No. of species: 15-18
Range: Mexico to Brazil

92 Puffbirds
Bucconidae
No. of species: 30-34
Range: Mexico to Brazil and Paraguay

93 Barbets
Capitonidae
No. of species: 68-81
Range: Middle & northern part of S. America, Africa, India, Burma, Thailand, Malaysia, Philippines
Plate No: 79

94 Honeyguides
Indicatoridae
No. of species: 14-17
Range: Africa, Himalayas, Burma, Thailand, Malaysia, Sumatra, Borneo

95 Toucans
Ramphastidae
No. of species: 33-55
Range: S. America
Plate No: 59, 81, 82

96 Woodpeckers, Wrynecks
Picidae
No. of species: 200-215
Range: Worldwide exc. Australia, Papuan Region and most oceanic islands
Plate No: 59

Passeriformes
97 Broadbills
Eurylaimidae
No. of species: 14
Range: Africa, Himalayas, China, Malaysia, Kampuchea, Sumatra, Java, Borneo, Philippines

98 Woodcreepers
Dendrocolaptidae
No. of species: 48-52
Range: Mexico to Argentina

99 Ovenbirds
Furnariidae
No. of species: 217-231
Range: Mexico and S. America

100 Antbirds
Formicariidae
No. of species: 228-244
Range: Mexico to Argentina

101 Gnateaters
Conopophagidae
No. of species: 8-11
Range: Tropical S. America

102 Tapaculos
Rhinocryptidae
No. of species: 28-30
Range: Costa Rica and southern S. America

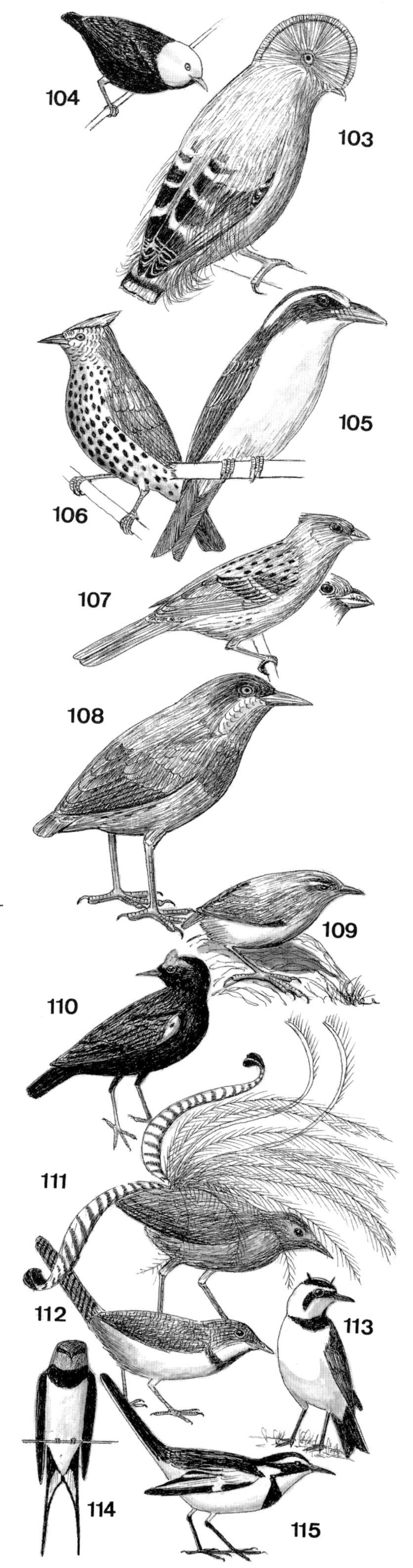

103 Cotingas
Cotingidae
No. of species: 65-84
Range: Texas, Mexico to S. America, Jamaica (very variable)
Plate No: 61, 79

104 Mannakins
Pipridae
No. of species: 52-55
Range: Mexico to Paraguay

105 Tyrant Flycatchers
Tyrannidae
No. of species: 362-412
Range: N. America (exc. extreme north) to S. America

106 Sharpbill
Oxyruncidae
No. of species: 1
Range: Costa Rica, Panama, Guyana, Venezuela, Paraguay, Brazil

107 Plantcutters
Phytotomidae
No. of species: 3
Range: Peru, Chile, Uruguay, Paraguay, Bolivia, Argentina, Falkland Islands

108 Pittas
Pittidae
No. of species: 24-31
Range: C. & S. Africa, India, Burma, Thailand, Kampuchea, China, Japan
Plate No: 62

109 New Zealand Wrens
Xenicidae
No. of species: 3-4
Range: New Zealand

110 Asities
Philepittidae
No. of species: 4
Range: Madagascar

111 Lyrebirds
Menuridae
No. of species: 2
Range: S.E. Australia

112 Scrubbirds
Atrichornithidae
No. of species: 2
Range: S.E. and S.W. Australia
Plate No: 63

113 Larks
Alaudidae
No. of species: 75-91
Range: Nearly worldwide

114 Swallows, Martins
Hirundinidae
No. of species: 74-89
Range: Worldwide
Plate No: 63

115 Wagtails, Pipits
Motacillidae
No. of species: 54-64
Range: Worldwide

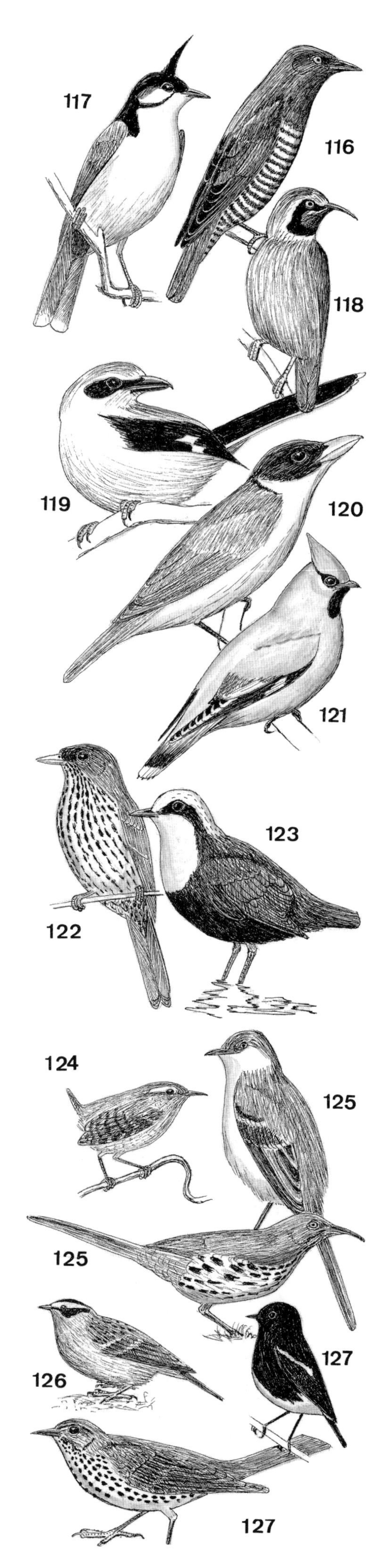

116 Cuckoo Shrikes
Campephagidae
No. of species: 67-72
Range: Africa, India to China, Japan, Philippines, Malaysia, Papuan Region, Australia to Samoa

117 Bulbuls
Pycnonotidae
No. of species: 118-137
Range: Africa, Madagascar, S. Asia, Malaysia, Philippines, Moluccas

118 Leafbirds, Ioras, Fairy-Bluebirds
Irenidae
No. of species: 14
Range: India, Burma, Thailand, Kampuchea, China, Malaysia, Philippines

119 Shrikes, Helmet Shrikes
Laniidae
No. of species: 74-81
Range: Eurasia, Africa, New Guinea, Timor, N. America

120 Vanga Shrikes
Vangidae
No. of species: 8-13
Range: Madagascar

121 Waxwings, Silky Flycatchers, Hypocolius
Bombycillidae
No. of species: 7
Range: Eurasia, N. America to Guatemala

122 Palmchat
Dulidae
No. of species: 1
Range: Hispaniola

123 Dippers
Cinclidae
No. of species: 4-5
Range: Eurasia, Japan, N. America to Argentina

124 Wrens
Troglodytidae
No. of species: 59-74
Range: Eurasia, N. Africa, N. and S. America

125 Mocking-Birds, Thrashers
Mimidae
No. of species: 30-34
Range: N. & S. America, West Indies

126 Accentors
Prunellidae
No. of species: 12-13
Range: Eurasia, N. Africa

127 Thrushes, Chats
Muscicapidae (1)
No. of species: 304-315
Range: Worldwide exc. New Zealand and extreme north.

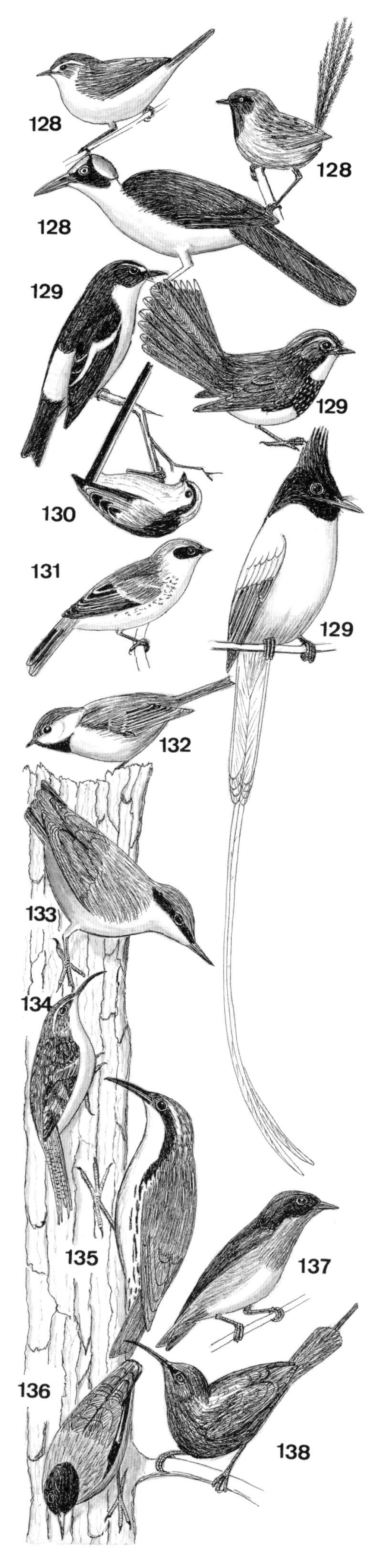

128 Old World Warblers, Logrunners, Warblers, Parrotbills, Rockfowl, Gnatwrens and Gnatcatchers
Muscicapidae (2)
No. of species: 718-757
Range: Worldwide
Plate No: 63, 64

129 Old World Flycatchers, Puffback, Wattle-eye Flycatchers, Australian Wrens, Thornbills, Monarch Flycatchers, Fantail Flycatchers, Australasian Robins, Whistlers
Muscicapidae (3)
No. of species: 353-358
Range: Old World east to Hawaiian Islands
Plate No: 64

130 Long-tailed Tits
Aegithalidae
No. of species: 7-8
Range: Eurasia, Java, N. America, Guatemala

131 Penduline Tits
Remizidae
No. of species: 9-12
Range: Eurasia, Africa, S. USA, Mexico
Plate No: 68

132 Tits, Chickadees
Paridae
No. of species: 46-53
Range: Eurasia, Canary Islands, Africa, Java Sumatra, Philippines, N. America

133 Nuthatches
Sittidae
No. of species: 21-25
Range: Eurasia, Japan, Formosa, New Guinea, Australia, Philippines, N. America

134 Treecreepers
Certhiidae
No. of species: 5-6
Range: Eurasia, Africa, N. America

135 Philippine Creepers
Rhabdornithidae
No. of species: 2-3
Range: Philippines

136 Australian Creepers
Climacteridae
No. of species: 6-8
Range: Australia, New Guinea

137 Flowerpeckers
Dicaeidae
No. of species: 44-58
Range: India, Burma, Thailand, Malaysia, Papuan Region, Kampuchea, China, Australia, Philippines, Solomons

138 Sunbirds
Nectariniidae
No. of species: 116-123
Range: Africa, Madagascar, Israel, Arabia, Iran, India, S.E. Asia to China, Malaysia, Australia, Papuan Region, Philippines

139 White Eyes
Zosteropidae
No. of species: 79-96
Range: Africa, Madagascar, India, Burma, Thailand, Kampuchea, China, Japan, Australia, New Guinea, Malaysia, Philippines
Plate No: 65

140 Honey Eaters
Meliphagidae
No. of species: 169-182
Range: Australia, Papuan Region, Bali, New Zealand, Samoa, Hawaii, South Africa
Plate No: 81

141 Buntings, Cardinals, Grosbeaks, Plush-capped Finch, Tanagers, Swallow Tanager
Emberizidae
No. of species: 553-569
Range: Worldwide exc. Australasia
Plate No: 64

142 New World Warblers
Parulidae
No. of species: 115-126
Range: New World, West Indies

143 Hawaiian Honeycreepers
Drepanididae
No. of species: 15-23
Range: Hawaiian Islands

144 Vireos
Vireonidae
No. of species: 39-51
Range: N. C. & S. America, West Indies

145 New World Blackbirds
Icteridae
No. of species: 92-95
Range: N. M. & S. America, West Indies
Plate No: 65

146 Finches
Fringillidae
No. of species: 122-125
Range: Nearly worldwide exc. Australia
Plate No: 65

147 Waxbills
Estrildidae
No. of species: 124-139
Range: Worldwide exc. New World
Plate No: 66, 66A, 67, 68, 69, 74

148 Weavers, Viduines, Sparrows
Ploceidae
No. of species: 143-157
Range: Worldwide exc. New World
Plate No: 70, 71, 72, 73, 74, 75

149 Starlings
Sturnidae
No. of species: 106-111
Range: Eurasia, Africa, Malaysia, Papuan Region, Australia
Plate No: 69, 81

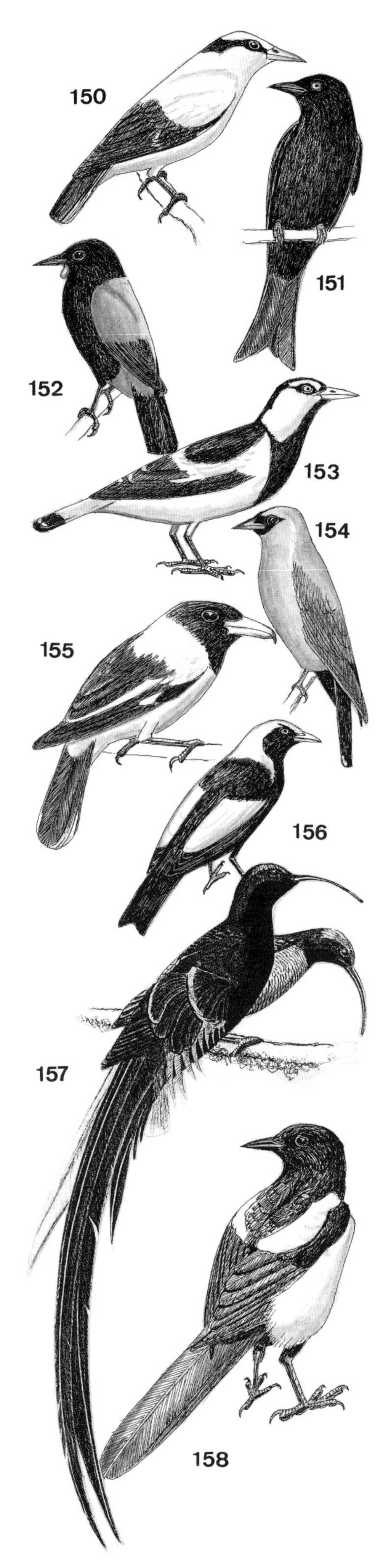

150 Orioles
Oriolidae
No. of species: 25-28
Range: Eurasia, Africa, Australia, Philippines, East Indies

151 Drongos
Dicruridae
No. of species: 20-24
Range: Africa, Madagascar, India, Burma, Thailand, Kampuchea, Malaysia, China, Australia, Papuan Region, Philippines, Solomons

152 Wattlebirds
Callaeidae
No. of species: 2-3
Range: New Zealand

153 Magpie Larks
Grallinidae
No. of species: 4
Range: Australia, New Guinea

154 Wood Swallows
Artamidae
No. of species: 10-11
Range: India, Burma, Thailand, Kampuchea, China, Philippines, Australia, Fiji Islands

155 Butcher Birds
Cracticidae
No. of species: 8-11
Range: Australia, New Guinea

156 Bowerbirds
Ptilonorhynchidae
No. of species: 17-20
Range: Australia, New Guinea

157 Birds of Paradise
Paradisaeidae
No. of species: 40-45
Range: New Guinea, Australia, Moluccas
Plate No: 76, 77, 78

158 Crows, Jays
Corvidae
No. of species: 103-117
Range: Worldwide exc. New Zealand

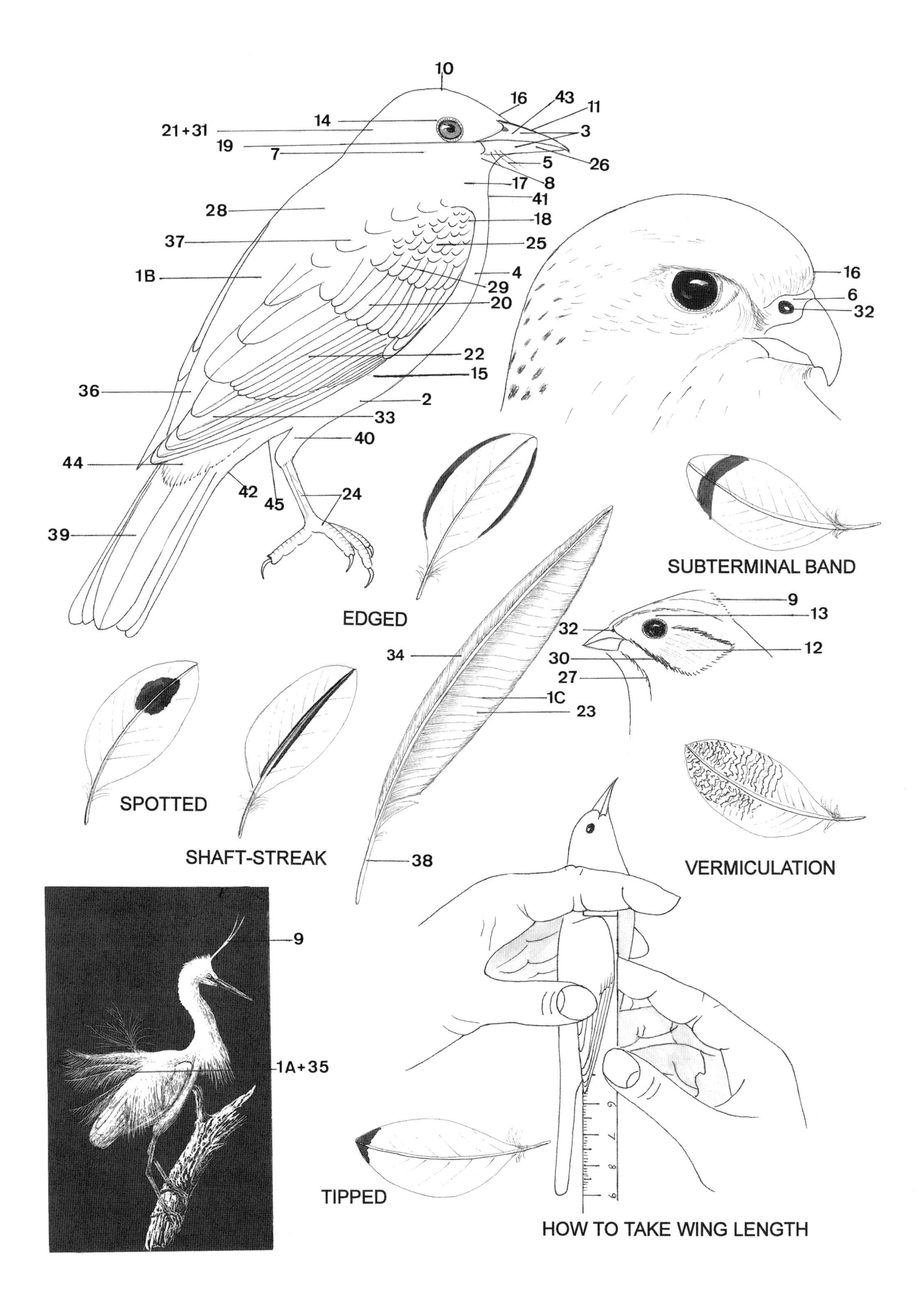
10
16
43
11
14
21+31
3
19
7
26
5
8
17
41
28
18
37
25
1B
4
16
29
6
20
32
22
36
15
2
33
40
44
42
45
24
39
SUBTERMINAL BAND
9
13
EDGED
32
34
12
30
27
1C
23
SPOTTED
VERMICULATION
SHAFT-STREAK
38
9
1A+35
TIPPED
HOW TO TAKE WING LENGTH

GLOSSARY

TO THE TOPOGRAPHY OF A BIRD
AND SOME IMPORTANT ORNITHOLOGICAL TERMS

Ornithological terms have been avoided to some extent in this book, in order that those without much knowledge of birds can understand the text.

1B. Back. The portion of the upperparts located behind the mantle and before the rump.
1C. Barb. A feather consists of a shaft on either side of which a long series of sidebranches (barbs) are attached.
2. Belly. Between breast and the under tail-coverts.
3. Bill. Consists of upper bill (upper mandible) and lower bill (lower mandible).
4. Breast. Between the foreneck and the belly. Also called chest.
Breeding plumage. Distinct plumage borne by an adult during the breeding season.
5. Bristle. Stiff, hairlike feather, usually with a few barbs at base of shaft.

Cap. Area of contrasting colour on top of the head.
Casque. An enlargement on the upper surface of the bill, in front of the head.
6. Cere. The fleshy covering of the proximal portion of the upper bill in some bird families e.g. birds of prey and pigeons.
7. Cheek. The side of the face.
Chest. Synonymous with breast.
8. Chin. The area below the base of lower bill.
Collar. A band of contrasting colour that runs across the foreneck or hindneck, or both.
Comb. Unfeathered flaps or appendages, usually of a fleshy texture and brightly coloured.
Coverts. Small feathers hiding the bases of larger ones, except for ear-coverts, which cover the ear-opening.
9. Crest. A tuft of feathers on the crown or nape of the head.
10. Crown. The upper surface of the head, between the eyebrows.
11. Culmen. The dorsal ridge of the upper bill from forehead to tip.

Distal. Pertaining to part of feather, leg, tail, etc. which is farthest from body.

12. Ear-coverts. Small, loose-webbed feathers on the check behind and below the eye, which cover the outer openings of the ears.
13. Eyebrow. A stripe above the eye.
14. Eye-ring. A fleshy or feathered ring around the eye.

Facial disc. A bird's face, disc-like in form, being well-defined and relatively flat, e. g. in owls, harriers.
15. Flanks. The sides of the mid and lower ventral surface.
22. + 33. Flight-feathers. The long, firm feathers of the wings.

16. Forehead. Just above the base of the upper bill.
17. Foreneck. The front or underside of the neck.
18. Forewing. The front of the wing when closed.

19. Gape. The fleshy corner of the bill which is often yellow in young birds.
20. Greater wing-coverts. A row of short feathers that covers the base of the inner flight-feathers.
Gular. Pertaining to expandable or inflatable throat area. Also called gular pouch.

21. + 31. Hindneck. The rear or upper surface of the neck, also called the nape.

22. Inner flight-feathers. The inner long, firm feathers of the wings, (secondaries). Innermost flight-feathers. The three innermost wing-feathers, (tertials).
23. Inner web. See: Vane.

Juvenile. The first covering of true feathers, usually of a somewhat looser texture than later plumages. In order to save space no difference is made in this work between juvenile or young birds and immature birds.

24. Leg. The paired hind limb including the toes.
Length. From tip of bill to end of tail with stretched neck.
25. Lesser wing-coverts. The short feathers that are arranged in several irregular rows near the shoulder on the front of the wing.
26. Lower bill. The under-section of the bill.

27. Malar. Area between the gape and lower cheeks.
28. Mantle. The upper back.
29. Median wing-coverts. The row of short feathers that cover the bases of the greater wing-coverts.
Mirror. A patch of distinctive colour on the extended wing of some birds, e. g. ducks. Also called speculum.
Morph. One of two or more well-defined forms belonging to the same population. An older term is phase.
30. Moustacial streak. A coloured streak running from the base of the bill backwards along the side of the throat.

Nail. The hooded tip of the upper bill of albatrosses, petrels and waterfowl.
31. + 21. Nape. The back of the head, including the hindneck.
32. Nostrils. The paired openings of the nasal cavities, placed mostly near base of upper bill.

33. Outer flight-feathers. The outermost, longest wing-feathers, also called primaries.
34. Outer web. See: Vane.

Phase. See: Morph.
35. + 1A. Plume. A long showy, display feather. Also called aigrette.
Population. A group of individuals of the same species inhabiting a given area.

Race. See: Subspecies.
Ruff. Collar of elongated feathers on the neck.
36. Rump. The area between the upper tail-coverts and the back.

37. Scapulars. The feathers of the upperparts at the side of the mantle that cover the area where the wing joins the body. Also called the shoulder-feathers.
Sedentary. Living locally, not travelling far.
38. Shaft. Central stiff axis of a feather.
Shaft-streak. Narrow streak of contrasting colour along the shaft of the feather.
Shoulder-feather. See: Scapulars.
Speculum. See: Mirror.
Spur. Sharp bony projection on the leg.
Streak. Pattern of colour oriented longitudinally on feather.
Subspecies. Geographically distinct populations of a species that have diverged from one another in the past but have not yet reached species status.

39. Tail-feathers. Usually 12 stiff long feathers, two central feathers and five lateral feathers on each side. Also called rectrices.
40. Thigh. The feathered portion of the upper leg.
41. Throat. The area of the underparts between the chin and the breast.
Tuft. A bunch of feathers held together at the base.

Underparts. The lower surface of the body including the chin, throat, breast, belly, sides and flanks, and under tail-coverts, excluding rest of head, neck, wings and tail.
42. Under tail-coverts. The small feathers that lie beneath and cover the bases of the tail-feathers.
Under wing-coverts. The small feathers that cover the underside of the long flight-feathers.
43. Upper bill. The upper section of the bill. Also called upper mandible.
Upperparts. The upper surface of the body, including the back, scapulars, rump and upper tail-coverts, excluding head, neck, wings and tail.
44. Upper tail-coverts. The small feathers that lie over the bases of the upper tail feathers. (Enormously long in peacocks).

23. + 34. Vane. The webbed part of the feather, the outer vane being the segment outside the shaft and the inner vane being the segment inside. Also called web.
45. Vent. Area around the cloaca.
Vermiculated. Marked with fine lines.

Web. a) The fleshy membrane between two toes, or b) the vane of a feather.
Wing-bar. Transverse band of contrasting colour in any part of the wing.

Young bird. See: Juvenile.

COLOUR GUIDE

4 Ashy-grey
11 black
14 blue
53 bronze-green
46 brown
41 buff
28 carmine
34 chestnut
43 cinnamon
6 creamy
29 crimson
49 earth-brown
36 flesh
52 green
8 grey
33 hazel
44 horn-brown
20 indigo-blue
42 isabelline
3 ivory
12 jet-black
lead-grey, see:plumbeous
17 lilac
32 mahogany-red
35 maroon
18 mauve-pink
45 mouse-brown
51 olivaceous-green
50 olive
38 orange
5 pearly-grey
21 pink
9 plumbeous
24 purple
27 red
23 rose
30 rufous
39 saffron-yellow
37 salmon
7 sandy
15 sapphire-blue
26 scarlet
2 silvery-white
13 sky-blue
10 slate
48 sooty-brown
31 tawny
54 turquoise
16 ultramarine
47 umber
vermilion, see: scarlet
22 vinaceous
19 violet
vineous, see: wine-red
1 white
25 wine-red

1	2	3	4	5	6
7	8	9	10	11	12
13	14	15	16	17	18
19	20	21	22	23	24
25	26	27	28	29	30
31	32	33	34	35	36
37	38	39	40	41	4
43	44	45	46	47	48
49	50	51	52	53	54

1. STRUTHIO C. CAMELUS

English: Ostrich; *French:* Autruche d'Afrique; *German:* Strauss; *Italian:* Struzzo; *Spanish:* Avestruz

Range: Widely distributed in Africa
Identification:
A. **Male:** Height fully erect 210-275cm, (largest living bird). Flightless. Head, neck, and thighs nearly naked. Body black except white collar, wings, and tail. All feathers loose, plume-like. Only two toes.
B. **Female:** Height 175-190cm. Naked skin and down dirty grey-brown. Body brown with pale feather edgings. Wings and tail whitish to brownish-white. Variable.
Young birds: Like female but darker. Males attain white wings and tail in 2nd year.
Bare parts: Bill yellowish horn, lower bill tinged red, female grey-brown bill. Legs pinkish, more red in front with yellow-margined shields. Bill and legs redder in males in breeding plumage. Eyes brown. Skin on head, neck and thighs pinkish.

Note! Only Ostriches from Algeria, Burkina Faso, Cameroon, Central African Republic, Chad, Mali, Mauritania, Morocco, Niger, Nigeria, Senegal, and Sudan are listed in:
List: I

All other populations are not listed:

Subspecies:

Struthio camelus molybdophanes
Range: Somalia, Kenya
Bare skin of head, neck, and thighs blue-grey. Tail white.
Geographical variation:
Birds from Southern Africa have bare skin grey, red in breeding plumage. Tail of male brown.
Lit.: 20, 37, 51, 129, 410, 428

2. PTEROCNEMIA P. PENNATA

Also known as: Rhea pennata; Rhea darwini

English: Lesser Rhea, Darwin's Rhea, Puna Rhea; *French:* Nandou de Darwin; *German:* Darwinstrauss, Darwinnandu; *Italian:* Nandu di Darwin; *Spanish:* Ñandú cordillerano, Ñandú petizo

Range: Peru, Bolivia, N. Chile, N.W. Argentina
Identification:
Length 92-100cm, height up to 160cm. Flightless.
Male: Face and throat fulvous, otherwise grey-brown often white-tipped on back and wings. Belly whitish. Thighs and upper part of legs grey feathered in front; lower part of front legs with 16-18 transverse shields. Three toes. Feathers long and plume-like on body.
Female: Duller with fewer white spots, like males in worn plumage.
Young birds: Fulvous on head and neck. Upperparts brown without white spots.
Bare parts: Bill brown. Legs light grey. Eyes brown.

Subspecies:

Pterocnemia pennata tarapacensis
Range: S. Chile and S. Argentina
Head and neck ash grey. Forewing dark chestnut-brown. 8-10 frontal shields on legs.
List: I
Lit.: 13, 37, 77, 129, 144, 410, 428

3. RHEA AMERICANA

English: Greater Rhea, Common Rhea; *French:* Nandou d'Amérique; *German:* Nandu; *Italian:* Nandu, *Portuguese:* Ema; *Spanish:* Ñandú, Ñandú común

Range: Brazil, Bolivia, Paraguay, Uruguay, Argentina
Identification:
A. Length 127-140cm. Flightless. **Male:** Crown, nape, lower neck, and upper back usually dark brown or black. Rest grey or grey-brown above, whitish below. Individuals with necks quite white or entire plumage white are not rare. Feathers long and plume-like. Three toes.
B. **Female:** Smaller and entirely grey.
Young birds: Buffish with chestnut-brown spots.
Bare parts: Bill and legs grey, juveniles have yellowish-brown legs. Eyes brown.

Geographical variation:
Birds from S. Brazil and Uruguay have sooty crown and lower foreneck tinged buff. Upper back more greyish.
Birds from Paraguay have upper half of neck orange-cinnamon, lower half black.
Birds from Bolivia and S. E. Brazil are larger. Upper two-thirds of neck pale buff, lower third black.
4. Birds from N. Argentina are larger. Almost entirely black neck and upper back.
List: II
Lit.: 13, 37, 129, 410, 428

PLATE 1

1. TINAMUS SOLITARIUS

English: Solitary Tinamou; *French:* Tinamou solitaire; *German:* Grausteisstinamu; *Spanish:* Tinamú macuco

Range: Brazil, Paraguay, Argentina
Identification:
Length 42-48cm. Sexes similar. Crown dark olive to rufous-brown. Yellow-brown stripe behind eye. Sides of head and lower neck yellow-brown with grey-brown spots; chin and throat white. Underparts and wings olive to rufous brown, narrowly barred black on lower back, rump, and wings. Underparts grey-brown, belly whitish or buffish often finely vermiculated. Flanks with buffish vermiculation. Under tail-coverts cinnamon with grey-brown spots. Tail rufous barred black.
Young birds: White spots on upperparts and wings.
Bare parts: Bill black. Legs grey. Eyes reddish brown.

Geographical variation:
Birds from E. Brazil are paler and more yellowish. Stripe behind eye lacking or not distinct.
List: I
Lit.: 13, 112, 307, 410, 428

2. RHYNCHOTUS R. RUFESCENS

English: Red-winged Tinamou; *French:* Tinamou isabelle; *German:* Pampahuhn; *Spanish:* Tinamú alirrojo

Range: E. Bolivia, Brazil and Uruguay
Identification:
Length 39-42cm. Sexes similar. Crown black edged buff. Rest of head and neck buffy-cinnamon with a creamy throat. Upperparts and wings grey-brown barred and spotted black and buff; outer parts of wings cinnamon-rufous. Underparts buffy-cinnamon, flanks variably barred black. Tail short and hidden.
Young birds: Like adult.
Bare parts: Long and decurved bill blackish. Legs dark yellowish to grey. Eyes yellow.

Subspecies:

3. RHYNCHOTUS RUFESCENS PALLESCENS
Range: N. Argentina
Bigger and greyer.

4. RHYNCHOTUS RUFESCENS MACULICOLLIS
Range: W. & S. Bolivia, W. Argentina
Black streaks on foreneck.
List: II

Note: Rhynchotus rufescens catingae
Range: Central Brazil
Similar to pallescens but neck much darker yellow-brown.
Not listed
(Sibley and Monroe have no subspecies)
Lit.: 13, 77, 97, 129, 157, 307, 410, 428

5. SPHENISCUS DEMERSUS
Formerly known as: Diomedea demersa

English: Jackass Penguin, African Penguin, Black-footed Penguin, Cape Penguin; *French:* Manchot du Cap, *German:* Brillenpinguin; *Italian:* Sfenisco del Capo; *Spanish:* Pingüino del Cabo

Range: South African coast and islands
Identification:
A. Length 70cm. Sexes similar. Head black with a white band from bill over eye and behind cheeks to white upper breast. Upperparts, wings and tail black. Underparts white with various black patches on breast and belly. Some birds have a partial or complete second breast band.
B. Flanks with a black band ending on belly.
C. Young birds: Blackish-grey without white bands on head. Lack breast band or incomplete.
Bare parts: Bill black with irregular grey bands. Legs blackish spotted pink; juvenile pale flesh. Eyes brown, around eyes pink.
List: II
Lit.: 20, 37, 112, 113, 129, 410, 428

6. SPHENISCUS HUMBOLDTI

English: Humboldt Penguin, Peruvian Penguin; *French:* Manchot de Humboldt; *German:* Humboldtpinguin; *Italian:* Sfenisco di Humboldt; *Spanish:* Pingüino de Humboldt

Range: West coast of S. America and adjacent islands
Identification:
A. Length 65cm. Sexes similar. Head black with a narrow white band on each side of crown and behind cheeks to white upper breast. Black stripe above eye. Upperparts and wings blackish-grey. Underparts white with a single black breast band continuous at sides of body to thigh. Tail black.
C. Young birds: Head grey and brown-grey without white stripe, underparts white.
Bare parts: Bill black with vertical grey bar and fleshy base. Legs blackish. Eyes reddish-brown, eye-ring pink.
List: I
Lit.: 13, 37, 112, 113, 116, 129, 410, 428

PLATE 2

1. PODILYMBUS GIGAS

English: Atitlán Grebe, Giant Grebe, Pied-billed Grebe; *French:* Grèbe du Lac Atitlan; *German:* Atitlantaucher; *Spanish:* Zampullín del Atitlán

Range: Lake Atitlán, Guatemala
Identification:
A. Length 48cm. Sexes alike. Head and neck blackish-grey, crown and nape darker; nape and sides of neck with a bottle-green wash. Chin and throat black. Upperparts blackish-brown faintly tinged green. Dark wings short and nearly flightless. Underparts black spotted silvery white, belly white. Tail blackish above, white below. Non breeding: Browner and whitish stripes on sides of head and neck.
C. Young birds: Like non-breeding adult.
Bare parts: Bill ivory with broad black band in adult. Legs olive-grey. Eyes black, skin around eye white.
List: I
Status: Recently extinct at the only known locality, Lake Atitlán.
Lit.: 100, 113, 129, 317, 410, 428

2. DIOMEDEA ALBATRUS

English: Short-tailed Albatross, Steller's Albatross; *French:* Albatros à queue courte; *German:* Kurzschwanzalbatros; *Japanese:* Ahodori; *Spanish:* Albatros colicorto

Range: North Pacific, Izu Islands, formerly Ryukyu Islands and Bonin Islands
Identification:
A. Length 94cm. (Wingspan 213-229cm). Sexes similar. General coloration white. Black line between bill and head. Head washed yellow, often extending to nape and throat. Wings white with variable black areas. Underwing mostly white with black margins. Short tail white with broad black tip.
C. Young birds: Blackish-brown with a narrow line between eye and paler chin. In successive four-five moultings juvenile birds become more and more white.
D. Bare parts: Bill distinctive pink with blue tip and a narrow black line from base through nostril to near tip; juvenile in first two plumages have pale flesh bill without blue tip. Legs bluish-white, juvenile in first two plumages flesh. Eyes blackish.
List: I
Lit.: 113, 129, 301, 410, 428

3. PELECANUS CRISPUS

English: Dalmatian Pelican; *French:* Pélecan frisé; *German:* Krauskopfpelikan; *Italian:* Pellicano riccio; *Russian:* Kudryavy Pelikan; *Spanish:* Pelicano ceñudo

Range: S. E. Europe to China
Identification:
A. Length 160-180cm. Sexes alike. Silvery-white except grey-brown outer wings and some feathers on mantle, scapulars, and tail which have black shafts. Breast often with a wash of yellow or rarely brown. Untidy crest on nape.
C. Young birds: Head, neck and upperparts mottled greyish-brown. Underparts dirty white. Tail grey-brown.
Bare parts: Bill pale yellow, basal half grey, cutting edges distally tinged red, nail at tip orange. Legs dark grey. Eyes pale yellow. Bare skin of face purple. Gular pouch deep orange. Non-breeding and juvenile have yellow skin on face and gular pouch.
List: I
Lit.: 51, 113, 129, 301, 410, 428

4. PAPASULA ABBOTTI

Also known as: Sula abbotti

English: Abbott's Booby; *French:* Fou d'Abbott; *German:* Abbott-Tölpel, *Spanish:* Piquero de Abbott

Range: Tropical Indian Ocean, Christmas Island, (Australia)
Identification:
A. Length 79cm. Sexes similar except for bill. Head, neck, narrow line on middle of back, rump, and upper tail-coverts white with large black spots in area around tail. Rest of upperparts black. Wings black with white on forewing and some white spots on wing-coverts; outer flight feathers with white inner webs. Underparts white, thighs spotted black. Tail black. Note: upperparts of faded birds are brown.
C. Young birds: Upperparts brown.
Bare parts: Bill blue-grey tinged pinkish, female pinkish, both sexes with black tip and cutting edges serrated; juvenile grey, tip black. Legs grey. Eyes brown. Naked skin of face blue-black. Gular pale yellow.
List: I
Lit.: 113, 129, 178, 200, 201, 237, 288, 301, 396, 410, 428

5. FREGATA ANDREWSI

English: Christmas Fregate Bird, Andrews' Frigatebird; *French:* Frégate d'Andrews; *German:* Weissbauch-Fregattvogel; *Spanish:* Rabihorcado de la Christmas

Range: East Indian Ocean, Christmas Island (Australia)
Identification:
A. Length 89-100cm. **Male:** Black glossed green on lanceolate feathers of upperparts. Forewing paler. Lower belly white.
B. **Female:** Larger. Blackish-brown with narrow white stripes on nape and hindneck. Collar on lower neck, breast, and belly white.
C. Young birds: Like female but head and neck pale tawny-yellow. Feathers on upperparts have a more white mottled appearance.
Bare parts: Bill blackish, female pink. Legs blackish, female flesh-white. Eyes brown, eye-ring black, female pink. Males during breeding have a scarlet throat pouch, which can be inflated as a balloon during courtship.
List: I
Lit.: 113, 129, 178, 301, 410, 428

PLATE 3

1. BULBULCUS IBIS

Also known as: Ardeola ibis; Ardea ibis; Agretta ibis

Dutch: Koereiger; *English:* Cattle Egret, Buff-backed Heron; *French:* Héron garde-boeufs; *German:* Kuhreiher; *Italian:* Airone guardabuoi; *Japanese:* Ama-sagi; *Malaya:* Burong apuh; *Spanish:* Garcilla bueyera

Range: Spain to Iran, N. and C. Africa, Madagascar, E. North America, N. and C. South America
Identification:
A. Length 48-53cm. Sexes similar in breeding dress. Crown, nape, mantle, and breast buffish to pale rufous, feathers elongated, loose, and hair-like; mantle feathers nearly reach tail tip. Rest of plumage white. Non breeding: Males have paler crown and nape feathers. Centre of mantle and breast washed with cream-buff. Feathers only slightly elongated on throat and mantle. Female: Entire plumage white and feathers not elongated.
Young birds: Like non-breeding female. Crown becomes slightly cinnamon tinged after 10 weeks.
Bare parts: Bill short (52-60mm) and red to orange during short period before egg-laying, otherwise yellow. Legs red for a short period, otherwise dark green, upper part and sole yellowish; juveniles have first black legs, later dark green. Eyes and bare skin between bill and eye red, non- breeding yellow.

Subspecies:

Bulbulcus ibis coromanda
English: Eastern Cattle Egret
Range: India to Japan, Philippines, Moluccas, Australia, New Zealand
Ornamental feathers more golden cinnamon. More of upper legs bare of feathers.
List: III Ghana
Lit.: 20, 37, 51, 108, 129, 173, 310, 410, 428

2. EGRETTA GARZETTA

Formerly known as: Ardea garzetta

Dutch: Kleine Zilverreiger; *English:* Little Egret; *French:* Aigrette garzette; *German:* Seidenreiher; *Italian:* Garzetta; *Japanese:* Ko-sagi; *Spanish:* Garceta común

Range: S. Europe to Japan, Philippines, to New Guinea, Australia, Sunda Islands, Africa
Identification:
A. Length 55-65cm. Sexes alike. Wholly snow-white. 2-3 narrow crest feathers 13-16cm long. Lanceolate plumes on foreneck, chest, mantle, and scapulars, last two forming loose train and reaching tip of tail, called aigrettes. Non breeding: Without long crest feathers. Plumes of upperparts shorter and fewer.
D. Dark variant: Wholly grey, or head white with undulating grey bars, and long flight feathers grey mottled white, shafts black.
Young birds: Like adult non breeding but without any elongated feathers.
Bare parts: Bill black, (67-93mm), juvenile brown. Legs black, feet yellow washed green, juveniles have grey-green feet. Eyes yellow. Bare skin of face grey, washed green, orange in breeding season, juvenile lead-grey.

Geographical variation:
Birds from Sunda Islands, Philippines to New Guinea and Australia have black feet.
List: III Ghana
Lit.: 20, 37, 51, 89, 108, 129, 173, 178, 410, 428

3. CASMERODIUS ALBUS

Also known as: Egretta alba; Ardea alba

American: Great Egret; *Brazilian:* Garça-branca-grande; *Dutch:* Grote Zilverreiger; *English*: Great White Egret; *French:* Grande aigrette; *German:* Silberreiher, *Italian:* Airone bianco (maggiore); *Japanese:* Dai-sagi; *Spanish:* Garceta grande

Range: S. E. Europe to China, Japan, Australia, New Zealand, Africa, N. & S. America
Identification:
Length 85-102cm. Sexes similar but male larger. Wholly snow-white. Nape feathers slightly elongated. Long aigrette plumes from scapulars extending far beyond tip of tail. Lanceolate feathers on lower foreneck and upper breast. Non breeding: Without aigrettes. A few scapulars with aigrette-like tips.
Young birds: Similar to non breeding adult but breast feathers not soft and loose, and any scapulars with aigrette-like tips are shorter.
Bare parts: Bill black, base variable amount of yellow, non breeding yellow. Legs black, upper legs pinkish-yellow with narrowing yellow stripes down sides to toes; non breeding upper legs yellow, rest dark grey-green, juvenile upper legs yellow-green, rest brown black, sides of lower legs and soles yellow. Eyes yellow. Bare skin around eye pale green.

Geographical variation:
Birds from Africa have bill black only for a very short time or missing altogether. Upper legs black during breeding.
Birds from N. & S. America have upper legs black in breeding. Bill orange-yellow in breeding, often with a blackish tip.
List: III Ghana
Lit.: 2, 20, 37, 51, 89, 129, 232, 410, 428, 431

4. ARDEA GOLIATH

Formerly known as: Ardea nobilis

Dutch: Goliathreiger; *English:* Goliath Heron; *French:* Héron goliath; *German:* Goliathreiher; *Italian:* Airone gigante; *Spanish:* Garza goliat

Range: Senegal to Sudan & Cape Province
Identification:
Length 135-150cm. Sexes alike. Like a gigantic Purple Heron. Top of head, short crest, nape, and sides of neck rich chestnut, darkest on head. Crest bushy. Chin, throat, front of neck, and upper breast white, last two streaked black. Upperparts and wings slaty-grey, mantle and scapulars with lanceolate feathers. Rest of underparts and under wing-coverts deep rufous-chestnut. No seasonal changes.
Young birds: General appearance paler. Head and neck duller and paler rufous. Upperparts browner. Underparts white streaked with brown.
Bare parts: Bill, upper black, lower paler. Legs black. Eyes yellow.
List: III Ghana
Lit.: 2, 7, 20, 37, 108, 129, 268, 325, 378, 410, 428

5. BALAENICEPS REX

English: Shoebill, Whale-headed Stork; *French:* Bec-en-sabot du nil; *German:* Schuhschnabel; *Spanish:* Picozapato

Range: Sudan to Zambia
Identification:
Length 120-153cm. Sexes alike. Plumage grey to blue-grey, slightly washed greenish on back. Plumes on lower neck with black shafts. Abdomen and undertail-coverts pale grey.
Young birds: Browner without green tinge on back.
Bare parts: Bill enormous and shoe-shaped, pinkish-grey. Cannot be confused with any other bird. Legs blackish. Eyes grey to bluish-white.
List: II
Lit.: 7, 20, 129, 171, 325, 410, 428

PLATE 4

1. CICONIA BOYCIANA

Also known as: Ciconia ciconia boyciana

English: Oriental White Stork, Eastern White Stork; *French:* Cigogne orientale; *German:* Schwarzschnabelstorch; *Italian:* Cicogna dal becco nero; *Spanish:* Cigüeña blanca coreana, Cigüeña oriental

Range: N. E. Asia
Identification:
Length 110cm. Sexes similar but female smaller. Like European White Stork but bill larger, slightly upturned, and a tendency to be open in middle. Inner flight feathers edged greyish, otherwise black and white.
Young birds: Like adult but wings brown, not black, and brown spots on shoulders.
Bare parts: Bill diagnostic black. Legs pink to dull orange-red. Eyes and bare skin on face red.
List: I
Lit.: 2, 37, 51, 64, 129, 145, 191, 196, 323, 360, 410, 428

2. CICONIA NIGRA

Dutch: Zwarte Ooievaar; *English:* Black Stork; *French:* Cigogne noire; *German:* Schwarzstorch; *Italian:* Cicogna nera; *Spanish:* Cigüeña negra

Range: Europe to China, India, Africa
Identification:
A. Length 95-100cm. Sexes similar but female smaller. Head, neck, upperparts and wings black with conspicuous purple and green gloss. Underparts white. Tail black with brown tinge.
C. Young birds: Browner with pale feather tips and plumage less glossy.
Bare parts: Bill and legs bright scarlet, juvenile olive-green. Eyes brown, juvenile grey-brown.
List: II
Lit.: 20, 37, 51, 129, 145, 410, 428

3. JABIRU MYCTERIA

Brazilian: Jabiru; *English:* Jabiru, Jabiru Stork, *French:* Jabiru d'Amérique; *German:* Jabiru; *Spanish:* Garzon soldado, Jabirú Americano

Range: Mexico to Argentina
Identification:
Length 120-130cm, female slightly smaller, sexes otherwise similar. Head and neck bare. Entire body including wings and tail white.
Young birds: Dark grey to brownish.
Bare parts: Large slightly upturned bill and legs black. Eyes brown. Bare skin of head and neck mainly black, lower neck bright orange-red.
List: I
Lit.: 13, 37, 129, 145, 192, 410, 428, 431

4. MYCTERIA CINEREA

Also known as: Ibis cinereus

English: Milky Stork; *French:* Tantale blanc; *German:* Milchstorch; *Spanish:* Tántalo malayo

Range: Malaya, Sumatra, Java, and Greater Sunda Islands
Identification:
Length 97cm. Sexes alike. Head naked. Rest of plumage white except glossy green-black, long flight-feathers and tail.
Young birds: Head and neck dull brown, upperparts greyish-brown with white rump and black flight-feathers. Breast grey, rest of underparts white.
Bare parts: Long, decurved, and rounded bill yellowish-orange, tip paler. Legs dark red, outside breeding grey. Eyes dark. Naked facial skin red and black.
List: I
Lit.: 129, 145, 150, 230, 277, 410, 428

5. EPHIPPIORHYNCHUS SENEGALENSIS

Also known as: Mycteria senegalensis

English: Saddle-billed Stork, Saddlebill; *French:* Jabiru du Sénégal; *German:* Sattelstorch; *Italian:* Mitteria del Senegal; *Spanish:* Jabirú Africano

Range: Senegal to Sudan & Transvaal
Identification:
Length c. 150cm, female 10-15% smaller. Sexes similar. Head and neck black iridescent with green; on base of neck purplish. Scapulars, upper tail-coverts, wing-coverts, and tail black glossed green. Rest including flight-feathers pure white.
Young birds: Grey where adults are white and black colour duller.
Bare parts: Bill large, slightly upturned, red with a black band and yellow frontal saddle is diagnostic; juvenile grey and lacking frontal saddle, 2nd-year birds have red and black bill, yellow saddle first occurs in 3rd-year birds. Legs black with joints and feet red. Eyes brown, female yellow.
List: III Ghana
Lit.: 20, 37, 129, 145, 173, 268, 410, 428

6. LEPTOPTILOS CRUMENIFERUS

Dutch: Afrikaanse Maraboe; *English:* Marabou Stork; *French:* Marabout d'Afrique; *German:* Marabu; *Italian:* Marabu d'Africa; *Spanish:* Marabú, Marabú Africano

Range: Senegal to Sudan & Transvaal
Identification:
Length c. 140cm, female smaller. Sexes similar. Bare head and neck with sparse black down and feathers. Two inflatable air sacs: one long on throat, one smaller and normally hidden in white ruff at base of neck. Back and wing-coverts waxy blue-grey. Wings and tail black, glossed green; inner flight-feathers edged white. Underparts white. Non breeding: Back and wing-coverts slate grey, otherwise like breeding dress.
Young birds: Head and neck covered in hair-like downs and feathers. Wings dark brown with little gloss. 2nd-year birds have black wings but only in 3rd-year completely black and glossy.
Bare parts: Bill horn to pinkish, mottled black. Legs black powdered with white. Eyes brown. Bare skin on head and neck dirty pinkish, side of face orange-red with blackish in front of bill and around ear. Large air sac reddish, small orange-red.
List: III Ghana
Lit.: 20, 37, 129, 145, 173, 268, 410, 428

PLATE 5

1. GERONTICUS CALVUS

African: Kalkoen; *English:* Bald Ibis, Southern Bald Ibis; *French:* Ibis du Cap, Ibis chauve de l'Afrique du Sud; *German:* Glattnackenrapp, Kahlkopfrapp; *Italian:* Ibis calvo; *Spanish:* Ibis calvo

Range: South Africa

Identification: A. Length 80cm. Sexes similar. Skin of head and throat naked. Feathers of lower neck form fluffy ruff. General colour black with bluish-green gloss and purple area on upper wing-coverts.

C. Young birds: Head and throat with light grey feathers. Body less iridescent and no purple on wings.

Bare parts: Bill long and decurved, red, juvenile pinkish flesh to brownish, base grey. Short legs dull red, juvenile brownish. Eyes red. Bare skin on head and throat flesh coloured with bright red crown, domed posteriorly; juveniles have no red on crown.

List: II

Lit.: 20, 37, 88, 129, 172, 174, 177, 205, 410, 428

2. GERONTICUS EREMITA

English: Waldrapp, Northern Bald Ibis; *French:* Ibis chauve; *German:* Waldrapp; *Italian:* Ibis eremita; *Spanish:* Ibis eremita

Range: Turkey (?), N. Africa, Ethiopia, Arabia

Identification: A. Length 70-80cm. Sexes similar. Head and throat naked. Crest of narrow, long glossy blue or purple feathers. Elongated ruff on neck. Plumage black with green metallic gloss. Upper wing-coverts glossed purple-red.

Young birds: Like adult but less iridescent. Grey feathers on head. Crest smaller and wings lacking purple-red gloss.

Bare parts: Bill long and decurved, red. Short legs dirty red. Eyes orange-red, juvenile yellow-grey. Bare skin on crown black with orange central streaks; rest of naked head and throat red.

List: I

Lit.: 20, 37, 51, 89, 126, 129, 159, 173, 255, 336, 410, 428

3. EUDOCIMUS RUBER

Brazilian: Guará; *English:* Scarlet Ibis; *French:* Ibis rouge; *German:* Scarlachsickler; *Spanish:* Corocoro Colorado, Corocoro rojo

Range: Northern S. America, Trinidad

Identification: A. Length 56-61cm. Sexes alike. Entire plumage scarlet except for black wing tips.

Young birds: Head and neck light brown streaked whitish. Upperparts, wings, and tail uniformly brown, rump and underparts white, rump often tinged pink.

Bare parts: Bill large, decurved, variably black to buffish, in non breeding pinkish to reddish. Legs scarlet, juvenile pinkish to dusky. Eyes dark brown. Bare facial skin scarlet, juvenile pinkish.

List: II

Lit.: 13, 125, 129, 192, 410, 428, 431

4. NIPPONIA NIPPON

English: Japanese Crested Ibis; *French:* Ibis nippon, Ibis blanc; *German:* Nipponibis; *Italian:* Ibis del Giappone; *Japanese:* Toki; *Spanish:* Ibis moñudo japonés, Ibis nipón

Range: N. E. Asia, Japan

Identification: A. Length 77cm. Sexes similar. Forehead naked. Rest of head, long crest feathers, and back grey. Wings pale grey, underside salmon-pink. Underparts and tail white.

D. Non Breeding: Plumage white with pale salmon-pink tinge.

Young birds: Brown-grey.

Bare parts: Bill long and decurved, black, tip red. Short legs reddish to brown. Eyes red, juvenile brownish-yellow. Bare skin on forehead red.

List: I

Status: In 1989 there were 46 birds estimated in China. In Japan extinct in 1981. Captive population in 1982: 5 birds.

Lit.: 37, 64, 79 129, 196, 323, 410, 428

5. BOSTRYCHIA HAGEDASH

Also known as: Hagedashia hagedash

English: Hadada Ibis, Hadada, Hadeda, Hadedah; *French:* Ibis hagedash; *German:* Hagedasch; *Spanish:* Ibis hadada

Range: Senegal to Ethiopia and Cape Provinces

Identification: Length 76cm. Sexes similar. Head and neck brownish-grey, white stripe on cheeks. Body dull olive-grey to brown with metallic sheen on upperparts and wings. Underparts brownish-grey, Tail blue-black.

Young birds: Duller and bill shorter.

Bare parts: Bill blackish, base of upper bill red. Legs greyish-black, top of toes dull red. Eyes brown.

Geographical variation:

Birds from Gambia to Zaire are darker below and have green gloss on wing-coverts. Bill longer, 126-163mm, against 117-153mm in birds from southern Africa.

Birds from Ethiopia to Uganda have longest bill, 152-174mm.

List: III Ghana

Lit.: 20, 88, 129, 174, 205, 268, 410, 428

6. BOSTRYCHIA RARA

Also known as: Lampribis rara

English: Spot-breasted Ibis; *French:* Ibis vermiculé; *German:* Fleckenibis, Fleckenbrustibis; *Spanish:* Ibis moteado

Range: Liberia to Zaire & Angola

Identification: Length 55-65cm. Male: Front of face black with a turquoise spot in front and behind eye. Rest of head brown with a narrow white line below ear-coverts. Crest black glossed green. Neck, breast, and belly rufous edged black giving a spotted appearance. Upperparts brown glossed blue, green, and bronze. Feathers on mantle edged buff. Under tail-coverts black with green gloss. Tail brown glossed dark blue-green.

Female: Like male but lacking turquoise spots on head.

Young birds: Duller with shorter bills and chest.

Bare parts: Bill long, decurved, red. Legs pinkish-brown. Eyes dark brown.

List: III Ghana

Lit.: 7, 20, 129, 173, 410, 428

7. THRESKIORNIS AETHIOPICUS

English: Sacred Ibis; *French:* Ibis sacré; *German:* Heiliger Ibis; *Spanish:* Ibis sagrado

Range: Mauritania, Senegal to Somalia and south to S. Africa

Identification: A. Length 65-75cm. Sexes similar. Head and upper neck naked. Body white except for black ornamental plumes over rump and wing tips, both with some purple-blue gloss. Underparts and greater wing-coverts often partly stained cinnamon or brown.

C. Young birds: Head and hind neck feathered, giving a mottled black, brownish, and white appearance. Without ornamental plumes.

Bare parts: Bill long, decurved, and glossy black, juvenile greyish-black. Legs black, tinged red, juvenile greyish-black. Eyes brown with red outer-ring, juveniles have narrow white outer- rings. Bare skin on head and upper neck black, in breeding season bare skin reaches base of neck.

List: III Ghana

Lit.: 7, 20, 34, 51, 88, 128, 129, 268, 410, 428

8. PLATALEA LEUCORODIA

Dutch: Lepelaar; English: Spoonbill, Eurasian Spoonbill; *French:* Spatule blanche; *German:* Löffler; *Italian:* Spatola; *Spanish:* Espátula blanca, Espátula común

Range: Netherlands, S. Europe to China, Japan, India, and Africa

Identification: A. Length 80-90cm. Sexes alike. All white except for yellow-buff collar round base of neck, variable extent. Loose and long crest feathers sometimes with a yellow-buff wash. Non breeding: Like breeding but lacking crest and yellow-buff collar.

C. Young birds: Like adult non-breeding but shafts of flight-feathers black.

Bare parts: Bill long and with spoon-like tip, black, barred black and grey, tip orange-yellow, edged black; juvenile pink. Legs black, juvenile pinkish-flesh. Eyes red, juvenile grey-brown. Bare skin between bill and eye black bordered white to pale yellow, bare chin and throat yellow, tinged orange. Around eyes yellow, juvenile blue.

Geographical variation:

Birds from Mauritania have bill completely black and no yellow-buff collar.

List: II

Lit.: 20, 37, 51, 129, 410, 428

PLATE 6

1. PHOENICOPTERUS ANDINUS

Also known as: Phoenicoparrus andinus

Brazilian: Flamingo-dos-Andes; *English:* Andean Flamingo; *French:* Flamant des Andes; *German:* Gelbfussflamingo, Andenflamingo; *Italian:* Fenicottero delle Ande; *Spanish:* Flamenco andino, Parina grande

Range: Peru, Chile, Bolivia, Argentina
Identification:
Length 110cm. Sexes similar. Plumage white with a wash of pink. Foreneck and breast often distinctly vinaceous. Wing-coverts rosy vermilion, flight-feathers black.
Young birds: Dull greyish with dark streaks above.
Bare parts: Bill yellowish-white, terminal half black; red spot between nostrils. Legs yellow (only flamingo with yellow legs). Lacks hind toe. Eyes dark brown.
List: II
Lit.: 13, 37, 77, 129, 211, 410, 428, 431

2. PHOENICOPTERUS JAMESI

Also known as: Phoenicoparrus jamesi

English: James' Flamingo, Puna Flamingo; *French:* Flamant de James; *German:* Kurzschnabelflamingo; *Italian:* Fenicottero di James, Fenicottero dal becco corto; *Spanish:* Flamenco andino chico, Parina chica, Flamenco de James

Range: Peru, Bolivia, Chile, Argentina
Identification:
Length 92cm. Sexes alike. Like Andean Flamingo but smaller and paler. Black in flight-feathers not so visible in folded wing, as rose-red shoulder feathers cover it. In breeding dress has carmine area on breast, rarely seen in captivity.
Young birds: Drab-coloured. Underparts with bold dark brown streaks. Flanks often with fine streaks.
Bare parts: Bill yellowish-orange, tip black; bare skin between bill and eye bright red. Legs red, hind toe absent. Eyes orange.
List: II
Lit.: 13, 37, 49, 77, 129, 211, 410, 428

3. PHOENICOPTERUS CHILENSIS

Formerly known as: Phoenicopterus ruber chilensis

Brazilian: Flamingo chileno; *English:* Chilean Flamingo; *French:* Flamant du Chili; *German:* Chileflamingo; *Italian:* Fenicottero del Cile, *Spanish:* Flamenco chileno

Range: Peru, Bolivia, Chile, Uruguay to Tierra del Fuego
Identification:
Length 105cm. Sexes alike. Plumage white faintly washed pink, feathers on lower neck and breast are washed bright pink. Long wing-coverts bright crimson. Flight-feathers black, hardly visible with folded wings. Non breeding: Lower neck and breast only washed faintly with pink.
Young birds: Greyish-white and brown. Wing-coverts pale pink at base.
Bare parts: Bill whitish tip black. Legs greenish-grey to light blue, joints and feet red. Eyes pale yellow.
List: II
Lit.: 13, 37, 129, 211, 410, 428, 431

4. PHOENICOPTERUS R. RUBER

Brazilian: Flamingo; *English:* American-, Caribbean-, Cuban-, Rosy- or West Indian Flamingo; *French:* Flamant rouge, Flamant de Cuba; *German:* Flamingo, Rosaflamingo; *Italian:* Fenicottero rosso, Fennicottero dei Caraibi; *Spanish:* Flamenco de Cuba, Flamenco rojo, Flamenco común

Range: Atlantic, C. & S. America, West Indies
Identification:
A. Length c. 120cm. Sexes alike but male larger. Plumage varies from white washed pink to vermilion, deepest colour on neck and wing-coverts. Flight-feathers black often not seen when wing is folded.
C. Young birds: Mostly greyish with wash of pink on underparts and wings. Back feathers with black shaft streaks. Older birds are light salmon with dusky streakings on upperparts.
Bare parts: Bill orange-pink, tip black, juveniles have brown bill. Between bill and eye pink, young birds lead-colour. Legs pinkish, joints ultramarine-violet to bright pink. Juveniles have lead coloured legs. Eyes yellow, juvenile brown.

Subspecies:

5. PHOENICOPTERUS RUBER ROSEUS

English: Greater Flamingo; *French:* Flamant rose; *Spanish:* Flamenco común
Range: S. Europe, S. Asia to India, Africa
Larger (125-145cm) and paler especially on head, neck, and breast. Black at tip of bill smaller, does not reach upwards to bend of culmen.
List: II
Lit.: 13, 20, 37, 51, 129, 211, 217, 222, 410, 428, 431

6. PHOENICOPTERUS MINOR

Also known as: Phoeniconaias minor
English: Lesser Flamingo; *French:* Petit Flamant, Flamant nain; *German:* Zwergflamingo; *Italian:* Fenicottero minore; *Spanish:* Flamenco enano
Range: Africa, Madagascar, India
Identification:
Length 80-90cm. Sexes similar but female smaller. General colour deep rose-pink but very variable. Wing-coverts pink and deep crimson. Flight-feathers black. Tail deep pink.
Young birds: Brownish-grey with dark brown streaks on back, wing-coverts and breast.
Bare parts: Bill dark crimson, tip black. Bare skin between bill and eye dark red; juveniles have purple-brown bills with black tips. Legs pale vermilion, darker at joints and feet; young birds dark grey. Eyes yellow to red, juveniles brown.
List: II
Lit.: 20, 37, 51, 129, 211, 410, 428

PLATE 7

1. CYGNUS MELANOCORYPHA

English: Black-necked Swan; *French:* Cygne á cou noir, *German:* Schwarzhalsschwan; *Italian:* Cigno dal collo nero; *Spanish:* Cisne cuellinegro

Range: Southern S. America, Falkland Islands
Identification:
A. Length 102-124cm. Sexes alike but female smaller. Head and neck unmistakably black except for a small white eye stripe. Rest white.
Young birds: Head and neck more brown than black. Body white and pale brown.
Bare parts: Bill blue-grey, tip pinkish white; frontal knob red, juvenile without knob. Legs pink, juvenile grey. Eyes dark brown.
List: II
Lit.: 37, 60, 129, 176, 297, 410, 428

2. CYGNUS COLUMBIANUS

English: Whistling Swan, Tundra Swan; *French:* Cygne siffleur; *German:* Pfeifschwanz; *Spanish:* Cisne chico

Range: Canada, USA
Identification:
A. Length 120-150cm. Sexes similar. Plumage white everywhere.
C. Young birds: Greyish-brown, paler below.
Bare parts: Bill black with a small yellow area between eye and base of bill but sometimes absent; juvenile pink. Legs black, juvenile fleshy-grey. Eyes dark brown.

Subspecies:

3. Cygnus columbianus bewickii
English: Bewick's Swan; *French:* Cygne de Bewick; *German:* Zwergschwan
Range: N. Europe, Siberia to China, Korea, Japan
A little smaller. Yellow between eye and bill with a rounded yet variable closing most often extends to nostrils.

Geographical variation:
4. Birds from N. E. Asia, China, Korea and Japan have a larger and higher bill. Yellow patch on bill more extensive.
NO MORE LISTED
Lit.: 60, 129, 146, 176, 195, 297, 410, 428

5. COSCOROBA COSCOROBA

Brazilian: Capororoca; *English:* Coscoroba Swan; *French:* Coscoroba blanc, Cygne coscoroba; *German:* Coscorobaschwan; *Italian:* Cigno coscoroba; *Spanish:* Cisne coscoroba, Ganso blanco

Range: Southern S. America
Identification:
A. Length 90-115cm. Sexes alike. Completely white except six outer flight-feather tips black.
C. Young birds: White and grey-brown on head and upperparts, underparts white.
Bare parts: Bill bright red, tip paler; form duck-like; juvenile bluish-grey. Legs pink, juvenile blue-grey. Eyes orange, female brown.
List: II
Lit.: 37, 60, 129, 146, 176, 297, 410, 428, 431

6. BRANTA SANDVICENSIS

Formerly known as: Nesochen sandvicensis

English: Nene, Hawaiian Goose; *French:* Bernache néné, Bernache d'Hawaii; *German:* Nene, Hawaiigans; *Italian:* Oca delle Hawaii; *Spanish:* Barnacla Hawaiana, Barnacla nené

Range: Hawaii
Identification:
Length 56-71cm. Sexes alike. Head and hindneck brownish-black. Sides of head and neck buff. Feathers on neck strongly furrowed. Upperparts grey-brown with transverse barring. Rump and tail black. Underparts grey-brown, tip of feathers grey. Lower belly and under tail-coverts white. Flanks like upperparts.
Young birds: Duller and more mottled on body.
Bare parts: Bill and legs black. Eyes dark brown.
List: I
Lit.: 37, 60, 129, 146, 176, 297, 410, 428

7. BRANTA CANADENSIS LEUCOPAREIA

English: Aleutian Canada Goose; *French:* Bernache des Aléoutiennes; *German:* Aleuten-Zwergkanadagans; *Italian:* Oca delle Aleutine; *Spanish:* Barnacla canadiense Aleutiana, Ganso canadiense Aleutiana

Range: Aleutian Islands, Japan & W. USA
Identification:
Length 55-70cm. Sexes similar. Head and neck black except for white patches on cheek and neck ending in a white ring, in front at least 2cm wide. Upperparts grey-brown with grey transverse barring. Rump and tail black. Underparts grey-brown. Lower belly and under tail-coverts white.
Young birds: Like adults but white neck ring smaller.
Bare parts: Bill and legs black. Eyes dark brown.
List: I
Note: all other races of Canada Goose are not listed.
Lit.: 37, 60, 129, 146, 176, 297, 410, 428

8. BRANTA RUFICOLLIS

Formerly known as: Anser ruficollis

English: Red-breasted Goose; *French:* Bernache à cou roux; *German:* Rothalsgans; *Italian:* Oca dal collo rosso; *Russian:* Krasnozobaya Kazarka; *Spanish:* Barnacla cuellirroja

Range: W. Siberia, S. Russia, S. E. Europe
Identification:
Length 53-55cm. Sexes similar. Crown, face, throat and hindneck black. Between eyes and bill a white patch. Rest of head, neck and forebreast chestnut with white border round ear region, and lateral white line between black hindneck and chestnut foreneck. Fore body encircled by a white band. Upperparts black with two narrow white wing bars. Underparts black, forebreast chestnut. Band on upper flanks, belly and under tail-coverts white.
Young birds: A little duller, ear area often whitish.
Bare parts: Bill and legs black. Eyes dark brown.
List: II
Lit.: 37, 129, 146, 176, 297, 410, 428

PLATE 8

1. DENDROCYGNA BICOLOR

Formerly known as: Dendrocygna fulva

Brazilian: Marreca-caneleira; *English:* Fulvous Whistling Duck, Fulvous Tree Duck; *French:* Dendrocygne fauve; *German:* Gelbbrustpfeifgans; *Malagasy:* Tahia, Tsoea; *Spanish:* Pato silbon, Yaguaso Colorado, Suirirí bicolor

Range: USA, Northern S. America, Hawaiian Islands, Africa, Madagascar, India, Burma
Identification:
Length 45-53cm. Sexes alike. Bright fulvous with dark brown upperparts and wings, mantle with rufous feather fringes. Upper tail-coverts white. Vent and under tail-coverts creamy-white. Flanks with elongated creamy-buff feathers with black outer webs. Sides of neck paler, almost whitish, forming "striations". Tail black.
Young birds: Duller and greyer, flank pattern reduced, upper tail-coverts greyish.
Bare parts: Bill and legs dark grey. Eyes dark brown.
List: III Ghana, Honduras
Lit.: 20, 60, 129, 176, 192, 297, 410, 428, 431

2. DENDROCYGNA VIDUATA

Brazilian: Irere; *English:* White-faced Whistling Duck, White-faced Tree Duck; *French:* Dendrocygne veuf; *German:* Witwenpfeifgans; *Malagasy:* Tsiriry, Vivy; *Spanish:* Yaguaso cariblanco, Suirirí cariblanco

Range: S. America, Africa, Madagascar
Identification:
Length 43-48cm. Sexes alike. Front of head and upper throat white. Nape and upper neck black. Upperparts pale brown with darker feather centres. Wings dark, forewing chestnut. Underparts and tail black. Foreneck and upper breast chestnut. Flanks finely barred black and buff.
Young birds: Duller, head greyish instead of white.
Bare parts: Bill black, a grey band near tip. Legs greyish-black. Eyes brown.
List: III Ghana
Lit.: 60, 129, 176, 192, 297, 410, 428, 431

3. DENDROCYGNA ARBOREA

English: West Indian Whistling-Duck, Black-billed Whistling Duck, West Indian Tree Duck; *French:* Dendrocygne à bec noir; *German:* Kubapfeifgans; *Italian:* Anatra arborea, Dendrocigna di Cuba; *Spanish:* Pato silbón de Cuba, Suirirí yaguaza

Range: West Indies
Identification:
Length 48-58cm. Sexes similar. Forecrown chestnut, top of head and nape blackish, rest of head and neck buffish white with fine streaking at side of neck. Upperparts blackish-brown, feathers with rufous brown edges. Underparts spotted black and white. Breast rufous-brown with dark feather centres. Tail blackish.
Young birds: Duller, less streaked on head and neck. Black spots on underparts less developed.
Bare parts: Bill and legs blackish. Eyes dark brown.
List: II
Lit.: 37, 60, 129, 176, 297, 335, 410, 428

4. DENDROCYGNA AUTUMNALIS

English: Black-bellied Whistling Duck, Red-billed Tree Duck; *French:* Dendrocygne à bec rouge; *German:* Rotschnabel-Pfeifgans, *Spanish:* Guirirí, Suirirí piquirrojo

Range: Texas to S. America, Trinidad
Identification:
Length 48-53cm. Sexes alike. Top of head, nape, and lower neck chestnut. Rest of head and upperneck pale buffish grey. Eye-ring white. Upperparts chestnut-brown. Wings blackish with a large white area. Underparts and tail black, breast chestnut.
Young birds: Duller.
Bare parts: Bill reddish-pink. Legs bright pink; juvenile yellowish-grey to blue-grey bill and legs. Eyes dark brown.

Geographical variation:
Birds from Panama to Ecuador and Argentina have a varying amounts of grey on lower breast.
List: III Honduras
Lit.: 37, 60, 129, 176, 297, 410, 428

5. ALOPOCHEN AEGYPTIACUS

Also known as: Alopochen aegypticus; Alopochen aegyptica

English: Egyptian Goose, Nile Goose; *French:* Ouette d'Egypte, Oie d'Egypte; *German:* Nilgans; *Italian:* Oca egiziana; *Spanish:* Oca del Nilo, Ganso del Nilo

Range: Africa south of Sahara
Identification:
A. Length 63-73cm. Sexes alike. Head and neck pale buff, around eye, base of bill, and ring on neck dark brown.Some brown mottling on crown and neck. Upperparts greyish brown, brown to red-brown. Back, rump and tail black. Wings black, white and metallic green. Underparts buffy to greyish-buff, belly and under tail-coverts whitish. Patch on lower breast dark brown. Note, birds very variable!
C. Young birds: Duller, lacking dark brown eye and breast spot.Crown and nape dusky-brown.
Bare parts: Bill pink, tip dusky. Legs pink; juveniles have bill and legs yellowish-grey. Eyes pale yellow, sometimes brown.
List: III Ghana
Lit.: 37, 60, 129, 176, 297, 410, 428

6. PLECTROPTERUS GAMBENSIS

Formerly known as: Anas gambensis

English: Spur-winged Goose; *French:* Plectroptère de Gambia, Canard armé, Oie-armée Gambie; *German:* Sporengans; *Italian:* Oca dallo sperone; *Spanish:* Ganso del Gambia, Ganso espolonado

Range: Gambia to Sudan and south to South Africa
Identification:
Length 75-100cm. Male: Large knob on bill. Plumage black glossed metallic green and bronze, except for white behind eye, throat, area on wings, breast, lower flanks and remainder underparts. Wings with prominent spur.
Female: Smaller, less iridescent and white area on forewing less extensive. Knob smaller or absent.
Young birds: Duller, white areas buffish brown. No bare skin on head.
Bare parts: Bill, knob, and legs fleshy red. Eyes dark brown. Bare skin in front of eyes and area on side of neck fleshy red.

Geographical variation:
Birds from southern Africa are a little smaller and have less white on face and underparts. Bare skin on head bluish-grey.
List: III Ghana
Lit.: 20, 37, 129, 176, 410, 428

PLATE 9

1. CAIRINA MOSCHATA

Brazilian: Pato-do-mato; *English:* Muscovy Duck, Musk Duck; *French:* Canard-musqué d'Amérique; *German*: Moschusente; *Spanish*: Pato real, Pato criollo

Range: Mexico to Peru and Uruguay to Argentina
Identification:
Length male 76-84cm, female 58-63cm. **Male:** Prominent crest on head and nape. Entire bird brownish-black, above glossed with green and purple except white wing patches. Under wing-coverts also white.
Female: Duller than male.
Young birds: Little or no white area on wings. Less iridescent.
Bare parts: Bill and knob blackish patterned pinkish white. Legs greyish-black. Eyes yellowish brown. Skin around eye black, seldom reddish (red more typical of domesticated birds) with small fleshy wattles, females have less naked skin and with few or no fleshy wattles.
List: III Honduras
Lit.: 13, 60, 129, 176, 192, 223, 297, 410, 428, 431

2. CAIRINA SCUTULATA
Also known as: Asarcornus scutulata

English: White-winged Duck, White-winged Wood Duck; *French:* Canard-musqué à ailes blanches, Canard à ailes blanches; *German:* Weissflügelente; *Italian:* Anatra della Malesia, *Spanish:* Pato almizclero aliblanco, Pato de Jungla

Range: Assam and Thailand to Sumatra and Java
Identification:
A. Length 66-81cm. Sexes alike. Head and neck white spotted dark grey.Upperparts blackish-brown, glossed green.Wings dark brown with white, black, and greyish-blue areas. Underparts dark brown tinged chestnut, breast glossy. Tail dark brown.
Young birds: Duller, head and neck pale brown.
Bare parts: Bill orange spotted dusky-grey. Legs orange-yellow. Eyes yellow-orange to red, female brownish.

Geographical variation:
D. A variant from Assam and Sumatra has entire forepart white.
List: I
Lit.: 2, 37, 60, 129, 146, 149, 175, 297, 410, 428

3. SARKIDIORNIS M. MELANOTOS
Also known as: Sarkidiornis melanota

Brazilian: Pato-de-crista; *Burmese:* Mauk-tin; *English:* Comb Duck, Knob-billed Goose; *French:* Canard-à-bosse bronzé, Sarcidiorne à crète; *German:* Glanzente, Höckerglanzente; *Italian*: Sarcidiorne dalla cresta, *Spanish:* Pato crestudo, Pato de moco

Range: Africa, India, China
Identification:
A. Length 56-76cm. **Male:** Large fleshy comb diagnostic. Head and neck white spotted black, often with a golden yellow tinge to sides of head and neck. Upperparts, tail and wings black strongly glossed blue-green or bronze. Underparts white, flanks pale grey. Black narrow bands on sides of upper breast and in front of under tail-coverts, which are often pale yellow.
Female: Smaller, less glossy. Underparts often with some brownish mottling and ill-defined black breast bands. Lacks comb.
C. Young birds: Crown and upperparts dark brown. Rest of head, neck and underparts buffish-brown, with scaling below. Dark eye-stripe.
Bare parts: Bill black, comb dark grey. Legs dark grey. Eyes dark brown.

Subspecies:

D. Sarkidiornis melanotos sylvicola
English: American Comb-Duck
Range: Panama, S. America east of the Andes, Trinidad
A little smaller. Males with glossy black flanks. Females with dark grey flanks.
List: II
Lit.: 13, 37, 60, 129, 176, 192, 276, 297, 410, 428, 431

4. PTERONETTA HARTLAUBII
Formerly known as: Cairina hartlaubi; Querquedula hartlaubi

English: Hartlaub's Duck; *French:* Ptéronette de Hartlaub, Canard de Hartlaub; *German:* Hartlaubente; *Italian:* Anatra di Hartlaub; *Spanish:* Pato de Hartlaub

Range: Liberia to Zaire and Sudan
Identification:
A. Length 56-58cm. **Male:** Head and neck black except for white forehead. Body chestnut, dark olive-brown on shoulders, rump and tail. Wings with a blue-grey area.
Female: Smaller and duller, often without white forehead.
Young birds: Duller than female, feathers pale-edged.
Bare parts: Male bill: Black with pale subterminal spot, swollen at base in breeding. Female: Black with paler pinkish spots and bands. Legs yellowish brown, webs dusky. Eyes reddish-brown.

Geographical variation:
D. Amount of white on forehead variable, most frequent in birds from Zaire.
List: III Ghana
*Lit.:*20, 37, 60, 129, 176, 297, 410, 428

5. NETTAPUS AURITUS
Formerly known as: Anas aurita

English: African Pygmy Goose; *French:* Anserelle naine, Sarcelle à oreillons; *German:* Rotbrust-Zwerggans, Afrikanische Zwergglanzente; *Italian:* Oca pigmea; *Spanish:* Patito africano, Gansito africano

Range: Gambia to Kenya and Cape Province, Madagascar
Identification:
A. Length 30-33cm. Male: Forehead, sides of head, throat and foreneck white. Crown, nape and hindneck iridescent greenish black. Oval greenish area on sides of upper neck bordered black. Above glossy dark green. Wings black with white area. Below rufous-chestnut, centre of breast and belly white. Tail blackish.
B. Female: Duller. Lacks oval greenish patch on neck. Dark eye-stripe.
Young birds: Like female but less distinctly marked except more distinct eye-stripe.
Bare parts: Bill bright yellow, tip black, females lack black tip. Legs dark grey. Eyes brown to red.
List: III Ghana
Lit.: 20, 37, 60, 129, 176, 297, 410, 428

PLATE 10

1. ANAS ACUTA

Dutch: Pijlstaart; *English:* Northern Pintail; *French:* Canard pilet; *German:* Spiessente; *Italian:* Godone; *Spanish:* Pato rabudo, Anade rabudo

Range: Eurasia, China, N. W. Africa, N. America
Identification:
A. Length 51-76cm. **Male:** Chocolate-brown head, hindneck and foreneck. Sides of neck, breast and belly white. Back and flanks vermiculated grey. Shoulder feathers elongated black and white. Speculum green. Long tail feathers and rump black.
Male non breeding: Like female but greyer without buff edges to feathers.
B. **Female:** Distinguished from other female ducks by long neck and brown speculum.
Young birds: Darker above than female. Underparts more heavily streaked. Tail feathers notched at tip.
Bare parts: Bill grey-blue, juveniles dark grey. Legs grey with dusky webs, female green-grey. Eyes yellow-brown.
List: III Ghana
Lit.: 51, 129, 297, 410, 428

2. ANAS CAPENSIS

Dutch: Kaapse Taling; *English:* Cape Teal; *French:* Sarcelle du Cap; *German:* Kaplöffelente, Fahlente; *Spanish:* Pardilla del Cabo, Cerceta del Cabo

Range: Ethiopia to Botswana, Cape Province
Identification:
Length 44-48cm. **Male:** Whole head and underparts greyish with dark brown spots. Throat white. Upperparts dark brown with red-buff feather edgings. Speculum green. Note: males have only one plumage.
Female: Like male but fewer and broader spots on breast.
Young birds: Spots on upperparts smaller and paler buff. Tail feathers notched at tip.
Bare parts: Bill pink tinged lilac, light blue at tip, juvenile paler. Legs buffish. Eyes variably brown, yellow, red or orange.
List: III Ghana
Lit.: 20, 51, 129, 297, 410, 428

3. ANAS CLYPEATA

Formerly known as: Spatula clypeata

Dutch: Slobeend; *English:* Common Shoveler, Northern Shoveler; *French:* Canard suchet; *German:* Löffelente; *Italian:* Mestolone; *Spanish:* Pato cuchara,Cuchara común

Range: Eurasia, India, China, Africa, N. America
Identification:
A. Length 44-52cm. Enormous spatulate bill. **Male:** Head dark green. Neck, breast, shoulder and stripe between belly and under tail-coverts white. Shoulder feathers elongated and pale blue, dark green, black and white. Forewing light blue. Speculum green. Belly and flanks chestnut.
Male non breeding: Like female but darker.
B. **Female:** Like other female dabbling ducks except bill and paler spots and feathers edged cinnamon-pink to pink-buff. Bluish forewing and green speculum. Tail whitish.
Young birds: Like female but upperparts more uniform and underparts more streaked. Tail feathers notched at tip.
Bare parts: Bill blackish, female and juvenile dark olive-brown. Legs orange-red, juvenile yellow-orange. Eyes orange, female and juvenile brown.
List: III Ghana
Lit.: 51, 129, 336, 410, 428

4. ANAS C. CRECCA

Dutch: Wintertaling; *English:* Common Teal, Teal; *French:* Sarcelle d'hiver; *German:* Krickente; *Italian:* Alzavola; *Spanish:* Cerceta común

Range: Eurasia, India, China, Africa, Aleutian Islands
Identification:
A. Length 34-38cm. **Male:** Head chestnut with broad green area from eye to nape encircled by a buff line. Upperparts and flanks grey vermiculated cream. Speculum metallic green. White horizontal line above wing. Breast creamy with dark spots. Belly white. Under tail-coverts black, sides cream-buff. Tail dark grey.
Male non-breeding: Like female but crown and hindneck duller. Light line over eye indistinct. Variable.
B. **Female:** Drab grey-brown with light eye-line. Speculum green. Buff triangular patch on dark under tail-coverts.
Young birds: Like female but tail feathers with notched tip.
Bare parts: Bill dark slate, juvenile pink-horn. Legs olive-grey to brown-grey. Eyes dark hazel.

Subspecies:

Anas crecca carolinensis
English: Green-winged Teal
Range: N. & C. America, West Indies
Male has a broad vertical white band on side of breast.
List: III Ghana
Lit.: 59, 70, 129, 197, 297, 410, 428

5. ANAS PENELOPE

Dutch: Smient; *English:* Eurasian Wigeon, European Wigeon, *French:* Canard siffleur; *German:* Pfeifente; *Italian:* Fischione; *Spanish:* Anade silbón, Silbón europeo

Range: Eurasia, India, Japan, Africa
Identification:
A. Length 45-51cm. **Male:** Pale yellow forehead, rest of head chestnut with a green stripe behind eye. Sometimes a metallic dark green spot behind eyes. Hindneck and forebreast pink-brown. Upperparts and flanks vermiculated grey. Elongated shoulder feathers grey. Speculum dark green. Forewing and belly white.
Male non breeding: Like female but forewing white and forecrown with large rounded black spots. Variable.
B. **Female:** Head and neck pink-brown with black spots and streaks. Mantle brown with pink-buff bars. Forewing grey-brown, speculum blackish-green. Forebreast and flanks brown with pink-buff bars. Belly white. Variable.
Young birds: Like female but speculum without green (female) or only with a slight green tinge towards base (male).
Bare parts: Bill slate-blue, female darker. Legs pale blue-grey to olive-grey. Eyes yellow brown to dark brown.
List: III Ghana
Lit.: 51, 129, 336, 410, 428

PLATE 11

1. ANAS BERNIERI

Also known as: Querquedula bernieri

English: Madagascar Teal, Bernier's Teal; *French:* Sarcelle de Bernier,Sarcelle de Madagascar; *German:* Bernierente; *Italian:* Anatra de Bernier; *Spanish:* Cerceta de Madagascar, Cerceta malgache, Pato Bernier

Range: Madagascar
Identification:
Length 40cm. Sexes alike. Almost entire plumage light reddish brown. Crown and nape darker. Chin and throat buff. Feathers of upperparts edged buffish brown. Wings blackish, mirror blackish between two white bands. Underparts with darker but indistinct feather centres, particularly on flanks.
Young birds: Undescribed.
Bare parts: Bill and legs reddish, female said to be duller brown. Eyes brown.
List: II
Status: Population very small, probably less than 200.
Lit.: 37, 60, 129, 162, 176, 196, 410, 428

2. ANAS A. AUCKLANDICA

Formerly known as: Nesonetta aucklandica

English: Brown Teal, New Zealand Teal, Auckland Island Teal, Flightless Teal; *French:* Sarcelle brune, Sarcelle terrestre des îles Auckland; *German:* Aucklandente, Auckland-Kastanienente; *Italian:* Anatra delle Auckland; *Spanish:* Cerceta alicorta de Auckland, Cerceta maori

Range: Auckland Island (New Zealand)
Identification:
A. Length 38cm. Flightless. **Male:** Head and neck dark brown glossed green. Upperparts and wing-coverts dark brown, feathers edged paler brown. Wings blackish tinged green. Mirror green and brown, hindmost edged white. Underparts deep chestnut, sides of belly grey. Flanks with buff vermiculation.
Male non breeding: Like female.
B. **Female:** Like male but head without green gloss. Underparts duller and browner with black spots, lack grey on sides of belly.
Young birds: Resembles female but underparts with more black spots.
Bare parts: Bill bluish-grey. Legs dark grey. Eyes dark brown.

Subspecies:

4. ANAS AUCKLANDICA CHLOROTIS

English: New Zealand Brown Teal, New Zealand Brown Duck; *French:* Sarcelle de la Nouvelle-Zélande; *German:* Neuseelandente, Grünohrente; *Italian:* Anatra della Nouva Zelanda; *Spanish:* Cerceta alicorta neozelandesa
Range: New Zealand
Length 43-48cm. Like Anas a. aucklandica but with a narrow white eye-ring, also often with a white collar on lower neck. Sides of belly white, not grey. Flight-feathers longer.
List: II

3. ANAS AUCKLANDICA NESIOTIS

English: Flightless Teal
Range: Campbell Island
Length 36cm. Flight-feathers shorter and therefore flightless. Darker and lacking any vermiculation below or a distinct mirror.
List: I
Status: Population on Campbell Island extinct, but on nearby islet there were 30-50 birds in 1975.
Lit.: 60, 69, 129, 146, 178, 195, 297, 302, 410, 428
(Sibley and Monroe have this race included in Anas a. aucklandica)

5. ANAS LAYSANENSIS

Also known as: Anas platyrhynchos laysanensis

English: Laysan Duck, Laysan Mallard, Laysan Teal; *French:* Canard de Laysan; *German:* Laysanente; *Italian:* Germano reale di Laysan; *Spanish:* Pato real de Laysan

Range: Laysan Island (resident)
Identification:
A. Length 35-41cm. **Male:** Perhaps only a subspecies of Mallard. Head and neck very dark brown, sometimes with green gloss, and marked irregularly with white. Body dark reddish-brown, each feather with dark brown subterminal markings. Wings brown with green mirror bordered black and white behind. Underwing whitish.
B. **Female:** Mirror dull dark brown.
Young birds: Like female.
Bare parts: Bill dull green, blackish on upperside and tip, female brownish-yellow. Legs orange. Eyes brown.
List: I
Status: 69 birds count in 1974.
Lit.: 60, 176, 222, 227, 250, 297, 318, 410, 428

ANAS OUSTALETI

Also known as: Anas p. platyrhynchos; Anas platyrhynchos oustalleti; Anas superciliosa pelewensis

English: Marianas Duck, Mariana Mallard, Oustalet's Duck; *French:* Canard d'Oustalet; *German:* Marianen-Stockente; *Italian:* Anatra di Oustalet, Germano reale delle Marianne, *Spanish:* Pato real marismeño

Range: S. Mariana Islands (USA) (resident)
Identification:
Two types occur, both probably of hybrid origin:

6. PLATYRHYNCHOS TYPE:

A. Length 52cm. **Male:** Head mostly dark green with buff mottling and faint neck-ring. Upperparts dark brown edged paler. Mirror purple-blue bordered black and white. Breast reddish-chestnut, strongly mottled brown. Rest of underparts brownish edged lighter. Tail grey and white. Tail- coverts black, central feathers on upperside slightly curled up.
Male non breeding: Like female.
B. **Female:** Like Mallard female.
Young birds: Like female.
Bare parts: Bill olive with extensive blackish over basal portion. Legs orange, webs darker. Eyes dark brown.

7. SUPERCILLIOSA TYPE:

Length 52cm. **Male:** Like platyrhynchos type, but head and upper neck buffish with dark brown stripes on top of head and nape, through eye and from bill to cheek. Mirror green bordered black and buff. Underparts dark brown edged paler, broadest on upper breast and flanks, Tail dark brown. Note: males have no non breeding dress!
Female: Like Mallard female.
Young birds: Like female.
Bare parts: Bill olive, tip and narrow stripe on upperside black. Legs orange. Eyes dark brown.
List: I
Status: May be extinct.
Note: Sibley and Monroe listed only Anas superciliosa, Pacific Black Duck, and regard Anas oustaleti as probably a hybrid between Anas platyrhynchos and Anas superciliosa.
Lit.: 5, 37, 60, 146. 176, 178, 179, 195, 329, 380, 410

PLATE 12

1. ANAS QUERQUEDULA

Dutch: Zomertaling; *English:* Garganey; *French:* Sarcelle d'été; *German:* Knäckente; *Italian:* Marzaiola; *Spanish*: Cerceta carretona

Range: W. Europe to Japan, India, Indonesia, Africa
Identification:
A. Length 37-41cm. **Male:** Upper head black-brown; rest of head and neck dark chestnut flecked with fine whitish streaks and a conspicuous broad white stripe from before and above eye to nape. Upperside black brown edged pale buff to olive-grey. Forewing pale blue-grey, shoulder feathers elongated and pointed black and white. Wings dark grey-brown, mirror dark green, black and white. Underside white, flanks finely vermiculated black. Breast chestnut with black scale-like spotting. Belly and under tail-coverts buffish with dark spots. Tail dark brown edged buff.
Male non breeding: Like female but a little darker.
B. **Female:** Two pale buffish bands, one over eye and one below eye. Plumage otherwise dark brown edged pale to dull buffish. Centre of underside white. Mirror dark grey often with a green gloss, edged white.
Young birds: Like female but buff feather-edges narrower.
Bare parts: Bill dark grey. Legs grey, webs darker grey, juvenile until autumn yellow-grey to brown-grey. Eyes dark brown, juvenile grey-brown to brown.
List: III Ghana
Lit.: 51, 60, 112, 129,176, 297, 410, 428

2. RHODONESSA CARYOPHYLLACEA

English: Pink-headed Duck; *French:* Nette à cou rose, Canard à tête rose; *German:* Nelkenente, Rosenkopfente; *Italian:* Anatra dalla testa rosa; *Spanish:* Pato cabecirrosa

Range: India, Burma
Identification:
A. Length 60cm. **Male:** Head and neck bright pink, slightly tufted on nape. A brownish black band runs from chin to breast. Body brownish black finely lined pale pink on scapulars, mantle, and breast. Mirror pinkish buff edged white. Underwing pale pink.
B. **Female:** Duller, rosy-white tinged dusky where male is bright pinkish. Without brownish black front neck. Mirror pale brownish buff.
Young birds: Paler. Head and neck pale rosy-white tinged grey, crown, nape and hindneck brown. Underside pale brown, feathers edged white.
Bare parts: Bill bright pink, female duller. Legs reddish black, female duller. Eyes red to orange, female brownish orange.
List: I
Status: Probably extinct. Last authentic sight record 1935. A few records from the sixties too doubtful to be believed.
Lit.: 1, 2, 28, 37, 60, 98, 112, 146, 176, 410

3. AYTHYA NYROCA

Formerly known as: Nyroca nyroca

Dutch: Witoogeend; *English:* Ferruginous Pochard, Ferruginous Duck, White-eyed Pochard; *French:* Fuligule nyroca; *German:* Moorente; *Italian:* Moretta tabaccata; *Spanish:* Porrón pardo

Range: Europe to China, Iran, N. Africa
Identification:
A. Length 38-42cm. **Male:** Head and neck rich chestnut-red. Small white spot on chin. Narrow black band round lower neck. Upperparts black finely spotted chestnut. Wings blackish brown with a broad wide band visible in flight. Underwing whitish. Underparts white, on centre of belly mottled grey-brown; breast and flanks rich chestnut-red. Tail dark grey-brown.
Male non breeding: Like female except for eye colour.
B. **Female:** Like male but duller with some white mottling on throat.
Young birds: Resembles female but paler and more buffish.
Bare parts: Bill grey, tip black. Legs olive-grey, joints and webs black. Eyes white, female and juvenile brown.
List: III Ghana
Lit.: 51, 60, 112, 129, 176, 297, 410, 428

4. OXYURA LEUCOCEPHALA

Formerly known as: Anas leucocephala

English: White-headed Duck, White-headed Stifftail; *French:* Erismature à tête blanche; *German*: Weisskopf-Ruderente; *Italian*: Gobbo rugginoso; *Spanish:* Malvasía, Malvasía cabeciblanca

Range: Mediterranean to C. Asia, India, Egypt
Identification:
A. Length 43-48cm. **Male:** White large head with dark crown and swollen-based bill distinctive. Upper neck black shading into chestnut. Upperparts finely barred grey-brown, uppertail-coverts chestnut. Wings grey-brown without any mirror. Underparts grey-brown with buffish wash. Breast deep chestnut, flanks paler chestnut finely barred black. Stiff tail black.
Male non breeding: More dusky areas in white head. Neck and breast greyish-buff.
B. **Female:** Crown, cheeks, and nape blackish-brown with rufous feather tips. Rest of head whitish. Note: also female and juvenile have swollen bill. Otherwise like male non breeding.
Young birds: Like female, but upperparts darker and underparts more buffish. Tips of tail feathers spiked.
Bare parts: Bill bright blue; in non breeding male, female and juvenile greyish. Legs greyish, sometimes reddish, juvenile pale cinnamon, webs blackish. Eyes yellow to orange-yellow, juvenile pale grey to yellow-brown.
List: II
Lit.: 37, 51, 60, 129, 176, 297, 410, 428

PLATE 13

1. GYMNOGYPS CALIFORNIANUS

English: California Condor; *French:* Condor de Californie; *German:* Kalifornischer Kondor; *Italian:* Condor della California; *Spanish:* Cóndor de California

Range: California
Identification:
Length 109-146cm. Sexes alike. Head, neck and crop area bare except for few short bristly black feathers on forehead, base of bill, and in front of eyes. Rest of plumage blackish, on back edged brown. Round neck a ruff of long pointed feathers with grey shafts. Wings black with greyish to white edges of inner flight-feathers and greater wing-coverts. Underwing-coverts white.
Young birds: Head and neck greyish. Little or no white or greyish on wings. Underwing white mottled with brown or quite black.
Bare parts: Bill whitish. Legs pink. Eyes red. Naked head skin orange, sometimes greyish-yellow to yellow. On lower side of neck a purplish-red area.
List: I
Status: Wild population extinct. 27 birds in captivity (1987)
Lit.: 19, 37, 101, 129, 196, 285, 313, 410

2. VULTUR GRYPHUS

English: Andean Condor; *French:* Condor des Andes; *German:* Kondor, Andenkondor; *Italian:* Condor delle Ande; *Spanish:* Cóndor Andino

Range: Andes from Venezuela to Tierra del Fuego
Identification:
A. Length 109-146cm. **Male:** Head and neck bare except for few bristles. Large fleshy "comb" on crown is characteristic. White fluffy down around lower neck, and a large white area on wings. Rest glossy black.
Female: Like male but smaller and lacks a comb.
C. Young birds: Dusky brown and without comb.
Bare parts: Bill ivory, base greyish-black. Legs greyish to black. Eyes greyish-brown, female red. Naked head and neck red to blackish-red.
List: I
Lit.: 13, 19, 37, 101, 129, 313, 410

3. SARCORAMPHUS PAPA

English: King Vulture; *French:* Sarcoramphe roi; *German:* Königsgeiher; *Spanish:* Rey zamuro

Range: Mexico to Argentina, Trinidad
Identification:
Length 76-81cm. Sexes similar. Sparse black bristle-like feathers on crown, ear region, and around eye. At base of bill a large wattle. Grey feathers round base of neck. Wings, rump, and tail black. Rest white.
Young birds: Sooty black-brown, base of feathers below white. Look like young California Condors but much smaller, average 69cm.
Bare parts: Bill orange to red, juvenile black with some red at tip. Legs greyish, juvenile blackish. Eyes white, juvenile yellow-grey. Large wattle at base of bill orange. Bare skin of head and neck yellow, orange, blue and purple.
List: III Honduras
Lit.: 13, 19, 101, 129, 313, 410

4. AQUILA HELIACA

Also known as: Aquila h. heliaca

Dutch: Keizerarend; *English:* Imperial Eagle; *French:* Aigle impérial; *German:* Kaiseradler; *Italian:* Aquila imperiale; *Spanish:* Aguila imperial

Range: Greece to Siberia, N. E. Africa, India, China
Identification:
A. Length 79-84cm. Sexes alike but female larger. Crown, sides of head and neck with lanceolate pale golden or rarely silvery feathers. Rest of plumage black-brown. Base of tail grey-brown. Shoulders with white patches.
Young birds: Paler yellow-brown and streaked. Tail and most wing-feathers black, wings with conspicuous buff bands. No white on shoulders.
Bare parts: Bill bluish, tip black. Cere and gape pale yellow to yellow-green. Legs yellow. Eyes yellow-grey to orange-brown. Young birds grey washed brown.
List: I
Lit.: 19, 51, 93, 101, 129, 195, 313, 410, 423

D. AQUILA ADALBERTI
Also known as: Aquila heliaca adalberti
Range: Spain
Identification:
Like previous but white patches on shoulders larger and often also with white upper wing-coverts.
List: I
Lit.: 19, 51, 93, 101, 129, 195, 313, 410, 423

5. CHONDROHIERAX U. UNCINATUS

Brazilian: Caracoleiro; *English:* Hook-billed Kite; *French:* Milan bec-en-croc; *German:* Langschnabelweih

Range: Mexico to Argentina, Trinidad
Identification:
Length 38-42cm. Sexes alike but plumage variable.
D. Grey-phase: Bluish-slate, underparts paler grey barred white or buff. Feathers on hind neck with white base of feathers. Rump spotted and barred white. Wings barred black, underside of wings barred black and white. Tail blackish, outermost tip and two broad bands white or grey.
E. Black phase: Leaden black. Tail with white tip and one broad white bar.
F. Brown phase: Head and upperparts bluish-grey. Hind neck rufous, lower back brown. Underparts white with cinnamon bars. Wings brown. Tail black, tipped white with two broad white or pale grey bars.
Young birds: Upperparts dark brown, hind neck white. Underparts creamy-white. Wings brownish-black edged rufous or white. Tail with three or four pale bands.
Bare parts: Bill black, lower bill, cere and facial skin greenish-yellow. Legs orange-yellow. Eyes white.
List: II

Subspecies:

Chondrohierax uncinatus wilsonii
English: Cuban Kite
Range: Cuba
Smaller and bill entirely yellow.
List: I
Lit.: 13, 19, 101, 129, 196, 313, 410, 431

PLATE 14

1. HALIAEETUS ALBICILLA

Dutch: Zeearend; *English:* White-tailed Sea Eagle; *French:* Pygargue à queue blanche, Pygargue commun; *German:* Seeadler; *Italian:* Aquila di mare; *Japanese:* Ojiirowashi; *Russian:* Orlan-belokhvost; *Spanish:* Pigargo coliblanco; *Swedish:* Havsörn

Range: Greenland, Europe, N. Asia, India, China, Japan
Identification:
A. Length 79-92cm. Sexes similar but female slightly larger. Head dark brown with lanceolate feathers tipped buffish-white on crown, nape, and neck; paler in worn plumage. Black bristles between bill and eye. Mantle, scapulars, and wing-coverts dark brown with purple gloss, contrasting to much paler head and neck. In worn plumage paler. Rump and short upper tail-coverts dark brown. Flight-feathers dark brown. Underparts dull brown, chin and throat buff, breast feathers paler with dark brown shaft-streaks. In worn plumage feathers on underparts are tipped buff-white. Tail short, wedge-shaped, and white, diagnostic for adult birds.
C. Young birds: Darker brown, on crown and nape tipped pale rufous or brown, bases white. Upperparts and wings dark brown mottled pale rufous-brown. Long upper tail-coverts mottled dark brown and white. Tail whitish strongly mottled dark brown, tip and outer webs of outer tail-feathers wholly black-brown. Requires at least 5 years to reach full adult plumage.
Bare parts: Huge bill and cere yellow, juvenile brown-black. Legs unfeathered yellow, juvenile green-yellow to grey-yellow. Eyes yellow, juvenile black-brown, later yellow-brown.
List: I
Lit.: 19, 26, 30, 37, 551, 65, 86, 89, 101, 119, 129, 217, 256, 410

2. HALIAEETUS LEUCOCEPHALUS

English: American Bald Eagle, Bald Eagle; *French:* Pygargue à tête blanche, Aigle à tête blanche; *German:* Weisskopf-Seeadler; *Italian:* Aquila di mare a testa bianca; *Spanish:* Pigargo cabeciblanco, Aguila cabeza blanca

Range: N. America
Identification:
A. Length 76-109cm. Sexes alike but female slightly larger. Head and neck white. Body and wings dark brown, most feathers above edged buffish. Tail and its coverts white.
C. Young birds: Head and neck dark brown, body and wings also dark brown but less black than in adults; rarely blotched with white or buffy. Underwings and tail have blotchy buffy-white patterns, tail otherwise dark brown. Requires 4-5 years to reach full adult plumage. Highly variable.
Bare parts: Bill and cere bright yellow, juvenile brownish horn. Legs unfeathered, yellow. Eyes bright yellow, juvenile pale yellowish-grey.

Geographical variation:
Birds from Alaska and W. Canada are larger.
List: I
Lit.: 19, 31, 37, 70, 101, 129, 197, 217, 273, 300, 410

3. HARPIA HARPYJA

Brazilian: Gravião-real; *English:* Harpy Eagle; *French:* Aigle harpie; *German:* Harpyie; *Spanish:* Aguila harpía

Range: Mexico to Bolivia and Argentina
Identification:
Length 86-100cm. Sexes alike. Head ashy grey with bushy divided blackish crest edged paler. Neck, upperparts, wings and breast black indistinctly edged grey. Bend of wings and underparts white, thighs barred black. Tail black with three bars and tips pale grey.
Young birds: Head, breast, and underparts white. Upperparts incl. divided crest brownish-grey to pale grey, crest tipped white. Tail marbled and barred black and dusky.
Bare parts: Bill and cere black, bill tinged blue. Legs yellow. Eyes light brown, mottled grey, blackish-brown or orange, juvenile dark brown.
List: I
Lit.: 13, 19, 31, 101, 125, 129, 192, 223, 238, 410, 431

4. PITHECOPHAGA JEFFERYI

English: Philippine Eagle, Great Philippine Eagle; *French:* Aigle des singes; *German:* Affenadler

Range: Luzon, Mindanao, Samar (Philippines)
Identification:
A. Length 91cm. Sexes similar. Feathers on crown and nape lanceolate, buff with dark brown shaft streaks forming a conspicuous crest. Upperparts and wings brown edged paler. Chin white finely streaked black, rest of underparts whitish-grey, thighs buffy streaked rufous. Tail dark brown with 4-5 paler bands, shafts white.
C. Young birds: Lack hair-like black streaks on chin and thighs uniformly buff.
Bare parts: Bill bluish, cere greenish-blue. Legs dull yellow. Eyes blue-grey.
List: I
Status: Estimated to be less than 200 individuals in the wild with 13 in captivity (1992)
Lit.: 19, 31, 61, 66, 101, 129, 148, 330, 410

PLATE 15

1. FALCO ARAEA

Créole: Katiti; *English:* Seychelles Kestrel; *French:* Emouchel des Seychelles, Crécerelle katitie;*German:* Seychellen Turmfalke;*Italian:* Gheppio delle Seychelles; *Spanish:*Cernícalo de Seychelles.

Range: Seychelles
Identification:
Length 20-23cm (smallest member of the genus Falco). A.**Male**: Crown and hind neck slate grey. Side of head and neck greyish with a faint moustachial streak. Upperparts and wing-coverts dark chestnut spotted with black. Upper tail-coverts grey. Wings dark brown, outermost edged white, innermost edged white and chestnut. Under wing-coverts pinkish-buff. Underparts rufous-buff, throat and under tail-coverts paler.Tail grey, broad subterminal and three or four narrower bars black, tips white.
Female: Like male but paler.
C.Young birds: Crown chestnut with black streaks. Black spotting on upperparts heavier. Flight-feathers edged pale rufous. Breast and abdomen rufous-buff spotted black. Tail greyish-rufous with same black bars as adult but tipped buff.
Bare parts: Bill: no information available but slate-grey in all illustrations. Cere, skin round eyes, and legs yellow. Eyes dark brown.
List: I
Lit.: 19, 31, 33, 37, 101, 221, 410, 415

2. FALCO PUNCTATUS

*English:*Mauritius Kestrel;*French:* Emouchel de l'ile Maurice, Crécerelle de Maurice; *German:* Mauritius Turmfalke; *Italian:* Gheppio delle Maurizio; *Spanish:* Cernícalo de Mauricio

Range: Mauritius
Identification:
Length 23-26cm, female larger, sexes otherwise similar. Like no. 1 it looks very much like adult female Common Kestrel, also with moustachial streak. Crown, nape, and upperparts are chestnut, crown streaked black, back and wing-coverts chestnut barred black. Long flight-feathers blackish tipped buffish. Underparts whitish with distinct black and V-shaped spots, smaller on breast. Tail chestnut with a subterminal band and six or seven narrower black bands.
Young birds: Paler above and more streaky below
Bare parts: Bill not described but slate-grey in all illustrations. Cere, skin round eyes, and legs yellow. Eyes dark brown.
List: I
Status: Only four birds in 1974. Then one pair changed from nesting primarily in tree cavities, where introduced macaques preyed heavily on eggs and nestlings, to nesting in holes and niches on cliffs. This change in nesting-place and enormous work on the part of the WWF and ICBP, now called Birdlife International, resulted in about 66 birds in 1985!
Lit.: 33, 37, 75, 101, 196, 287, 295, 410

3. FALCO N. NEWTONI

Créole: Katiti;*English:* Madagascar Kestrel;*French:* Emouchel de Madagascar, Crécerelle Malgache; *German:* Malegassenturmfalke; *Italian:* Gheppio del Madagascar; *Malagasy:* Hitikitike, Hitsikitsike; *Spanish:* Cernícalo de Madagascar

Range: Madagascar
Identification:
A. Length 30cm. **Male**: Crown and nape greyish-rufous streaked black. Thin black moustache. Upperparts and wing-coverts chestnut, spotted black, upper tail-coverts grey, spotted black. Long flight-feathers blackish brown, inner webs with white and chestnut spots. Short flight-feathers chestnut with black bars. Underparts whitish; breast, belly and under wing-coverts spotted black. Tail grey, broad subterminal and six or seven narrower black bars, all feathers tipped white.
D. **Male red phase:** Crown and nape nearly black. Body and under wing-coverts dark chestnut with black streaks and spots, throat buffish-white and under tail-coverts greyish-white spotted black.
Female: Like male but with more chestnut on crown, underparts more heavily spotted and tail brown with black bars.
Young birds: Similar to female.
Bare parts: Bill slate-grey, tip blackish. Cere and legs yellow, in red phase bright orange. Eyes dark brown.
List: II

Subspecies:

3. FALCO NEWTONI ALDABRANUS

English: Aldabra Kestrel
Range: Aldabra Island, Anjouan Island (last Comoros)
Average smaller (wings male 170-183mm, female 177-186mm against male 180-195mm, female 188-203mm in Falco n. newtoni).
B. Some females have entire underparts white.
List: I
Status: About 100 birds in 1977.
Lit.: 19, 33, 37, 102, 129, 152, 162, 195, 222

4. FALCO JUGGER

English: Lagger Falcon; *French:* Faucon laggar; *German:* Laggerfalke; *Hindi:* Lagudu

Range: Baluchistan, Himalayas, India, Burma
Identification:
Length 43-46cm. Sexes alike but female larger. Crown and nape red-brown with black shaft streaks. Forehead, eyebrow, and sides of head white to pale rufous, with a distinct brown moustachial stripe. Upperparts dark brown to grey-brown on lower back, rump, and upper tail-coverts, each feather edged paler. Wings dark brown, edged paler and inner webs broadly barred on outermost feathers. Wing-coverts dark brown, each feather edged paler. Underparts white washed pinkish to pale fulvous with blackish streaks on abdomen, flanks and thighs; thighs brownish. Tail grey-brown barred and tipped buffish, except for central unbarred feathers.
Young birds: Upperparts dark brown edged paler. Underparts dark brown with some white mottling. Chin and throat white.
Bare parts: Bill bluish-slate, tip darker. Cere and legs yellow, juvenile greyish-green cere and legs pale grey, greenish-grey or dull slate. Eyes dark brown.
List: I
*Lit.:*2, 19, 31, 33, 37, 101, 410

PLATE 16

1. FALCO PEREGRINUS

Brazilian: Falcão-peregrino; *Dutch:* Slechtvalk; *English:* Peregrine Falcon; *French:* Faucon pèlerin; *German:* Wanderfalke; *Italian:* Falcone pellegrino; *Spanish:* Halcón viajero, Halcón real, Halcón común; *Swedish:* Pilgrimsfalk

Range: Worldwide
Identification:
Length 36-50cm. **Male:** Top of head and hindneck blackish tinged blue-grey, forehead paler. Sides of head black with a broad moustachial streak and a whitish area behind malar stripe. Upperparts and wing-coverts blue-grey barred with ill-defined darker bars and feathers with blackish shafts. Flight-feathers blackish edged narrowly whitish, shorter flight-feathers blue-grey with tooth-like dark bars on inner webs. Under wing-coverts white, narrowly barred black. Chin and throat white, rest of underparts whitish with a pale buff or creamy tinge and with fine black streaks and spots on breast, being more narrowly barred on rest of lower part. Tail blue-grey, paler at base and narrowly barred darker; tip of tail whitish.
B. **Female:** Like male but larger, darker above and more buffish to pale rufous below, barring often heavier.
C. Young birds: Like adult but browner on head and upperparts, most feathers with paler rufous edging. Forehead and area on sides of face creamy, last mentioned with brown shaft-streaks. Chin white, rest of underparts cream to rufous-buff, heavily marked with dark brown streaks and spots forming bars on flanks; under tail-coverts have more wavy narrow bars. Wings dark brown tipped narrowly buff. Under wing-coverts broadly barred dark brown and cream. Juveniles with blue-grey upperparts are rare.
Bare parts: Bill slate-blue, tip of upper bill black, cere and round eyes yellow, cere and eye-ring in juveniles blue-white to blue-green. Legs yellow, rarely blue-grey to blue-green in juvenile. Eyes dark brown.

Geographical variation:
Birds from Canada and W. USA are very large. Upperparts dark slaty blue.
Birds from N. C. & S. America are larger, paler grey than above, and upper breast narrower barred and spotted, underparts more rufous and juvenile darker especially below.
Birds from Chile, Tierra del Fuego and Falkland Islands have sides of head entirely black. Darker grey on upperparts with broader black bars. More heavily spotted below than last mentioned. Juvenile darker below and more spotted and streaked. Rare colour phase very pale, formerly known as Falco kryenborgi.
Birds from Russia, Northern Siberia to Southern Africa and New Guinea are larger and paler. Crown grey-blue, malar stripe narrower. Underparts whiter and less heavily barred.
Birds from Eastern Siberia, Japan and Taiwan are darker than above but more barred on underparts, abdomen dark buff, grey on sides.
Birds from Mediterranean and Asia Minor are smaller, darker above and more rufous below with denser barring.
3. Birds from Iraq to Mongolia and Northern India are small and palest of all. Crown and nape rufous, underparts sparingly barred.
4. Birds from India, Sri Lanka to Southern China are small and generally much darker than last and heavily barred below. Breast and belly normally deep rufous.
Birds from Ghana to Ethiopia to Cape Province are small and darker than birds from Mediterranean with indication of rufous on crown and nape.
5. Birds from Australia (except South-West) are very dark above, below with a strong buff or grey tinge.
Birds from South-West Australia are like last but pale rufous on foreneck and upper breast, and lower breast and belly deep rufous barred black.
Birds from Cape Verde Islands: Male has a brown tinge on head and upperparts, underparts dull pinkish-buff; female much more brown.
Birds from Madagascar and Comoro Islands are smaller and darker than birds from C. and S. Africa, and underparts more heavily barred.
Birds from Volcano Island like birds from Canada but darker and tail more black and not so clearly barred.
6. Birds from Philippine Islands and New Guinea are darkest of all, head nearly entirely black, upperparts, flanks and thighs also nearly black. Breast with rufous tinge and underparts heavily barred black.
Birds from New Hebrides, Loyalty Island and New Caledonia are very dark on head and upperparts, underparts less barred and more rufous than last. Grey tinge on flanks and thighs.
List: I
Birds from North Africa, the Canaries and Sudan are separated by Sibley and Monroe as a different species:

2 FALCO PELEGRINOIDES
English: Barbary Falcon
Smaller and much paler than Peregrine Falcons from Mediterranean. Crown heavily washed rufous. Underparts generally more rufous.
List: I
Lit.: 2, 8, 13, 19, 20, 33, 37, 51, 63, 64, 66, 68, 70, 76, 86, 89, 122, 129, 162, 163, 191, 197, 222, 233, 260, 278, 313, 323, 410, 431

7. FALCO RUSTICOLUS

English: Gyrfalcon; *French:* Falcon gerfaut, Le Gerfaut; *German*: Gerfalke, Jagdfalke; *Italian:* Girfalco; *Spanish:* Halcón gerifalte; *Swedish:* Jaktfalk

Range: Arctic Europe, Asia and N. America
Identification:
Length 50-60cm. Sexes alike but female larger. There are great differences in colour, white Gyrfalcons occurring mostly in north, darker Gyrfalcons further south, but almost any variant between
these two stages found everywhere.

A. + B. + C. White phase: Head and nape white often with some blackish shaft-streaks and arrow-like streaks, especially on ear-coverts. Upperparts and wing-coverts white with varying amounts of grey-brown bars, arrowheads, or hearts. Long flight-feathers white with black-brown shafts and irregularly shaped broad subterminal band, shorter flight-feathers white with irregular brown bars, near tip like arrowheads. Underparts white, in some birds with a few black-brown shaft-streaks or arrowheads, most on flanks. Tail white, often with some incomplete dark bars or spots, but individuals with barred upperparts always have barred tail.

D. Dark phase: Forehead cream with dark brown streaks. Crown and hindneck like forehead or dark brown edged cream or uniformly grey-brown. Sides of head dark brown streaked white, often with a faint white stripe over eye and a black-brown moustachial streak. Upperparts dark brown to brown-grey with paler feather-edgings and white bars. Long flight-feathers dark brown narrowly tipped white with white bars on inner webs and white spots on outer webs. Under wing-coverts white with dark bars or arrow-like spots. Underparts white to cream with black-brown markings like drops, hearts, arrowheads, and streaks or bars on under tail-coverts. Chin and throat white or white with fine streaks. Tail dark brown or dark grey barred and tipped white.

Young birds, white phase: Like adult but markings more often like streaks and pattern paler brown. Underside more heavily streaked.

E. Young birds, dark phase: Crown dark brown with variable amounts of whitish streaks, more prominent on nape. Upperparts and wing-coverts dark brown to dark grey-brown, most feathers edged paler, rarely with white bars. Flight-feathers like adults. Under wing-coverts brown with white spots. Underparts whitish with prominent dark brown streakings, except for white chin, which has fine streaks. Tail brown barred whitish.
Bare parts: Bill, white birds: Slate, tip darker, base and cere yellow. Bill, dark birds: Creamy, in both phases juveniles have green-grey to slate cere. Legs yellow, juvenile slate to green-grey. Eyes dark brown.
List: I
Lit.: 19, 33, 37, 51, 64, 70, 86, 89, 129, 158, 197, 226, 256, 257, 260, 313, 410

Scale applies to all birds except No. 1C, 2, 3, 4, 5, 6, 7A, 7C

PLATE 17

FALCONIFORMES *

(EXCEPT CATHARTIDAE)

BIRDS OF PREY except New World Vulture

Family: **PANDIONIDAE**

English: Osprey; *French:* Balbuzard pêcheur; *German:* Fischadler; *Italian:* Falco pescatore; *Spanish:* Águila pescadora

Identification:
Length c. 60cm. Sexes alike. Head white, crown and stripe through eye black. Rest of upperparts brownish black with white edgings. Underparts white. Legs pale blue with spines on soles.

Only one member of this family:	English name:
1. Pandion haliaetus	Osprey

List: II

PLATE 18

SPARROW HAWK 36

Family: **ACCIPITRIDAE**

English: Hawks and Eagles; *German:* Greife

Identification:
Length 20 to 114cm. Sexes usually alike, but female larger. Colour brown, grey, and white, often with parts barred or streaked. Wings large and rather rounded. Legs strong. Bill short and hooked. Naked skin (cere) between bill and forehead usually brightly coloured.

	All other Hawks and Eagles:	English names:
	Aviceda cuculoides	African Baza
	Aviceda madagascariensis	Madagascar Baza
	Aviceda jerdoni	Jerdon's Baza
	Aviceda subcristata	Pacific Baza
2.	Aviceda leuphotes	Black Baza
3.	Leptodon cayanensis	Grey-headed Kite
	Leptodon forbesi	White-collared Kite
4.	Chondrohierax uncinatus except: wilsonii	Hook-billed Kite
	Henicopernis longicauda	Long-tailed Honey-Buzzard
	Henicopernis infuscatus	Black Honey-Buzzard
5.	Pernis apivoruus	European Honey-Buzzard
	Pernis ptilorhyncus	Oriental Honey-Buzzard
	Pernis celebensis	Barred Honey-Buzzard
14.	Lophoictinia isura	Square-tailed Kite
15.	Hamirostra melanosternon	Black-breasted Buzzard
	Elanoides forficatus	Swallow-tailed Kite
7.	Macheiramphus alcinus	Bat Hawk
8.	Gampsonyx swainsonii	Pearl Kite
9.	Elanus caeruleus	Black-winged Kite
	Elanus axillaris	Black-shouldered Kite
	Elanus leucurus	White-tailed Kite
	Elanus scriptus	Letter-winged Kite
10.	Chelictinia riocourii	Scissor-tailed Kite
11.	Rostrhamus sociabilis	Snail Kite
	Rostrhamus hamatus	Slender-billed Kite
12.	Harpagus bidentatus	Double-toothed Kite
	Harpagus diodon	Rufous-thighed Kite
13.	Ictinia mississippiensis plumbea	Mississippi Kite Ictinia Plumbeous Kite
	Milvus milvus	Red Kite
16.	Milvus migrans	Black Kite
	Milvus lineatus	Black-eared Kite
	Haliastur sphenurus	Whistling Kite
17.	Haliastur indus	Brahminy Kite
	Haliaeetus leucogaster	White-bellied Fish-Eagle
	Haliaeetus sanfordi	Solomon Fish-Eagle
	Haliaeetus vocifer	African Fish-Eagle
	Haliaeetus vociferoides	Madagascar Fish-Eagle
	Haliaeetus leucoryphus	Pallas's Sea-Eagle
	Haliaeetus pelagicus	Steller's Sea-Eagle
	Ichthyophaga humilis	Lesser Fish-Eagle
18.	Ichthyophaga ichthyaetus	Grey-headed Fish-Eagle
19.	Gypohierax angolensis	Palm-nut Vulture
21.	Gypaetus barbatus	Lammergeier
20.	Neophron percnopterus	Egyptian Vulture
22.	Necrosyrtes monachus	Hooded Vulture
23.	Gyps africanus	White-backed Vulture
	Gyps bengalensis	White-rumped Vulture
	Gyps indicus	Long-billed Vulture
	Gyps rueppellii	Rüppell's Griffon
	Gyps himalayensis	Himalayan Griffon
	Gyps fulvus	Eurasian Griffon
	Gyps coprotheres	Cape Griffon
25.	Aegypius monachus	Black Vulture
24.	Torgos tracheliotus	Lappet-faced Vulture
	Trigonoceps occipitalis	White-headed Vulture
26.	Sarcogyps calvus	Red-headed Vulture
	Circaetus gallicus	Short-toed Snake-Eagle
	Circaetus pectoralis	Black-chested Snake-Eagle
	Circaetus cinereus	Brown Snake-Eagle
	Circaetus fasciolatus	Fasciated Snake-Eagle
27.	Circaetus cinerascens	Banded Snake-Eagle
28.	Terathopius ecaudatus	Bateleur
	Spilornis cheela	Crested Serpent-Eagle
	Spilornis minimus	Nicobar Serpent-Eagle
	Spilornis kinabaluensis	Mountain Serpent-Eagle
29.	Spilornis rufipectus	Sulawesi Serpent-Eagle
	Spilornis holospilus	Philippine Serpent-Eagle
	Spilornis elgini	Andaman Serpent-Eagle
30.	Dryotriorchis spectabilis	Congo Serpent-Eagle
31.	Eutriorchis astur	Madagascar Serpent-Eagle
34.	Circus aeruginosus	Western Marsh-Harrier
	Circus ranivorus	African Marsh-Harrier
	Circus spilonotus	Eastern Marsh-Harrier
	Circus approximans	Swamp Harrier
	Circus maillardi	Madagascar Marsh-Harrier
	Circus buffoni	Long-winged Harrier
	Circus assimilis	Spotted Harrier
	Circus maurus	Black Harrier
	Circus cyaneus	Northern Harrier
	Circus cinereus	Cinereous Harrier
	Circus macrourus	Pallid Harrier
	Circus melanoleucos	Pied Harrier
	Circus pygargus	Montagu's Harrier
	Polyboroides typus	African Harrier-Hawk
32.	Polyboroides radiatuus	Madagascar Harrier-Hawk
38.	Kaupifalco monogrammicus	Lizard Buzzard
35.	Melierax metabates	Dark Chanting-Goshawk
	Melierax poliopterus	Eastern Chanting-Goshawk
	Melierax canorus	Pale Chanting-Goshawk
	Melierax gabar	Gabar Goshawk
	Accipiter poliogaster	Grey-bellied Goshawk
	Accipiter trivirgatus	Chested Goshawk
	Accipiter griseiceps	Sulawesi Goshawk
	Accipiter toussenelii	Red-chested Goshawk
	Accipiter tachiro	African Goshawk
	Accipiter castanilius	Chestnut-flankedSparrowhawk
	Accipiter badius	Shikra

PLATE 19

No.	Species	Common name
	Accipiter butleri	Nicobar Sparrowhawk
	Accipiter brevipes	Levant Sparrowhawk
	Accipiter soloensis	Chinese Goshawk
	Accipiter francesii	Frances's Goshawk
	Accipiter trinotatus	Spot-tailed Goshawk
	Accipiter novaehollandiae	Grey Goshawk
	Accipiter fasciatus	Brown Goshawk
	Accipiter melanochlamys	Black-mantled Goshawk
	Accipiter albogularis	Pied Goshawk
	Accipiter haplochrous	White-bellied Goshawk
	Accipiter henicogrammus	Moluccan Goshawk
	Accipiter luteoschistaceus	Slaty-mantled Sparrowhawk
	Accipiter imitator	Imitator Sparrowhawk
	Accipiter poliocephalus	Grey-headed Goshawk
	Accipiter princeps	New Britain Goshawk
	Accipiter superciliosus	Tiny Hawk
	Accipiter collaris	Semicollared Hawk
	Accipiter erythropus	Red-thighed Sparrowhawk
	Accipiter minullus	Little Sparrowhawk
	Accipiter gularis	Japanese Sparrowhawk
	Accipiter virgatus	Besra
	Accipiter nanus	Small Sparrowhawk
	Accipiter erythrauchen	Rufous-necked Sparrowhawk
	Accipiter cirrocephalus	Collared Sparrowhawk
	Accipiter brachyurus	New Britain Sparrowhawk
	Accipiter rhodogaster	Vinous-Breasted Sparrowhawk
	Accipiter madagascariensis	Madagascar Sparrowhawk
	Accipiter ovampensis	Ovampo Sparrowhawk
36.	Accipiter nisus	Eurasian Sparrowhawk
	Accipiter rufiventris	Rufous-chested Sparrowhawk
	Accipiter striatus	Sharp-shinned Hawk
	Accipiter chionogaster	White-breasted Hawk
	Accipiter ventralis	Plain-breasted Hawk
	Accipiter erythronemius	Rufous-thighed Hawk
	Accipiter cooperii	Cooper's Hawk
	Accipiter gundlachi	Gundlach's Hawk
	Accipiter bicolor	Bicolored Hawk
	Accipiter melanoleucus	Black Goshawk
	Accipiter henstii	Henst's Goshawk
	Accipiter gentilis	Northern Goshawk
	Accipiter meyerianus	Meyer's Goshawk
	Erythrotriorchis buergersi	Chestnut-shouldered Goshawk
	Erythrotriorchis radiatus	Red Goshawk
	Megatriorchis doriae	Doria's Goshawk
	Urotriorchis macrourus	Long-tailed Hawk
	Butastur rufipennis	Grasshopper Buzzard
	Butastur teesa	White-eyed Buzzard
	Butastur liventer	Rufous-winged Buzzard
37.	Butastur indicus	Grey-faced Buzzard
33.	Geranospiza caerulescens	Crane Hawk
	Leucopternis plumbea	Plumbeous Hawk
	Leucopternis schistacea	Slate-colored Hawk
	Leucopternis princeps	Barred Hawk
	Leucopternis melanops	Black-faced Hawk
	Leucopternis kuhli	White-browed Hawk
	Leucopternis lacernulata	White-necked Hawk
	Leucopternis semiplumbea	Semiplumbeous Hawk
39.	Leucopternis albicollis	White Hawk
	Leucopternis occidentalis	Grey-backed Hawk
	Leucopternis polionota	Mantled Hawk
40.	Buteogallus aequinoctialis	Rufous Crab-Hawk
	Buteogallus anthracinus	Common Black-Hawk
	Buteogallus subtilis	Mangrove Black-Hawk
	Buteogallus urubitinga	Great Black-Hawk
	Buteogallus meridionalis	Savanna Hawk

No.	Species	Common name
44.	Parabuteo unicinctus	Harris's Hawk
42.	Busarellus nigricollis	Black-collared Hawk
43.	Geranoaetus melanoleucus	Black-chested Buzzard-Eagle
	Harpyhaliaetus solitarius	Solitary Eagle
41.	Harpyhaliaetus coronatus	Crowned Eagle
	Asturina plagiata	Grey Hawk
45.	Asturina nitida	Grey-lined Hawk
	Buteo magnirostris	Roadside Hawk
	Buteo lineatus	Red-shouldered Hawk
	Buteo ridgwayi	Ridgway's Hawk
	Buteo platypterus	Broad-winged Hawk
	Buteo leucorrhous	White-rumped Hawk
	Buteo brachyurus	Short-tailed Hawk
	Buteo albigula	White-throated Hawk
	Buteo swainsoni	Swainson's Hawk
46.	Buteo albicaudatus	White-tailed Hawk
	Buteo galapagoensis	Galapagos Hawk
	Buteo polyosoma	Red-backed Hawk
	Buteo poecilochrous	Puna Hawk
	Buteo albonotatus	Zone-tailed Hawk
	Buteo solitarius	Hawaiian Hawk
	Buteo jamaicensis	Red-tailed Hawk
	Buteo ventralis	Rufous-tailed Hawk
	Buteo buteo	Common Buzzard
	Buteo oreophilus	Mountain Buzzard
	Buteo brachypterus	Madagascar Buzzard
	Buteo rufinus	Long-legged Buzzard
	Buteo hemilasius	Upland Buzzzard
	Buteo regalis	Ferruginous Hawk
	Buteo lagopus	Rough-legged Hawk
	Buteo auguralis	Red-necked Buzzard
	Buteo augur	Augur Buzzard
	Buteo archeri	Archer's Buzzard
	Buteo rufofuscus	Jackal Buzzard
47.	Morphnus guianensis	Crested Eagle
48.	Harpyopsis novaeguineae	New Guinea Eagle
49.	Ictinaetus malayensis	Black Eagle
	Aquila pomarina	Lesser Spotted Eagle
	Aquila clanga	Greater Spotted Eagle
	Aquila rapax	African Tawny Eagle
	Aquila vindhiana	Eurasian Tawny Eagle
	Aquila nipalensis	Steppe Eagle
	Aquila wahlbergi	Wahlberg's Eagle
	Aquila gurneyi	Gurney's Eagle
	Aquila chrysaetos	Golden Eagle
	Aquila audax	Wedge-tailed Eagle
	Aquila verreauxii	Verreaux's Eagle
50.	Hieraaetus fasciatus	Bonelli's Eagle
	Hieraaetus spilogaster	African Hawk-Eagle
	Hieraaetus pennatus	Booted Eagle
	Hieraaetus morphnoides	Little Eagle
	Hieraaetus ayresii	Ayres's Hawk-Eagle
	Hieraaetus kienerii	Rufous-bellied Eagle
56.	Polemaetus bellicosus	Martial Eagle
51.	Spizatur melanoleucus	Black-and-white Hawk-Eagle
52.	Lophaetus occipitalis	Long-crested Eagle
53.	Spizaetus africanus	Cassin's Hawk-Eagle
	Spizaetus cirrhatus	Changeable Hawk-Eagle

PLATE 20

Spizaetus nipalensis	Mountain Hawk-Eagle
Spizaetus alboniger	Blyth's Hawk-Eagle
Spizaetus bartelsi	Javan Hawk-Eagle
Spizaetus lanceolatus	Sulawesi Hawk-Eagle
Spizaetus philippensis	Philippine Hawk-Eagle
Spizaetus nanus	Wallace's Hawk-Eagle
Spizaetus tyrannus	Black Hawk-Eagle
Spizaetus ornatus	Ornate Hawk-Eagle
54. Stephanoaetus coronatus	Crowned Hawk-Eagle
55. Oroaetus isidori	Black-and-chestnut Eagle

List: II

Family:**SAGITTARIIDAE**

English: Secretary-Bird;*French:* Messager serpentaire, *German:* Sekretäre

Identification:
Length 127-150cm. Sexes alike but female larger. Colour pale grey and black with a little white on rump and tail. Elongated chest and central tail feathers. Extremely long legs with short toes.
Only one member of this family:

57. Sagittarius serpentarius	Secretary-Bird

List: II

Family:**FALCONIDAE**

English: Falcons, Caracaras;*French:* Faucons, Caracaras;*German:* Falken, Geierfalken

Identification: Length 15-64cm. Sexes usually alike, but female larger. Colour grey to brown and white. Below often barred or streaked. Wings in most species long and pointed. Legs strong, toes long. Bill short, hooked and in most species toothed. Naked skin between bill and forehead usually brightly coloured.

All other Caracaras and Falcons:

Daptrius ater	Black Caracara
58. Daptrius americanus	Red-throated Caracara
Phalcoboenus carunculatus	Carunculated Caracara
Phalcoboenus megalopterus	Mountain Caracara
59. Phalcoboenus albogularis	White-throated Caracara
Phalcoboenus australis	Striated Caracara
60. Polyborus plancus	Crested Caracara
61. Milvago chimachima	Yellow-headed Caracara
Milvago chimango	Chimango Caracara
62. Herpetotheres cachinnans	Laughing Falcon
Micrastur ruficollis	Barred Forest-Falcon
Micrastur plumbeus	Plumbeous Forest-Falcon
Micrastur gilvicollis	Lined Forest-Falcon
Micrastur mirandollei	Slaty-backed Forest-Falcon
Micrastur semitorquatus	Collared Forest-Falcon
63. Micrastur buckleyi	Buckley's Forest-Falcon
64. Spiziapteryx circumcinctus	Spot-winged Falconet
65. Polihierax semitorquatus	Pygmy Falcon
Polihierax insignis	White-rumped Falcon
66. Microhierax caerulescens	Collared Falconet
Microhierax fringillarius	Black-thighed Falconet
Microhierax latifrons	White-fronted Falconet
Microhierax erythrogenys	Philippine Falconet
Microhierax melanoleucus	Pied Falconet
Falco berigora	Brown Falcon
Falco naumanni	Lesser Kestrel
Falco tinnunculus	Common Kestrel
Falco moluccensis	Spotted Kestrel
Falco cenchroides	Australian Kestrel
Falco sparverius	American Kestrel
Falco rupicoloides	Greater Kestrel
Falco alopex	Fox Kestrel
Falco ardosiaceus	Grey Kestrel
Falco dickinsoni	Dickinson's Kestrel
Falco zoniventris	Barred Kestrel
Falco chicquera	Red-necked Falcon
Falco vespertinus	Red-footed Falcon
Falco amurensis	Amur Falcon
Falco eleonorae	Eleonora's Falcon
Falco concolor	Sooty Falcon
Falco femoralis	Aplomado Falcon
Falco columbarius	Merlin
Falco rufigularis	Bat Falcon
Falco subbuteo	Eurasian Hobby
Falco cuvieri	African Hobby
Falco severus	Oriental Hobby
Falco longipennis	Australian Hobby
Falco novaezeelandiae	New Zealand Falcon
Falco hypoleucos	Grey Falcon
Falco subniger	Black Falcon
Falco biarmicus	Lanner Falcon
Falco cherrug	Saker Falcon
Falco altaicus	Altai Falcon
Falco mexicanus	Prairie Falcon
Falco pelegrinoides	Barbary Falcon
Falco deiroleucus	Orange-breasted Falcon
Falco fasciinucha	Taita Falcon

List: II
Lit.: 2, 3, 7, 9, 13, 19, 20, 33, 51, 63, 64, 66, 70, 73, 77, 79, 86, 101, 117, 125, 129, 150, 189, 190, 191, 192, 195, 223, 233, 260, 295, 313, 410

Note:
The following members of the family CATHARTIDAE, New World Vultures, **are not listed:**

Cathartes aura	Turkey Vulture
Cathartes burrovianus	Lesser Yellow-headed Vulture
Coragyps atratus	American Black Vulture

(Sibley and Ahlquist have recently discovered that New World Vultures are more closely related to Storks, Ciconidae, than to Birds of Prey).

Reading List:
Brown, L. & D. Amadon, 1968:*Eagles, Hawks and Falcons of the World.* Vol. 1-2. Hamlyn, Feltham, UK.

Cade, T. J. 1982: *The Falcons of the World.* Collins, London.

Weick, F. 1980: *Birds of Prey of the World.* Paul Parey. Hamburg & Berlin.

PLATE 21

1. MACROCEPHALON MALEO

English: Maleo, Maleo Fowl, Malee Fowl; *French:* Mégapode maléo; *German:* Hammerhuhn

Range: Sulawesi, (Celebes)
Identification:
Male: Length 52cm. Head bare with a prominent blue-black casque. Neck sparse covered with down. Lower neck, rump, wings and tail blackish brown. Breast and belly white suffused with buffish pink. Thighs and undertail-coverts blackish-brown.
Female: Like male but a little smaller.
Young birds: Undescribed.
Bare parts: Bill yellowish. Legs yellow green. Eyes dark brown.
List: I
Lit.: 34, 129, 229, 312, 321, 332, 391, 410

2. ORTALIS VETULA

English: Plain Chachalaca; *French:* Ortalide chacamel; *German:* Braunflügelguan

Range: From Texas and Mexico to Nicaragua
Identification:
Length 56cm. Sexes alike. Olive brown and grey with bare throat and slight crest. Tail with greenish gloss, tipped white, central tail feathers tipped grey.
Young birds: Browner.
Bare parts: Bill and legs light horn blue. Eyes light brown. Naked throat skin reddish.

Geographical variation:
Birds from Texas and Mexico are paler and birds from Yucatan are palest of all populations.
Birds from Utila (Honduras) have shorter tail feathers and tipped grey or cinnamon.
List: III Guatemala, Honduras
Status: Population from Utila very rare, if not extinct.
Lit.: 62, 129, 152, 223, 312, 410

3. PENELOPE ALBIPENNIS

English: White-winged Guan; *French:* Pénélope à ailes blanches; *German:* Weisschwingenguan

Range: Peru
Identification:
Length 60cm. Sexes probably alike. Small crest. Brownish-olive, on wings and tail with a bronzy wash. Large flight-feathers white. Underparts more rufescent.
Young birds: Undescribed.
Bare parts: Bill blackish. Legs red-brown. Eyes pale brown.
Bare skin on face violet, throat orange.
List: I
Status: Long thought extinct, but a few hundred birds found in 1977.
Lit.: 62, 67, 129, 152, 190, 196, 213, 312, 410

4. PENELOPE PURPURASCENS

English: Crested Guan; *French:* Pénélope huppée; *German:* Rostbauchguan, *Spanish:* Pava culirroja

Range: Mexico to Nicaragua
Identification:
Length 89cm. Sexes alike. Short, bushy crest, head and neck brown. Body dusky olive-brown, underparts, shoulders and forewing edged white. Central tail feathers brown with a green wash, rest darker with a bluish wash.
Young birds: Undescribed.
Bare parts: Bill black. Legs red. Eyes red to red-brown. Bare skin around eye blue, throat bright orange-red.

Geographical variation:
Birds from Honduras to Colombia are smaller and shoulder not edged white. More rufous.
Birds from Colombia and Venezuela are the smallest and even more rufous.
List: III Honduras
Lit.: 62, 129, 223, 312, 410

5. PIPILE JACUTINGA
Also known as: Aburria jacutinga

Brazilian: Jacutinga; *English:* Black-fronted Piping-Guan; *French:* Pénélope à front noir; *German:* Schakutinga

Range: Brazil, Paraguay, Argentina
Identification:
Length 60cm. Sexes similar. Forehead, side of head, neck and body velvety black. Long crown feathers white with black shafts. Forewings white. Underparts glossed violet-blue. Breast narrowly edged white.
Young birds: Not described.
Bare parts: Bill blue, tip black. Legs dark red. Eyes brown. Skin around eye white. Dewlap red.
List: I
Lit.: 62, 129, 190, 312, 410, 431

6. PIPILE P. PIPILE
Also known as: Aburria pipile

English: Trinidad Piping Guan; *French:* Pénélope siffleuse; *German:* Blaukehlguan; *Spanish:* Pava rajadora

Range: Trinidad
Identification:
Length 60cm. Sexes alike. Differs from number 5 by having more blackish-brown colour washed with purplish-brown. Crest black with white edges.
Young birds: Not described.
Bare parts: Bill blue, tip black. Legs dark red. Eyes brown. Dewlap and bare skin on head said to be dark blue in life.
List: I
Status: May be extinct.
Lit.: 62, 129, 152, 190, 192, 312, 410

Note:
Pipile (pipile) cumanensis
Blue-throated Piping-Guan
Range: Guianas to Colombia & Peru, W. Brazil

Pipile grayi
Gray's Piping-Guan
Range: Bolivia, Paraguay, S. W. Brazil

Pipile natteri
Natterer's Piping-Guan
Range: S. & W. Amazonia

Pipile (pipile) cujubi
Red-throated Piping-Guan
Range: N. C. Brazil

are not listed

PLATE 22

1. PENELOPINA NIGRA

English: Highland Guan; *French:* Pénélope pajuil; *German:* Schluchtenguan

Range: Mexico to Nicaragua
Identification:
A. Length 53-63cm. **Male:** Entire bird glossy black, except rump and belly, which are dark blue to green.
B. **Female:** Brownish-black heavily barred rufous on head, neck and upperparts. Underside barred but paler. Tail chestnut barred black.
Young birds: Male like female but darker. Female like adult female but not so distinctly barred.
Bare parts: Bill red, female dull brown. Legs orange-red, female brownish-red. Eyes dark maroon-red, female reddish-brown. Bare skin around eye brownish-red, dewlap crimson.
List: III Guatemala
Lit.: 62, 129, 223, 312, 410

2. OREOPHASIS DERBIANUS

English: Horned Guan; *French:* Oréophase cornu; *German:* Zapfenguan

Range: Mexico, Guatemala
Identification:
Length 81-91cm. Sexes alike. Characterised by erect casque on top of head. Plumage glossy green-black, or on wings and tail blue-black. Foreneck, breast and a broad band on tail white. Breast finely streaked black.
Young birds: Not described.
Bare parts: Bill horn to straw-yellow. Legs red. Eyes white, seldom yellowish. Casque on top of head red. Naked area on throat red.
List: I
Status: Last seen in 1978.
Lit.: 62, 129, 196, 223, 312, 410

3. CRAX BLUMENBACHII
Formerly known as: Crax rubirostris

Brazilian: Mutum-do-sudeste; *English:* Red-billed Curassow; *French:* Hocco de Blumenbach; *German:* Blumenbachhokko

Range: Brazil
Identification:
A. Length c. 80cm. **Male:** Large, curly, black crest. Bill knob and wattles conspicuous. Belly and undertail-coverts white. Rest black, glossed green-blue on body and head.
B. **Female:** Like male but crest with narrow white spots. Wings with rufous to chestnut markings. Belly and undertail-coverts rufous. Without knob and wattles.
Young birds: Undescribed.
Bare parts: Bill blackish. Legs blackish, female dull flesh to dull orange-rose. Eyes dark brown, female paler. Knob on bill and wattles red in male
List: I
Status: Close to extinction.
Lit.: 13, 62, 129, 196, 312, 410, 431

4. CRAX RUBRA

English: Great Curassow; *French:* Grand Hocco; *German:* Tuberkelhokko

Range: Mexico to Colombia, Ecuador
Identification:
A. Length 76-96cm. **Male:** Round knob above bill. Large curly crest. Abdomen and undertail-coverts white. Rest glossy black.
D. **Female** has three phases: 1) Red: Head and neck barred black and white. Rest chestnut, tail with seven or eight buffy and black bars.
E. 2) Dark: Like red phase but lower neck and breast black, upperparts suffused with black. Tail unbarred blackish mottled rufous.
F.3) Barred: (rare) Entire head, neck, upperparts and tail barred, underparts pale rufous. All three phases lack bill knob and are smaller than male.
Young birds: Male like dark phase of female. Female like adult female, all three phases occur.
Bare parts: Bill blackish, female yellowish. Legs greyish-brown, female red phase have flesh colour. Eyes brownish-red. Knob yellow.
List: III Colombia, Costa Rica
Lit.: 13, 27, 62, 129, 223, 312, 410

5. MITU M. MITU
Also known as: Crax m. mitu

Brazilian: Mutum-do-nordeste; *English:* Alagoas Curassow; *French:* Hocco mitou; *German:* Mitu

Range: E. Brazil
Identification:
Length 83-89cm. Sexes alike. Casque-like bill, swollen at base. Crest broadens out distally to rounded tip. Entire bird black glossed dark blue, except chestnut belly and undertail-coverts, and white tips to tail-feathers.
Young birds: Without keel on bill.
Bare parts: Bill and casque bright red. Legs red. Eyes reddish-brown.
List: I
Status: May be extinct.
Note: Mitu tuberosa, Razor-billed Curassow, South Amazonia, has more elevated upper-bill and broader white tip of tail, and is not listed.
Lit.: 13, 62, 129, 195, 312, 410, 431

6. AGRIOCHARIS OCELLATA

English: Ocellated Turkey; *French:* Dindon ocellé; *German:* Pfauentruthuhn

Range: Mexico, Belize, Guatemala
Identification:
Length 71-91cm. **Male:** Bare skin on head with frontal tubercule and many knobs. Plumage dark, glossed bronze and blue-green. Tail grey, barred and mottled black, tip of feathers black, blue-green, and outermost chestnut eye-like spots.
Female: Smaller and duller. Tubercule, knobs and spur lacking or very small.
Young birds: Undescribed.
Bare parts: Bill dull red. Legs red, male with spur. Eyes dark brown. Bare skin on head of male blue with many orange-red knobs.
List: III Guatemala
Lit.: 3, 13, 129, 223, 312, 410
Note: Scale applies to all birds except no. 4D, 4E and 4F

Crax alberti, Crax daubentoni, Crax globulosa and Crax pauxi listed 1989 on List III. See plate 80.

PLATE 23

1. ITHAGINIS CRUENTUS

English: Blood Pheasant;*French:* Ithagine ensanglantée;*German:* Blutfasan; *Italian:* Fagiano insanguinato;*Nepalese:* Chilime;*Spanish:* Faisán sanguineo; *Tibetan:* Semo

Range: Nepal, Sikkim, Bhutan, Burma, China

*Identification:*A. Length 40-45cm. **Male:** Forehead, between bill and eye, and below eye black. Crest buff and grey. Neck yellowish-white, streaked black, except on sides. Upperparts grey with prominent white streaks more or less bordered black. Upper breast yellowish-green with red streaks, lower breast and flanks yellowish bordered darker green. Belly ashy buff. Upper- and under tail-coverts crimson. Visible parts of tail whitish. B. **Female:** Mainly brown to greyish-brown. Crest grey. Upperparts finely vermiculated.

Young birds: Not as distinctly coloured as adult.

Bare parts: Bill black, around nostril red; juvenile males orange-red. Legs crimson. Eyes red-brown. Skin around eye and throat red in males, juvenile fleshy-grey.

Geographical variation:

Birds from Sikkim have red on breast much reduced and flanks greener. Birds from E. Bhutan and S. E. Tibet have red between bill and eye instead of black. Birds from E. Assam to Yunnan have broad black band on upper breast, breast solid red. Birds from W. Yunnan have little red streaking on breast and forehead red. Throat and side of head more or less red streaked pale buff. Birds from N. W. Szechwan have upper wing-coverts reddish-brown. Birds from N. Kansu have three inner flight-feathers chestnut, centered and margined green. Upper wing-coverts reddish-brown. Birds from N. E. Burma have broad black band of upper breast, broken in middle. Breast solid red. Eyebrow black. Birds from S. E. Tibet to W. Szechwan are without red on head and chin . Birds from S. Kansu and N. Szechwan have three inner flight-feathers green. Upper wing-coverts reddish-brown. Without red on breast. Birds from Shensi have tail pale grey, fringed crimson. Upper wing-coverts reddish-brown. Birds from W. Kansu have upper wing-coverts reddish-brown.

List: II

Lit.: 4, 37, 59, 129, 142, 229, 410

2. RHIZOTHERA LONGIROSTRIS

English: Long-billed Partridge, Long-billed Wood Partridge;*French:*Roulroul à long bec; *German:* Langschnabelwachtel; *Spanish:* Perdiz de bosque de pico largo

Range: Malaya, Sumatra, Borneo

*Identification:*A. Length 36cm. **Male:** Long and slightly decurved bill diagnostic, (at least 35mm). Head rufous-chestnut. Neck and forebreast blue- grey. Rest brown and rufous-chestnut with white stripes on shoulders.

B. **Female:** Like male but neck and breast rufous-chestnut.

Young birds: Very similar to adult female.

Bare parts: Bill black. Legs pale brown. Eyes brown.

List: III Malaysia

Lit.: 37, 129, 143, 229, 281, 410

3. MELANOPERDIX NIGRA

English: Black Partridge, Black Wood Partridge;*French:* Roulroul noir;*German:* Schwarzwachtel; *Italian:* Melanoperdice; *Spanish:* Perdiz de bosque negra

Range: Malaya, Sumatra, Borneo

Identification: A. Length 25cm. **Male:** Entire bird glossy black, except dark brown wings. B. **Female:** Dark chestnut. Throat and underparts paler.

Young birds: Males like adult females but with some black feathers. Females have paler upperparts than adult females, with irregular blackish vermiculation and usually with pale buffy feather tips. Bare parts: Bill very short and black, female brown. Legs pale blue. Eyes dark brown.

Geographical variation:

Birds from Borneo are paler with a slight slaty-greenish tinge. Females have deeper chestnut-brown breast.

List: III Malaysia

Lit.:37, 129, 143, 229, 251, 410

4. ARBOROPHILA BRUNNEOPECTUS

English: Bar-backed Partridge, Brown-breasted Hill-Partridge; *French:* Torquéole à poitrine brune;*German:*Rotbauch-Buschwachtel:*Spanish:*Perdiz de bosque de pecho pardo.

Range: Assam, Yunnan to Thailand and Vietnam

*Identification:*Length 28cm. **Male:** Forehead pale brown. Crown and nape dark brown in middle, sides buffish. Chin, throat, ear-coverts, and neck buffish, side of neck and throat with black spots. Upperparts brown barred black. Breast brownish. Flank-feathers white edged black. Belly whitish with buff markings. **Female:** Like male but probably with less colourful bare parts.

Young birds: Like adult except colour of legs.

Bare parts: Bill black. Legs red, juvenile pink to orange. Eyes brown. Naked skin round eyes and throat red forming red lines on side of neck.

List: III Malaysia

Lit.: 37, 129, 143, 229, 410

ARBOROPHILA ORIENTALIS (not depicted)

Formerly known as: Arborophila brunneopectus orientalis

English: Grey-breasted Partridge; *French:* Torquéole de Sumatra; *German:* Sumatrabuschwachtel; *Malaysia:* San serok gunong

Range: Sumatra, Java

Male: Like A. brunneopectus but eyebrow stripe, chin, upper throat and ear-coverts white, (great variation), more black on head extending down side of neck, and head without buffish colour. Breast greyish-brown, flanks greyish black with white spots. **Female:** Less black on head.

Young birds: Like adult but pale shaft-stripes, underparts more rufous and breast finely barred. Bare parts: Resembles last one.

List: III Malaysia

Lit.: 143, 170 229, 410

5. ARBOROPHILA CHARLTONII

Formerly known as: Tropicoperdix charltonii

English: Chestnut-necklaced Partridge, Chestnut-breasted Tree Partridge;*French:* Torquéole à poitrine chataine, Torquéole des bois, Torquéole de Merlin;*German:* Grünfuss-Buschwachtel; *Spanish:* Perdiz de bosque de Charlton

Range: Thailand, Malaya, Sumatra, Borneo

*Identification:*Length 30cm. **Male:** forehead and crown brown. Face, throat, and upper neck white with black spots on neck. Upperparts brown edged black. Lower neck and upper breast with an unmarked rufous band. Lower breast chestnut edged black. Without spurs. **Female:** Colours of bare parts probably less bright.

Young birds: Breast with white markings. Bare parts: Bill black. Legs yellowish-green to yellow. Eyes brown.

Geographical variation:

Birds from Vietnam have ear-coverts mingled grey and chestnut. Throat without black.

Birds from Borneo have deep red legs and yellow eyes.

List: III Malaysia

Lit.: 37, 58, 129, 143, 229, 272, 410

6. CALOPERDIX OCULEA

English: Ferruginous Partridge, Ferruginous Wood Partridge;*French:* Roulroul ocellé; *German:* Augenwachtel; *Malaysia:* Sang serok; *Spanish:* Perdiz de bosque ferruginosa.

Range: Thailand, Malaya, Sumatra, Borneo

Identification: Length 26cm. **Male:** Bright rusty brown head, neck, breast, and flanks; flanks with black and white spots. Upperparts blackish brown edged white, rump and upper tail-coverts edged rufous. One or rarely two spurs.

Female: Like male but only one short spur.

Young birds: Nape with black bars and breast sparsely spotted or barred black.

Bare parts: Bill black. Legs greenish. Eyes brown.

Geographical variation:

Birds from Sumatra have paler upperparts with irregular rufous colour. Birds from Borneo have more red-brown on head and foreneck.

List: III Malaysia

Lit.: 37, 129, 143, 208, 209, 229, 410

7. ROLLULUS ROULOUL

English: Crested Partridge, Crested Wood Partridge; *French:* Roulroul couronné, Roulroul;*German:* Strausswachtel;*Italian:*Quaglia crestata; *Spanish:* Perdiz de bosque crestada; *Malaysia:* Burong siul

Range: Thailand, Malaya, Sumatra, Borneo

*Identification:*A. Length 25cm. **Male:** Large bushy, chestnut crest unmistakable, rest of plumage blackish glossed green and blue, wings brown. B. **Female:** Crest more diffuse. Head blackish grey. Body green, wings chestnut.

Young birds: Wing-coverts broad tipped buffy-white. Bare parts: Bill black, base red only in male. Legs reddish. Eyes brown. Bare skin round eyes red.

List: III Malaysia

Lit.: 37, 129, 143, 150, 222, 229, 410

PLATE 24

1. COLINUS VIRGINIANUS RIDGWAYI

English: Masked Bobwhite; *French:* Colin de Virginie; *German:* Virginiawachtel; *Italian:* Quaglia della Virginia; *Spanish:* Codorniz común; Colín de Virginia

Range: Sonora (Mexico)
Identification:
Length 21-27cm. **Male:** Forehead, forecrown, face and throat black. Hindcrown and nape brown, white and black. Broad white band from bill over eye to side of neck. Upperparts brown, black and pale buff. Underparts rich chestnut-brown.
Female: No black or very little on head and throat. Underparts buffy, barred black.
Young birds: Like female.
Bare parts: Bill black. Legs pale brown. Eyes dark brown.
List: I
Lit.: 37, 129, 143, 195, 223, 229, 295, 410

Note: All other subspecies are not listed.
(Sibley and Monroe have no races)

2. TETRAOGALLUS CASPIUS

English: Caspian Snowcock; *French:* Tétraogalle de Perse, Tétraogalle de la Caspienne; *German:* Kaspi-Königshuhn; *Italian:* Gallo cedrone caspio; *Russian:* Kaspiysky Ular; *Spanish:* Faisán nival del Caspio, Gallo salvaje del Caspio.

Range: Taurus Mts. to Iran
Identification:
Length 58-62cm. **Male:** Forehead, crown and nape grey with buff wash. Narrow white eye-stripe. Rest of head grey. Broad white streak from ear-coverts downwards. Upperparts grey and black with chestnut and cinnamon-pink spots. Underparts grey tinged white or pale buff. Breast with black spots. Flanks with some fine buff vermiculation. Under tail-coverts white. Tail grey mottled cinnamon. Large white area in flight-feathers visible when wings are spread.
Female: Smaller and duller. Neck-stripes and breast grey faintly marked white and black. Without spurs.
Young birds: Less distinctly marked.
Bare parts: Bill yellowish, tip and base blackish. Legs orange. Eyes dark brown.

Geographical variation:
Birds from Zagros Mts. (Iran) are paler, more buffy.
List: I
Lit.: 37, 51, 118, 129, 140, 229, 310, 410

3. TETRAOGALLUS TIBETANUS

English: Tibetan Snowcock; *French:* Tétraogalle du Tibet; *German:* Tibetaner-Königshuhn, Tibet Königshuhn; *Italian:* Gallo cedrone Tibetano: *Mongol:* Hailik; *Russian:* Tibetsky Ular; *Spanish:* Gallo salvaje del Tibet

Range: India, Tibet, China
Identification:
Length 48-50cm. **Male:** Differs from Caspian Snowcock by having grey-lilac head. Below white with blackish and grey streaks on sides and flanks.
Female: Smaller and without spurs.
Young birds: Paler, breast grey, mottled brown and buff. Flanks and belly without black streaks.
Bare parts: Bill and legs reddish to orange. Eyes brown. Naked skin around eye red.
List: I
Lit.: 37, 70, 129, 143, 229, 310, 410

4. TYMPANUCHUS CUPIDO ATTWATERI

English: Attwater Prairie Chicken; *French:* Cupidon des prairies, Poule des prairies; *German:* Präriehuhn

Range: Texas and Louisiana
Identification:
A. Length 43cm. **Male:** Large bare skin on neck that can be exposed during courtship. On upper-neck about ten elongated dark and rounded feathers that can be erected. Heavily barred brown, cinnamon and pale buff. Back of legs unfeathered diagnostic.
B. **Female:** Like male but long feathers on neck shorter.
Young birds: Undescribed.
Bare parts: Bill dark horn, base paler. Legs feathered in front, hind legs and toes pale brown. Eyes brown. Comb yellow. Bare skin on neck yellow-orange, female without comb and bare orange skin on neck.
List: I
Lit.: 70, 98, 129, 140, 195, 197, 229

Note: **Tympanuchus cupido pinnatus**
Range: Canada to Texas
Tympanuchus cupido pallidicinctus
Range: Kansas to New Mexico
are not listed
Tympannuchus cupido extinct

5. CYRTONYX M. MONTEZUMAE

English: Montezuma's Quail

Range: N. & C. Mexico

6. CYRTONYX MONTEZUMAE MEARNSI

Range: South of U.S.A. and Mexico

5. + 6. are no longer listed

7. AGELASTES MELEAGRIDES

English: White-breasted Guineafowl; *French:* Pintade à poitrine blanche; *German:* Weissbrust-Perlhuhn; *Spanish:* Pintado de pecho blanco

Range: Liberia to Ghana
Identification:
Length 46cm. Sexes alike. Head and upper neck naked. Lower neck and breast white. Rest black finely vermiculated white.
Young birds: Brown with white belly.
Bare parts: Bill greenish-brown. Legs greyish-brown to greyish-black. Eyes brown. Bare skin on head and upper neck red.
List: III Ghana
Perhaps one of the most threatened birds in Africa.
Lit.: 7, 37, 45, 129, 229, 305, 410

PLATE 25

1. TRAGOPAN MELANOCEPHALUS

English: Western Tragopan; *French:* Tragopan de Hastings; *German:* Schwarzkopf-Traagopan; *Himalayas:* Singmoonal

Range: Himalaya
Identification:
A. **Male:** Length 68-74cm. Crown black. Crest lies flat on neck and is red-tipped. Neck and breast red. Upperside spotted buffy, grey and black with black-bordered white spots. Rest of underside black with white and red spots.
B. **Female:** Length 60cm. Ground colour dark brownish-grey, spotted black and white. Flanks and belly paler. Tail feathers except central pair have a broad black band.
Young birds: Like female but duller.
Bare parts: Bill black. Legs pinkish. Eyes brown. Naked skin around eyes bright red with fine blue spots below eyes. Throat skin of male during display is deep blue, cheeks blue-green, corner purplish-blue down centre pink margins, pale blue grooves.
List: I
Lit.: 59, 129, 142, 229, 410

2. TRAGOPAN SATYRA

English: Satyr Tragopan; *French:* Tragopan satyre; *German:* Satyr-Tragopan, Rot-Satyrhuhn; *Italian:* Tragopano satiro; *Nepalese:* Monal; *Spanish:* Tragopan sátiro; *Tibetan/Chinese:* See-agea

Range: Himalaya
Identification:
A. Length 67-72cm. **Male:** Head, crest, and a border around throat skin black. Stripe over eye to hindneck, and side of neck orange-crimson. Upper back and underparts dark orange-crimson with white black-edged spots. Rest of upperparts and wings olive-brown with rufous-buff vermiculations and white black-edged spots. Tail black vermiculated buff at base.
B. **Female:** Much like female of number 1 but more rufous-buff to rufous-brown. Tail rufous-brown with broken buff and black bars. Variable in colours.
Young birds: Duller than female.
Bare parts: Bill black. Legs pinkish, female fleshy grey-brown. Eyes brown. Horns and bare facial skin of male blue, throat skin blue and bright grey-green edged with four or five triangular scarlet patches. Only males during display have horn and throat skin.
List: III Nepal
Lit.: 37, 59, 129, 142, 229, 410

3. TRAGOPAN BLYTHII

Assamese: Hurr-hurrea; *English:* Blyth's Tragopan; *French:* Tragopan de Blyth; *German:* Graubauch-Tragopan, Blyth-Satyrhuhn; *Italian:* Tragopano di Blyth; *Spanish:* Tragopan de Blyth

Range: Assam, Tibet, Burma, China
Identification:
A. **Male:** Length 65-70cm. Forehead, crown, border around throat skin, and broad ear-stripe black. Rest of head, neck, breast, upper mantle, and shoulders bright crimson. Rest of upperparts brown, spotted with red, maroon and white. Lower breast and belly smoky grey with paler centres. Flanks and vent mottled crimson, buff and black.
B. **Female:** Length 58cm. Like female of number 2 but paler. Centre of belly and vent more uniform grey. Edges of eyelids lemon-yellow.
Young birds: Like female. Males attain red neck in spring.
Bare parts: Bill brown. Legs pinkish, female brownish. Eyes brown. Horns of male during display bright blue. Throat skin yellow bordered blue.

Geographical variation:
Birds from Tibet are smaller (length of male 53cm). Upperparts darker. White spots smaller but more numerous. Below paler, almost without contrast between light centres and surroundings.
List: I
Lit.: 37, 59, 129, 142, 229, 410

4. TRAGOPAN CABOTI

English: Cabot's Tragopan, Yellow-bellied Tragopan; *French:* Tragopan de Cabot; *German:* Braunbauch-Tragopan, Cabot Satyrhuhn; *Italian:* Tragopano de Cabot; *Spanish:* Tragopan de Cabot

Range: China
Identification:
A. **Male:** Length 61cm. Head black and orange-red. Upperparts black and red with many large buff spots. Underparts pale buff; flanks, lower rump and thighs bordered red and black.
B. **Female:** Length 50cm. Like the other Tragopans except skin around eye and greyish-brown underparts with large white spots.
Young birds: Like female.
Bare parts: Bill horny, female pinkish brown. Legs pink. Eyes brown. Facial skin orange-yellow, female orange. Bare throat skin of male during display pale blue, centre orange with bristly purple spots, surrounded by cobalt blue, bordered with up to nine big pale greenish-grey patches.
List: I
Lit.: 37, 59, 129, 142, 229, 410

5. GALLUS SONNERATII

English: Grey Jungle-fowl, Grey Sonnerat's Junglefowl; *French:* Coq de Sonnerat; *German:* Sonnnerathuhn; *Italian:* Gallo de Sonnerat; *Spanish:* Gallo de Sonnerat

Range: India
Identification:
A. Length 70-80cm. **Male:** Unmistakably black hind head, neck, and foremantle, with grey borders and many golden-yellow bars at ends of feathers. Rest of back black with white borders and shaft-streaks. Elongated rump feathers fringed chestnut, glossed purple and spotted with pale yellow or white. Forewings orange-yellow. Underparts dark brown to black with white to grey shaft-streaks and borders.
B. **Female:** A little smaller than male. Little bare skin around eye and comb. Upperparts sandy-brown with fine white streaks and vermiculated dull black. Underparts white edged and speckled with dark brown. Tail dull rufous-black, central feathers mottled with rufous.
Young birds: Like female.
Bare parts: Bill yellowish, culmen black. Legs yellowish, juvenile yellowish-brown. Eyes yellow to bright red, juvenile yellowish-brown. Comb, facial skin and appendages red, juvenile pink.
List: II
Lit.: 37, 59, 129, 142, 229, 410, 425

PLATE 26

1. LOPHURA EDWARDSI

Formerly known as: Hierophasis edwardsi

English: Edward's Pheasant; *French:* Faisan d'Edwards; *German:* Edwardsfasan; *Italian:* Fagiano di Edward; *Spanish:* Faisán de Edwards

Range: Vietnam
Identification:
A. **Male:** Length 58-65cm. Crest white. Wings blackish to blue with green borders of wing-coverts. Rest dark blue.
B. **Female:** Length c.55cm. No crest. Three central tail feathers dark brown, other tail feathers black. Rest of plumage chestnut-brown, finely vermiculated with black and shaft of feathers pale brown.
Young birds: Have two subterminal dark spots on mantle. Otherwise like female but more greyish on head and neck.
Bare parts: Bill whitish-green, darker at base, female brown. Legs crimson. Eyes reddish brown. Face scarlet.
List: I
Lit.: 37, 59, 129, 142, 222, 229, 410

2. LOPHURA IMPERIALIS

Formerly known as: Hierophasis imperialis

English: Imperial Pheasant; *French:* Faisan impérial; *German:* Kaiserfasan; *Italian:* Lofura imperiale; *Spanish:* Faisán imperial

Range: Indochina
Identification:
A. **Male:** Length 75cm. General colour dark blue. Small crest blue-black.
B. **Female:** Length 60cm. Crown feathers long and often erect. Head pale greyish-brown. Two central tail feathers chestnut, rest of tail and wings blackish. Upperparts chestnut-brown. Underparts more greyish-chestnut.
Young birds: Like female, but more olive-brown. Tail light chestnut.
Bare parts: Bill pale yellowish-green, base blackish. Legs crimson. Eyes red to orange. Face wattles scarlet.
List: I
Lit.: 37, 59, 129, 142, 222, 229, 410

3. LOPHOPHORUS IMPEJANUS

Also known as: Lophophorus impeyanus

English: Himalayan Monal Pheasant; *French:* Lophophore resplendissant; *German:* Gelbschwanz-Glanzfasan; *Italian:* Loforo slendente; *Spanish:* Faisán monal del Himalaya

Range: Himalayas
Identification:
A. **Male:** Length 70cm. Long metallic green crest of spatulate feathers and cinnamon tail distinguish this species. Rest of plumage metallic blue, green, purple, and bronze.
B. **Female:** Length 63cm. Short crest black and rufous-buff.Chin and throat white to behind ear-coverts. Feathers on mantle black, with two buff streaks and buff edges to each feather.Tail black, barred rufous, tip white.
Young birds: Like female but not so distinctly marked.
Bare parts: Bill brown. Legs variably yellowish, brownish-green to grey. Eyes brown. Naked skin around eyes blue.
List: I
Lit.: 37, 59, 129, 142, 229, 410

4. LOPHOPHORUS SCLATERI

English: Sclater's Monal Pheasant, Crestless Monal; *French:* Lophophore de Sclater; *German:* Weisschwanz-Glanzfasan; *Spanish:* Faisán monal de Sclater

Range: Assam, Burma, Yunnan
Identification:
A. **Male:** Length 68cm. No real crest but short, curly, blue-green feathers on crown. Lower back, rump, and upper tail-coverts white. Body black, with metallic blue-green and purple gloss on upperparts. Tail white with rufous band in middle.
B. **Female:** Length 63cm. No crest. Chin and throat white. Back, shoulders and forewing brown with buff streaks. Underparts dull brown with wavy bars of dull yellow-brown. Tail with six or seven white bars.
Young birds: Undescribed.
Bare parts: Bill dirty white, female pale yellow. Legs pale greenish. Eyes dark brown. Skin around eyes extending to bill blue.

Geographical variation:
Birds from Myitkyina (Burma) have terminal white tail band narrower in male. Female paler.
List: I
Lit.: 37, 55, 59, 129, 142, 229, 410

5. LOPHOPHORUS IHUYSII

Chinese: Impeyan, monal; *English:* Chinese Monal Pheasant; *French:* Lophophore de Lhuys; *German:* Grünschwanz-Glanzfasan; *Spanish:* Faisán monal Chino

Range: China
Identification:
A. **Male:** Length 80cm. Long hanging crest from hind crown purple to bronze. Body black with metallic blue, green, red copper, and bronze gloss on upperparts. Rump and upper tail-coverts white, last with triangular black spots at tip. Tail blue-green.
B. **Female:** Length 76cm. No crest. Back and rump white. Tail barred rufous and dark brown.
Young birds: Undescribed.
Bare parts: Bill brown, female grey. Legs blackish, female yellowish-grey. Eyes brown. Skin around eyes blue.
List: I
Status: Rare and likely to disappear before long.
Lit.: 37, 59, 129, 142, 152, 169, 229, 410

PLATE 27

1. CROSSOPTILON MANTCHURICUM

Chinese: Hoki; *English:* Brown-eared Pheasant, Manchurian eared Pheasant; *French:* Hoki brun, Faisan oreillard brun; *German:* Brauner Ohrfasan; *Italian:* Fagiano orecchiuto bruno; *Spanish:* Faisán orejudo pardo

Range: China
Identification:
Length 96cm. Sexes alike. Ear-coverts and tufts white. Neck black. Mantle, wings and underparts brown, wings with purple wash. Lower back and rump silvery white.
Young birds: Mottled brown with buff tips to feathers. Tail shorter.
Bare parts: Bill pale reddish. Legs crimson. Eyes pale reddish-brown. Facial wattles red.
List: I
Lit.: 37, 59, 129, 142, 189, 229, 410

2. CROSSOPTILON CROSSOPTILON

English: White-eared Pheasant; *French:* Hoki blanc, Faisan orellard blanc; *German:* Schmalschwanz-Ohrfasan, Weisser Ohrfasan; *Italian:* Fagiano orecchiuto bianco; *Spanish:* Faisán orejudo blanco; *Tibetan:* Sharkar

Range: S. & C. China, S. E. Tibet
Identification:
Male: Length 96cm. Crown black, ear-coverts and tufts white. Neck and body white with pale grey in wings and upper tail-coverts.Tail purplish bronze, tip greenish-blue.
Female: No spurs.
Young birds: Grey with rusty buff streaks and bands on head and upperparts. Wings brownish-grey with some dark brown. Underparts grey.
Bare parts: Bill reddish. Legs red. Eyes orange-yellow.

Geographical variation:
Birds from Koko Nor have hind neck and upperparts pale ashy-grey. Feathers hairlike.
Birds from Yunnan have lighter grey wings. Four outer tail feathers mostly with a distinct whitish-grey outer border.
List: I
Lit.: 37, 59, 129, 142, 189, 229, 410

CROSSOPTILON HARMANI (not depicted)
Formerly known as: Crossoptilon crossoptilon harmani
English: Tibetan Eared-Pheasant
Range: S. W.& S. C. Tibet, N. E. India
Like last one but pure white except black crown, tail, and usually inner web of outer wing-feathers, which are mottled brownish grey. (Highly variable).
List: I
Lit.: 59, 129, 142, 189, 229, 410

3. LOPHURA I. IGNITA

Borneo: Sempidan; *English:* Crested Fireback, Viellot's Fireback; *French:* Faisan noble; *German:* Feuerrückenfasan; *Italian:* Fagiano nobile; *Malaya:* Ayam pĕgar, Ayam suil; *Spanish:* Faisán de carúncula azul crestado

Range: Borneo, Sumatra, Greater Sunda Islands
Identification:
A. **Male:** Length 65-67cm. Large purplish-blue crest. Rump coppery maroon. Lower breast and flanks coppery chestnut. Three central tail-feathers cinnamon-buff, rest purplish-blue.
B. **Female:** Length 56cm. Whole upperparts rufous-chestnut, chin and throat white. Breast and flanks blackish-brown with white borders.Tail black vermiculated with dark chestnut. No spurs.
Young birds: Like female, but wing coverts with large dark spots.
Male darker than female.
Bare parts: Bill horn white. Legs greyish white to flesh. Eyes red. Face wattles blue.

Subspecies:

Lophura ignita rufa
English: Malaysian Fireback
Range: Thailand, Malaysia, Sumatra
Male: Central tail-feathers white. Breast and flanks blue, flanks with white stripes. Legs crimson.
Female: Tail chestnut and legs red.
List: III Malaysia
Lit.: 37, 59, 129, 142, 189, 229, 410

4. LOPHURA ERYTHROPHTHALMA

Formerly known as: Houppifer erythrophthalmus

Borneo: Singgier; *English:* Crestless Fireback, Rufous-tailed Pheasant; *French:* Faisan à queue rousse; *German:* Gabelschwanzfasan; *Italian:* Fagiano dalla coda gialla; *Malaya:* Kuang-bestam; *Spanish:* Faisán de carúncula azul

Range: Malaya, Sumatra, Borneo
Identification:
A. **Male:** Length 47-50cm. Head crestless. Middle back metallic copper passing to silky maroon on rump. Tail cinnamon, middle tail-feathers a little shorter than rest. Rest purplish black with fine silvery grey spots. Legs with a long spur.
B. **Female:** Length 42-44cm. Brownish head. Rest steel blue.
Young birds: Black with rufous tips to feathers.
Bare parts: Bill greenish-white, female black. Legs bluish-grey. Eyes reddish-brown, female brown. Face wattles scarlet.

Geographical variation:
Birds from Borneo have neck and upper back light grey marbled with black and white stripes. Breast and flanks with wide white stripes.
List: III Malaysia
Lit.: 37, 59, 129, 142, 189, 222, 229, 410

5. LOPHURA SWINHOII

Formerly known as: Hierophasis swinhoii

Chinese: Wa-köe; *English:* Swinhoe's Pheasant; *French:* Faisan de Swinhoe; *German:* Swinhoefasan; *Italian:* Fagiano de Swinhoe; *Spanish:* Faisán de Swinhoe

Range: Taiwan
Identification:
A. **Male:** Length 79cm. Crest, upper back and two central tail-feathers white. Shoulders crimson-maroon. Wings black with green borders. Rest black with blue borders.
B. **Female:** Length 50cm. Without crest. Mantle and shoulder chestnut-brown with buff and black arrow-shaped marks. Underparts rusty buff with V shaped marks. Central tail-feathers black and brown with buff bars, rest of tail chestnut-red.
Young birds: Like female but duller.
Bare parts: Bill yellowish. Legs crimson. Eyes red-brown, female brown. Face wattles scarlet.
List: I
Lit.: 37, 59, 129, 142, 189, 222, 229, 269, 270, 410

PLATE 28

1. CATREUS WALLICHII

English: Cheer Pheasant, Wallich's Pheasant; *French:* Faisan de Wallich; *German:* Schopffasan; *Italian:* Catreo wallichi; *Kumaon and Garwhal:* Tshi-er; *Nepalese:* Chihir; *Spanish:* Faisán de Wallich

Range: Himalayas
Identification:
A. **Male:** Length 95-100cm. Long brown crest. Upperparts pale buff barred black. Rump rusty buff with black bars. Underparts paler and barred black. Tail buffy-grey with wide bars mixed with black and dark grey.
B. **Female:** Much like male but a little smaller.
Young birds: Like female but duller, and no crest.
Bare parts: Bill yellowish. Legs plumbeous to grey-brown. Eyes reddish to buff. Naked face skin red.
List: I
Status: No estimates have been made, but may be on the way to extinction.
Lit.: 37, 59, 84, 85, 142, 152, 155, 165, 229, 410

2. SYRMATICUS ELLIOTI

Chinese: Han-Ky; *English:* Elliot's Pheasant, Chinese Barred-backed Pheasant; *French:* Faisan d'Elliot; *German:* Elliotfasan; *Italian:* Fagiano di Elliot; *Spanish:* Faisán de Elliot

Range: China
Identification:
A. **Male:** Length 80cm. Sides of neck and lower part of underside white. Mantle and upper breast bright rufous-chestnut edged black and coppery red. Lower back and rump black barred white. Tail pale grey barred with chestnut and black irregular lines.
B. **Female:** Length 50cm. Chin, throat, and foreneck black. Central tail-feathers pale brown with indistinct chestnut bars at tip, other tail-feathers with broad white tips.
Young birds: Duller than female. Throat white.
Bare parts: Bill yellowish, female pale brown. Legs grey. Eyes brown.
List: I
Status: Increasingly rare in wild state.
Lit.: 37, 59, 129, 142, 152, 229, 410

3. SYRMATICUS HUMIAE

Burmese: Yit-min; *English:* Hume's Bar-tailed Pheasant; *French:* Faisan de Hume; *German:* Burmafasan, Hume Fasan; *Italian:* Fagiano di Burma; *Manipur:* Loe-nin-koi; *Spanish:* Faisán de Hume

Range: India, Burma, Thailand, China
Identification:
A. **Male:** Length 90cm. Black neck, deep chestnut on mantle, wings and entire underparts. Lower back and rump steel blue with narrow white fringes.
B. **Female:** Length 60cm. Throat pale fulvous brown. Tail grey with black and rufous bars.
Young birds: Like female but duller.
Bare parts: Bill horn yellow. Legs grey. Eyes brown to orange. Bare face skin red.

Geographical variation:
Birds from Burma have lower back and rump black with a wider white fringe, (5mm against 3mm).
List: I
Lit.: 37, 59, 129, 229, 410

4. SYRMATICUS MIKADO

English: Mikado Pheasant; *French*: Faisan mikado; *German*:Mikadofasan; *Italian:* Fagiano mikado; *Spanish:* Faisán mikado

Range: Taiwan
Identification:
A. **Male:** Length 88cm. Dull black with bluish-purple to metallic blue border to each feather. Long tail black with irregular white bars.
B. **Female:** Length 53cm. Like last, Hume's Bar-tailed Pheasant, but duller, more olive-brown; white spots on mantle usually larger. Tail more chestnut and black and white barring narrower.
Young birds: Like female, but head and neck with buffy white markings.
Bare parts: Bill blackish, Legs dark grey. Eyes reddish-brown, female light brown. Bare facial skin red.
List: I
Lit.: 37, 59, 129, 142, 229, 410

PLATE 29

1. POLYPLECTRON EMPHANUM

English: Palawan Peacock-Pheasant; *French:* Eperonnier Napoléon; *German:* Napoleonfasan; *Palawan:* Sulu maläk, Dusan bërtik

Range: Palawan (Philippines)
Identification:
A. **Male:** Length 50cm. Crown and long crest dark metallic green. Ear-coverts white. Upperparts metallic blue and green, hind back and rump spotted chestnut and black. Wings without ocelli.Underparts blackish. Tail blackish spotted buff with two bands of ocelli.
B. **Female:** Length 40cm. Crest long, dark brown and curved backwards. Mantle, back, and wings with tawny-buff bars and without ocelli.
Young birds: Male: Between bill and eye black. A few black or blue feathers on throat, mantle and wings. Female: Black spots on wings.
Bare parts: Bill and legs black, female dark grey. Eyes brown. Bare facial skin red.
List: I
Lit.: 59, 61, 129, 142, 229, 410

2. POLYPLECTRON MALACENSE

Also known as: Polyplectron malacensis

English: Malayan Peacock-Pheasant; *French:* Eperonnier de Hardwicke, Eperonnier Malais; *German:* Malaienpfaufasan; *Italian:* Speroniere malese; Malaya: Kuan, Kuang; *Spanish:* Faisán de cola ocelada malayo

Range: Thailand. Malay Peninsula
Identification:
A. **Male:** Length 50cm. Long metallic blue-green crest. General coloration buffish-brown spotted black. Hindneck violet. Many blue-green ocelli on wings. Tail with two rows of metallic green ocelli.
B. **Female:** Length 40cm. Crest short. Dull black triangular ocelli on mantle and wings.
Young birds: Like female.
Bare parts: Bill and legs dark grey, lower bill yellowish. Eyes bluish-white, female brown. Bare facial skin red.
List: II
Lit.: 37, 59, 129, 142, 222, 229, 410, 425

POLYPLECTRON SCHLEIERMACHERI (not depicted)
Also known as: Polyplectron malacense schleiermacheri

English: Bornean Peacock-Pheasant; *French:* Eperonnier de Borneo; *German:* Borneo-Spiegelpfau

Range: Borneo
Identification:
Male: Like No. 2, Malayan Peacock-Pheasant, but grey and black crest shorter and only centre glossed with green. Ear-coverts black. Large black and pale grey ruff with glossy violet-blue tips. Above darker, ocelli smaller and greener. Throat, upper breast and middle of lower breast white, breast sides contrasting glossy blue-green. Belly black.
Female: Like Malayan Peacock-Pheasant but general coloration more reddish. Belly darker.
Young birds: Undescribed.
Bare Parts: Like previous except female has brown eyes.
List II
Lit.: 59, 129, 142, 222, 229, 410, 425

3. POLYPLECTRON BICALCARATUM

Burmese: Daung-min, Daung-kala; *Chinese:* Tshinquis, Chin-tchienkhi; *Chainghpaw:* U-gaw; *English:* Grey Peacock-Pheasant; *French:* Eperonnier chinquis; *German:* Grauer Pfaufasan; *Italian:* Speroniere grigio; *Spanish:* Faisán de cola ocelada gris

Range: India to China, Burma to Laos and Vietnam
Identification:
A. **Male:** Length 65-75cm. Primary colour grey-brown. Upper mantleand wings with many violet green-blue ocelli surrounded by narrow brown and a broader white band.Tail with two bands of metallic green ocelli.
B. **Female:** Length 55cm. Duller than male. Ocelli on upper back and wings not round but broken by white bars. Only one band of ocelli on tail.
Young birds: Like female.
Bare parts: Bill creamy flesh, tip black. Legs dark slaty (variable). Eyes white or pearl-grey, female brown to grey.
Bare facial skin yellowish flesh.

Geographical variation:
Birds from Sikkim to Assam have rings and markings whitish buff or white.
Birds from Thailand are greyer and darker. Ocelli pure white.
Birds from Vietnam have ocelli of tail-feathers surrounded by wide buffy-grey border.
Birds from Hainan Island are smaller and all markings are very bright. Ocelli on mantle and wings blue and green.
List: II
Lit.: 37, 59, 129, 142, 229, 410

4. POLYPLECTRON GERMAINI

English: Germain's Peacock-Pheasant; *French:* Eperonnier de Germain; *German:* Brauner Pfaufasan, Germain Pfaufasan; *Italian:* Speroniere di Germain; *Spanish:* Faisán de cola ocelada de Germain

Range: Vietnam
Identification:
A. **Male:** Length 55cm. Like number 3 but smaller and with red facial skin. Ocelli on upperparts surrounded by black and buff border.
B. **Female:** Length 48cm. Duller than male. Ocelli on back and wings triangular. Like male, two bands of green ocelli on tail.
Young birds: Like female but ocelli dull black.
Bare parts: Bill and legs blackish-brown. Eyes brown. Facial skin red.
List: II
Lit.: 37, 59, 129, 142, 229, 410

5. POLYPLECTRON INOPINATUM

Formerly known as: Chalcurus inopinatus

English: Rothschild's Peacock-Pheasant, Mountain Peacock Pheasant; *French:* Eperonnier de Rothschild; *German:* Rothschild-Pfaufasan; *Spanish:* Faisán de cola ocelada indómito

Range: Malaya
Identification:
A. **Male:** Length 65cm. Unmistakably grey, chestnut, and greyish-black with fine white and pale grey spots on head, rump, and tail. Fine white lines along flanks. Small ocelli on back and forewing metallic blue. Outer tail-feathers with metallic green ocelli.
B. **Female:** Length 46cm. Resembles male but ocelli on mantle and wings duller.
Young birds: Similar to female.
Bare parts: Bill and legs grey. Eyes brown.
List: III Malaysia
Lit.: 37, 59, 129, 142, 229, 410

PLATE 30

1. PAVO MUTICUS

Burma: Oodoung; *Burmese:* Daung; *Chinghpaw:* U-daung; *English:* Green-necked Peafowl, Green Peafowl; *French:* Paon spicifère; *German:* Ährenträgerpfau; *Italian:* Pavone mutico; *Malaya:* Burong merak; *Spanish:* Pavo real verde

Range: India, Burma, Thailand, Indochina, Malaya, Java, China
Identification:
A. **Male:** Length 180-300cm. Differs chiefly from Common Peafowl by having a golden green neck and breast, where colour in Common Peafowl is blue.
B. **Female:** Length 100-110cm. Differs in same way as female from Common Peafowl male.
Young birds: Like female but lower back mostly bronze-green.
Bare parts: Bill horn grey. Legs dark brownish grey. Eyes brown. Skin around eye bluish-green.

Geographical variation:
Birds from Assam and W. Burma are duller and more blue.
Male birds from E. Burma, Thailand and Indochina have more coppery, not so golden green, fringes on neck, upper back and breast. Females are more heavily marked with buff on breast.
List: II
Lit.: 37, 59, 123, 129, 142, 229, 410

2. RHEINARDIA O. OCELLATA

Formerly known as: Rheinartia nigrescens

English: Crested Argus, Rheinard's Pheasant; *French:* Rheinarte ocellé; *German:* Perlenfasan; *Italian:* Rainardo ocellato; *Spanish:* Faisán de Rheinard; *Annam, Vietnam:* Tri

Range: Vietnam, Laos
Identification:
Male: Length 195-235cm. Crown dark brown.Crest white in middle, rest reddish-brown, 6cm long. Rest of plumage dark brown, red and chestnut with white spots and markings on each feather.Enormously long tail, central tail-feathers 150-173cm distinguishing this species.
Female: Length 74-75cm. Crest shorter and less full. Tail-feathers transversely mottled with black and buff.
Young birds: Duller and shorter tails.
Bare parts: Bill pink. Legs brown. Eyes brown.

Subspecies:

3. Rheinartia ocellata nigrescens
English: Malaysian Argus
Range: Malaysia
A. **Male:** Crest longer, 8,5cm and white except for a few black feathers in front. Darker and more white spots.
B. **Female:** Slightly brighter and more closely marked with black.
List: I
Lit.: 37, 59, 129, 142, 229, 410

4. ARGUSIANUS ARGUS

Borneo Dutch: Keee; *English:* Great Argus, Great Argus Pheasant; *French:* Argus géant; *German:* Argusfasan; *Iban:* Ruai; *Italian:* Fagiano argo, Argo; *N. Malaya:* Kuang raya; *Spanish:* Faisán argos; *Sumatra:* Koeweau

Range: Thailand, Malaya, Sumatra, Borneo
Identification:
A. **Male:** Length 170-200cm. Crown and crest black. Rest of head and neck bare with sparse fine shaft-like feathers. Body dark brown, barred and edged buffish and dark rufous. Brown and purple wing- and tail-feathers spotted white, rows of iridescent green-grey, purple and buffish ocelli surrounded by black and pale buff on inner flight-feathers. Tail 110cm long.
B. **Female:** Length 74-76cm. Crown dull rufous-buff edged black. Crest dark grey. Nape chestnut-rufous. Body black mottled, barred, and vermiculated rufous, chestnut and buff.
Young birds: Like female but more reddish and boldly marked.
Bare parts: Bill light bluish, female whitish to greyish-white. Legs reddish. Eyes hazel to greenish-brown. Bare skin on head and neck pale slate to bluish-grey.

Geographical variation:
Birds from Borneo are smaller, male 160-180cm, female 72-73cm. Male greyer and more reddish-orange below. Female have a brighter reddish orange-chestnut neck and breast. Underparts lighter brown.
List: II
Lit.: 37, 59, 129, 229, 410

PLATE 31

1. GRUS AMERICANA

English: Whooping Crane; *French:* Grue d'Amérique, Grue blanche d'Amérique; *German:* Schneekranich, Schreikranich; *Italian:* Gru Americana, Gru urlatrice; *Spanish:* Grulla trompetera, Grulla gritona, Grulla Americana
Range: North America
Identification: Length 127-151cm. Sexes alike. Forehead and little area on nape black. Rest of head feathers, neck and body white except for black long flight feathers. Innermost flight feathers elongated and tail completely hidden.
Young birds: Whitish, scattered reddish-brown, long flight-feathers black. Entire head feathered.
Bare parts: Bill wax yellow mixed with dull greenish to yellowish at tip, juveniles have darker bill. Legs black. Eyes yellow. Bare warty and red skin between bill and eye, crown, nape, malar and throat.
List: I
Status: Total population in wild 1965: 44 birds. 1987: 110 birds.
Lit.: 37, 129, 141, 152, 173, 175, 196, 410

GRUS C. CANADENSIS
English: Sandhill Crane; *French:* Grue canadienne, Grue du Canada; *German:* Kanadakranich, Sandhügelkranich; *Italian:* Gru Canadese, *Japanese:* Kanada-Zuru; *Russian:* Kanadskij shurawl; *Spanish:* Grulla cenicienta, Grulla del Canada
Range: North America, E. Siberia
Identification: Length 51-76cm. Sexes alike but female smaller. Naked forehead and crown. Rest of head and upper throat whitish. Wings grey to dark slate, innermost flight feathers elongated and bushy. Rest grey, often with red-brown spots from mud preening.
Young birds: Head completely feathered. Elongated flight-feathers shorter than adult's.
Bare parts: Bill grey to olive-grey. Legs blackish. Eyes red to reddish-brown. Bare skin between bill and eye, forehead and crown red with few black bristles.
Wing measurements: Male: 418-510mm, average 470mm; Female: 420-500mm, average 448mm. Weight: Male: under 3.75kg; Female under 3.5kg.
List: II

Subspecies:
Grus canadensis tabida
English: Greater Sandhill Crane
Range: Canada, USA
Length 102-127cm. Wing measurements: Male: 460-598mm, average 540mm; Female: 467-510mm, average 485mm. Weight: Male: over 4.9kg; Female: over 4.3kg.
List: II
Grus canadensis pratensis
English: Florida Sandhill Crane
Range: South-East USA (sedentary)
Wing measurements: Male: 460-533mm, average 501mm. Female: 445-517mm, average 478mm.
List: II
2. GRUS CANADENSIS PULLA
English: Mississippi Sandhill Crane
Range: Gulf Coast, South USA (sedentary)
Darker. Measurements: No data.
List: I
Status: Total population 1979: 40-50 birds.
GRUS CANADENSIS NESIOTES
English: Cuban Sandhill Crane
Range: Cuba and Isle of Pines (sedentary)
Length 51-76cm. Darker.
List: I
Status: About 200 birds in 1977.
Lit.: 37, 129, 133, 141, 152, 195, 291, 410
(Sibley and Monroe have no subspecies)

3. GRUS JAPONENSIS

Chinese: Tan-ting ho, Hsien-ho; *English:* Red-crowned Crane, Japanese Crane, Manchurian Crane; *French:* Grue du Japon, Grue de Mandchourie; *German:* Mandschurenkranich; *Italian:* Gru della Manciuria, Gru dell Giappone; *Japanese:* Taucho, Tozuro; *Russian:* Japonskij shurawl; *Spanish:* Grulla de Manchuria
Range: Siberia, Manchuria, Mongolia, China, Japan
Identification: Length 152cm. **Male:** Bare skin on crown. Between bill and eye, cheeks, throat and front and sides of neck black. Lower wings black, innermost flight-feathers elongated and decurved. Rest of plumage white.
Female: Like male but cheeks, throat, and neck pearly-grey.
Young birds: Head completely feathered and brownish, collar on neck grey to grey-brown. Scattered white, grey, and brown on rest of plumage.
Bare parts: Bill olive-green. Legs blackish. Eyes dark brown.
Bare skin on crown of adult red.
List: I
Lit.: 37, 64, 129, 133, 141, 180, 410

4. GRUS LEUCOGERANUS

Formerly known as: Bugeranus leucogeranus
English: Siberian Crane, Siberian White Crane, Great White Crane; *French:* Grue blanche, Grue nonne, Grue de Sibérie; *German:* Nonnenkranich, Schneekranich; *Italian:* Gru bianca asiatica; *Japanese:* Sod-egura-zuru, Sode guro-zuru; *Russian:* Sterch; *Spanish:* Grulla blanca asiática
Range: Russia, Siberia, India, China, Japan
Identification: Length 133cm. Sexes alike. Forehead naked. Long flight-feathers black. Rest of plumage snow white.
Young birds: Fully feathered head. Scattered rusty-cinnamon-brown and dull white. Long flight-feathers black.
Bare parts: Bill rufous, juvenile duller. Legs reddish, juvenile darker. Eyes reddish to bright yellow, juvenile hazel to pale yellow. Bare forehead red.
List: I
Lit.: 37, 64, 129, 133, 141, 410

5. GRUS MONACHA

Chinese: Huan-has; *English:* Hooded Crane; *French:* Grue moine; *German:* Mönchskranich; *Italian:* Gru monaca, Gru dal cappuccio; *Japanese:* Nabe-zuru; *Russian:* Tschernyj shurawl; *Spanish:* Grulla monjita
Range: Siberia, China, Japan
Identification: Length 92cm. Sexes alike. Forepart of crown sparsely covered with black bristles. Rest of head and neck white, often with some grey. Lower wings black, innermost flight-feathers elongated and decurved. Rest dark slate-grey, upperparts with brownish tinge.
Young birds: Head fully feathered, crown black and white.
Bare parts: Bill yellowish horn. Legs black. Eyes yellowish-brown, juvenile brown. Bare skin on forepart of crown red.
List: I
Lit.: 37, 64, 129, 133, 141, 410

6. GRUS NIGRICOLLIS

English: Black-necked Crane; *French:* Grue à cou noir; *German:* Schwarzhalskranich; *Italian:* Gru collonero; *Spanish:* Grulla cuellinegra

Range: Asia to China and India
Identification: Length 152cm. Sexes alike. Nearly naked between bill and eye and crown, only with a few black hair-like feathers. Little white spot behind eye. Rest of head and upper part of neck black. Lower wings and tail black, innermost flight-feathers elongated and decurved. Rest ashy-grey, more whitish on underparts.
Young birds: Head buffish and fully feathered. Neck black and white.
Bare parts: Bill greenish horn, tip yellow. Legs black. Eyes yellow. Bare skin on crown and between bill and eye red.
List: I
Lit.: 37, 64, 129, 141, 410

7. GRUS VIPIO

Chinese: Ts-ang-kua; *English:* White-necked Crane, Japanese White-necked Crane; *French:* Grue à cou blanc; *German:* Weissnackenkranich; *Italian:* Gru collobianco; *Japanese:* Mana-zuru; *Russian:* Daurskij shurawl; *Spanish:* Grulla cuelliblanca
Range: Siberia, Mongolia, China, Japan
Identification: Length 152cm. Sexes alike. Naked skin with thin hair-like feathers on face and forehead. Ear-coverts light grey. Rest of head white. Neck white with blackish-grey sides. Forewing light grey, innermost flight-feathers elongated, decurved and white, rest of flight-feathers black. Body slaty-grey, tail tipped black.
Young birds: Entire head feathered and cinnamon-brown.
Bare parts: Bill yellowish-green. Legs pinkish to blue-pink. Eyes orange-yellow. Bare skin on face and forehead red.
List: I
Lit.: 16, 37, 129, 133, 141, 410

PLATE 32

Family: **GRUIDAE ***

English: Cranes; *French:* Grues; *German:* Kraniche

Identification:
Large birds. Colour grey, brown, or white, many with bare skin on crown. Bill rather long and straight. Wings usually with curled or decurved and elongated feathers. Short tails. Legs long with hind toe.

PLATE 33

COMMON CRANE 5

All other cranes:

That is:

4.	Balearica pavonina	Black Crowned-Crane
	Balearica regulorum	Grey Crowned-Crane
	Grus antigone	Sarus Crane
	Grus rubicunda	Brolga
	Grus canadensis (- nesiotes & pulla)	Sandhill Crane
2.	Grus virgo	Demoiselle Crane
3.	Grus paradisea	Blue Crane
1.	Grus carunculatus	Wattled Crane
5.	Grus grus	Common Crane

List: II

PLATE 34

1. TURNIX MELANOGASTER

English: Black-breasted Button Quail; *French:* Turnix à poitrine noire; *German:* Schwarzbrust-Laufhühnchen

Range: Queensland, New South Wales, (Australia)
Identification:
A. Length 15-18cm. female larger. **Male:** Crown pale brown mottled dusky. Rest of head whitish mottled pale brown. Upperparts rufous with black bars and cream-white stripes. Wings grey-brown,wing-coverts dull chestnut with creamy spots. Underparts whitish edged black on breast and flank feathers. Abdomen and under tail-coverts buffish with fainter spotting. Tail short and hidden.
B. **Female:** Like male but head and breast black with white spots over eyes, cheeks, and across breast. Abdomen darker than male.
Young birds: Like male.
Bare parts: Bill grey. Legs (with 3 toes) dull yellow, juvenile greyish. Eyes cream-white, juvenile dark.
List: II
Lit.: 36, 129, 225, 232, 278, 307, 410

2. PEDIONOMUS TORQUATUS

English: Plains Wanderer; *French:* Pédionome errant; *German:* Trappenlaufhühnchen

Range: South Australia, New South Wales, Victoria (Australia)
Identification:
A. Length 15-17cm, female largest. **Male:** Crown and upperparts pale brown mottled black and edged white, nape paler with black and brown markings. Wings dull brown. Side of head buffish with black spots. Underparts pale buff, front of neck and breast buff with black bands, sparser on flanks. Tail short.
B. **Female:** Like male but colour brighter. Collar of white and black feathers. Upper breast and nape chestnut, yellow to orange during breeding.
Young birds: Like male but upperparts more heavily mottled black and breast and flanks with dusky, not black, bands.
Bare parts: Bill pale yellow, area round nostrils browner. Legs (4 toes) pale yellow. Eyes cream-white.
List: II
Lit.: 3, 11, 129, 225, 233, 278, 307, 410

3. GALLIRALLUS SYLVESTRIS

Also known as: Rallus sylvestris; Tricholimnas sylvestris

English: Lord Howe Rail, Lord Howe Wood Rail; *French:* Râle sylvestre, Râle de Lord Howe; *German:* Waldralle; *Spanish:* Rascón de Isla Lord Howe

Range: Lord Howe Island (Australia)
Identification:
Length 46cm. Sexes similar. Top of head and upperparts olive-brown. Wings chestnut regularly barred blackish-brown, only noticeable when wings are outstretched; wing-coverts dull chestnut barred dark brown. Side of head, eyebrow, chin, and throat whitish-grey. Breast, flanks, belly, and under tail-coverts olive-brown, latter barred dull chestnut. Tail olive-brown.
Young birds: Paler. Long flight-feathers (with pointed tips) and their coverts barred with black. Adult has square tips to flight-feathers.
Bare parts: Bill (length 47-50mm) pinkish-grey. Legs dark grey. Eyes red.
List: I
Status: 1984 only 18-20 birds!
Lit.: 37, 98, 129, 196, 241, 380, 410

4. GALLIRALLUS AUSTRALIS HECTORI

English: Buff Weka, Eastern Weka Rail; *French:* Râle wéka; German: Wekaralle; *Italian:* Rallo Weka; *Spanish:* Rascón Weka

Range: Eastern South Island, (New Zealand)
Identification:
Length 53cm. Sexes similar but male slightly larger. Flightless. Head yellowish-buff tinged red and spotted brownish-black. Eyebrow grey. Between bill and eye and band under eye rufous. Upperparts yellowish-buff longitudinally spotted dark brown. Wings and wing-coverts red-brown narrowly barred black and partly edged with buff. Throat grey, breast, flanks, and under tail-coverts yellow-brown, flanks and tail-coverts barred dusky. Belly ashy-brown. Tail reddish with dusky bands.
Young birds: Undescribed.
Bare parts: Bill reddish-brown. Legs dark pinkish. Eyes red to red-brown.
List: II

Note:
Gallirallus a. australis from N. & W. South Island
Gallirallus australis greyi from North Island and
Gallirallus australis scotti from Stewart Island
are not listed
They are all three much darker, general colour on upperparts reddish-brown.
Lit.: 37, 69, 129, 131, 195, 232, 241, 243

5. RHYNOCHETUS JUBATA

Also known as: Rhynochetos jubatus

English: Kagu; *French:* Kagou huppé, Kagou; *German:* Kagu; *Italian:* Kagu; *Spanish:* Kagú

Range: New Caledonia
Identification:
Length cc. 55-61cm. Sexes alike. Flightless. Head ashy-grey with a pronounced crest. Upperparts light ashy-grey with a brownish tinge on back and wings. Flight-feathers with broad black, white, and rufous bars, only visible when wings are spread. Underparts pale buffy-grey. Tail ashy-grey with black and brown bands.
Young birds: Plumage more rufous.
Bare parts: Bill and legs orange-red. Eyes red.
List: I
Lit.: 3, 37, 98, 129, 132, 183, 307, 343, 410

PLATE 35

1. CHLAMYDOTIS UNDULATA

Formerly known as: Psophia undulata; Otis houbara

English: Houbara Bustard; *French:* Outarde houbara, Houbara ondulé; *German:* Kragentrappe; *Italian:* Ubara Africana; *Spanish:* Hubara

Range: Canary Islands, Sahara to Nile Valley, Syria to Asia and India
Identification:
Length 65cm. **Male:** Forehead and sides of crown buff, fine vermiculated black. Crown with elongated narrow and erectile white feathers. Face buffish with fine brown streaking. Upper neck with erectile long black filamentous feathers on sides. Lower neck feathers white and longer. Upperparts orange-brown, vermiculated fine brown. Wings spread black with two large white bands. Forewing and tail like upperparts, tail with four pale blue-grey bands, and white tips, except central pair. Underparts white, on foreneck and breast with a buff wash.
Female: Smaller with shorter elongated feathers on head and neck. Only three bars on tail.
Young birds: Like female but crest smaller and more dark spots on upperparts. White areas in wings buffish.
Bare parts: Bill dark lead grey. Legs greyish to yellowish. Eyes pale yellow.

Geographical variation:
Birds from Canary Islands are smaller and more heavily vermiculated. Birds from Syria to Asia and India have a black and white crown. Foreneck white, vermiculated brown.
List: I
Lit.: 37, 51, 81, 89, 129, 133, 209, 305, 382, 410, 424

2. ARDEOTIS NIGRICEPS

Also known as: Choriotis nigriceps; Otis nigriceps; Otis edwardsii

English: Indian Bustard, Great Indian Bustard; *French:* Outarde à tête noire, Grande outarde de l'Inde; *German:* Hindutrappe, Indische Trappe; *Hindi:* Sohan, Huukna; *Italian:* Grande otarda dell'India; *Spanish:* Avutarda Indica

Range: India, Pakistan
Identification:
A. **Male:** Length 122cm. Forehead and crown black, feathers on hind crown elongated. Rest of head and neck white, finely striated with black. Upperparts raw umber, finely vermiculated buff. Wings black, raw umber, grey and with some white patches. Underparts white, a black band across hind breast.
B. **Female:** Length 92cm. White eyebrow. White on head, throat and foreneck heavily vermiculated black. Black breast band often absent or incomplete.
Young birds: Like female.
Bare parts: Bill greyish-brown above, yellow below. Legs yellowish to greyish. Eyes yellow.
List: I
Lit.: 2, 37, 129, 209, 231, 308, 382, 410, 424

3. EUPODOTIS BENGALENSIS

Also known as: Houbaropsis bengalensis; Otis bengalensis

Assamese: Ulu moira; *English:* Bengal Florican, Bengal Bustard; *French:* Outarde du Bengale, Outarde de l'Inde; *German:* Barttrappe, Indische Trappe; *Hindi:* Chards, Charg, Charat; *Italian:* Otarda del Bengala; *Spanish:* Avutarda Bengali

Range: India, Himalayas, Cambodia, Vietnam
Identification:
A. Length 66cm. **Male:** Entire head, neck, and underparts velvety jet black, on head and neck with a blue tinge. Crown with a black crest. Upperparts buffy with dark brown arrowed marks. Wings with a large white area. Tail with three dark brown bars.
B. **Female:** Slightly larger. Entire bird rufous-buff and sandy-buff mixed with dark brown arrowed marks. Forehead and crown dark brown and buff. No white area on wings.
Young birds: Like female.
Bare parts: Bill dark horn brown, lower yellowish. Legs yellow, sometimes washed with green or grey. Eyes yellow to brown.

Geographical variation:
Birds from Cambodia are smaller and darker on upperparts, female darker on neck and upperparts.
List: I
Lit.: 2, 37, 129, 135, 136, 138, 209, 308, 382, 410

4. OTIS TARDA

English: Great Bustard; *French:* Grande outarde, Outarde barbue; *German:* Grosstrappe; *Hungarian:* Túzok; *Italian:* Otarda comune; *Spanish:* Avutarda

Range: Central and South Europe, Asia to China
Identification:
A. **Male:** Length 105cm. Head and foreneck pale blue-grey with long white moustaches. Colour on neck gradually darkening to cinnamon then chestnut. Upperparts and wings black with cinnamon bars, wings with blackish tips to flight-feathers and a large white area. Tail cinnamon with a black subterminal band, outermost tail feathers white. Male non breeding: Without moustache and breast grey.
B. **Female:** Length 74cm. Like non breeding male, but a few females have short moustaches.
Young birds: Like female.
Bare parts: Bill grey, tip dark brown. Legs dark olive-brown to dark grey. Eyes dark brown.
List: II
Lit.: 37, 51, 87, 89, 133, 209, 382, 410

PLATE 36

Family: **OTIDIDAE** *

English: Bustards; *French:* Outardes; *German:* Trappen

Identification:
Length 36-132cm. Sexes unlike. Colour buff to grey on upperparts vermiculated black. Underparts pale buff, white or black. Some species crested, some with long bristly feathers on head or neck. Neck long. Bill rather short and flattened. Legs long and strong, only three toes.

PLATE 37

KORI BUSTARD (AND CARMINE BEE EATER NOT LISTED)

All other bustards
That is:

1.	Tetrax tetrax	Little Bustard
	Neotis denhami	Stanley Bustard
	Neotis ludwigii	Ludwig's Bustard
2.	Neotis nuba	Nubian Bustard
	Neotis heuglinii	Heuglin's Bustard
	Ardeotis arabs	Arabian Bustard
3.	Ardeotis kori	Kori Bustard
	Ardeotis australis	Australian Bustard
	Eupodotis savilei	Savile's Bustard
	Eupodotis gindiana	Buff-crested Bustard
4.	Eupodotis ruficrista	Red-crested Bustard
	Eupodotis afraoides	White-quilled Bustard
5.	Eupodotis afra	Black Bustard
	Eupodotis rueppellii	Rüppell's Bustard
	Eupodotis vigorsii	Karoo Bustard
	Eupodotis humilis	Little Brown Bustard
	Eupodotis senegalensis	White-bellied Bustard
6.	Eupodotis caerulescens	Blue Bustard
7.	Eupodotis melanogaster	Black-bellied Bustard
	Eupodotis hartlaubii	Hartlaub's Bustard
8.	Eupodotis indica	Lesser Florican

List: II
Lit.: 307, 308, 382, 408, 410

PLATE 38

1. BURHINUS BISTRIATUS

Brazilian: Téu-téu-da-savana; *English:* Double-striped Thick-Knee, Double-striped Stone Curlew; *French:* Oedicnème vocifère; *German:* Dominikanertriel

Range: Mexico to N. Brazil, Hispaniola
Identification:
Length 43-48cm. Sexes similar. Sides of crown black and long and broad eye-stripe. Side of face, neck and upperparts dark brown with narrow tawny-buff edges. Breast grey-brown. Throat and belly white. Border between grey-brown breast and white belly clear-cut. Graduated tail with white subterminal band.
Young birds: Head, neck, and breast more buff and feathers of upperparts have broader buff edges.
Bare parts: Bill black, base greenish-yellow. Legs greenish-yellow. Large eyes yellow.

Geographical variation:
Birds from Colombia to Guyana and Brazil are darker and breast more brownish.
Some birds from Colombia are paler.
Birds from Hispaniola are smaller.
List: III Guatemala
Lit.: 13, 115, 125, 129, 223, 410, 431

2. NUMENIUS BOREALIS

Brazilian: Maçarico-esquimó; *English:* Eskimo Curlew; *French:* Courlis esquimau; *German:* Eskimobrachvogel; *Italian:* Chiurlo piccolo eschimese; *Spanish:* Zarapito esquimal, Zatapito polar

Range: Canada to South America
Identification:
Length 29-34cm. Sexes alike but female a little larger. Buff cinnamon with dark brown streaks and spots.Very similar to Whimbrel (Numenius phaeopus) but differs in following respects: About 25% smaller. Bill 42-60mm long. Head less distinctly striped on crown. Flanks with broad y-shaped, dark markings. Under wing-coverts cinnamon, long flight-feathers uniformly grey, against white narrowly barred brown in Whimbrel and Bristle-thighed Curlew (Numenius tahitiensis). Eskimo Curlew can also be confused with Little Curlew (Numenius minutus), length 29-32cm, but Little Curlew has a distinctive dark eye-stripe and flanks with pale brown barring, and uniform scaling of front- and hind legs. Eskimo Curlew has much smaller scaling on hind legs.
Young birds: Underparts more buffish.
Bare parts: Decurved bill blackish, base fleshy brown. Legs bluish-grey. Eyes dark brown.
List: I
Status: Population recently estimated at about 20 birds.
Lit.: 92, 98, 115, 129, 139, 196, 410, 431

3. NUMENIUS TENUIROSTRIS

English: Slender-billed Curlew; *French:* Courlis à bec grêle; *German:* Dünnschnabel-Brachvogel; *Italian:* Chiurlottello; *Russian:* Tonkoklyuvy Kronshnep; *Spanish:* Zarapito fino

Range: Siberia, Mediterranean region, Iran, W. Morocco
Identification:
Length 36-41cm. Sexes alike but female larger. General appearance more whitish (particularly on breast and tail) than Whimbrel (Numenius phaeopus) and Curlew (Numenius arquata). Without strongly marked crown-stripe like Whimbrel. Differs from Curlew by its slenderer bill and flank feathers with bold black markings. Wing length 243-275mm, against at least 268mm in Curlew.
Young birds: Flanks have brown streaks rather than bold black spots, but become heart-shaped spots the following Spring.
Bare parts: Bill slender and decurved with no lateral expansion at tip, dark brown, nearly black at tip, pinkish-brown at base of lower bill. (January birds are said to have no pinkish at base?) Legs bluish to greyish. Eyes brown.
List: I
Status: Population believed to be 100-400 birds.
Lit.: 115, 129, 139, 410, 414, 427

4. TRINGA GUTTIFER

Formerly known as:
Pseudototanus guttifer

English: Spotted Greenshank, Nordmann's Greenshank; *French:* Chevalier tacheté; *German:* Kurzfuss-Wasserläufer, Fleckengrünschenkel; *Russian:* Okhotsky Ulit

Range: Siberia, India, Malaysia
Identification:
D. Length 29-32cm. Sexes alike. Breeding: Head and upperneck white heavily streaked dark brown. Upperparts blackish brown spotted and fringed white. Rump white. Tail white with pale grey barrings. Underparts white with blackish brown spotting on lower neck, breast and flanks.
E. Non-breeding: Much paler, tinged greyish and lacking dark spotting on underparts.
Young birds: Like adult non-breeding but tinged brownish, not grey.
Bare parts: Bill (almost straight) blackish, inner half greenish or brownish-yellow. Legs yellow to greenish or brownish yellow.(Partial webbing between three front toes). Eyes dark brown.

Similar Species:
Greenshank (Tringa nebularia) has 40% longer legs and bill is bent upwards.
Marsh Sandpiper (Tringa stagnatilis) has slender bill and longer legs.
Tattlers (Heteroscelus sp.) have slender bills and no white on rump and tail.
Great Knot (Calidris tenuirostris) has some red-brown feathers on shoulder in breeding dress.
List: I
Lit.: 64, 115, 129, 130, 139, 410

5. LARUS RELICTUS

Formerly known as: Larus melanocephalus relictus

English: Relict Gull; *French:* Goéland relique; *German:* Lönnbergmöwe, Reliktmöwe; *Russian:* Reliktovaya Chayka

Range: Asia, China
Identification:
Length 44cm. Sexes alike. Head and nearly whole neck black, except pale brown between bill and eye. Above and below eye a large white spot. Upperparts pale grey, rump white. Wings pale grey with black subterminal band on longest flight-feathers. Underparts and tail white. Winter plumage: Head white or white with varied dark markings.
Young birds: Head white mottled brown. Dark band on tail.
Bare parts: Bill dark red, juvenile first dark brown, base of lower bill paler, later orange-red. Legs dark red, juvenile dark grey. Note: Thicker bill and longer legs than other closely related gulls. Eyes dark red-brown, eye-ring red, juvenile have black eye-ring.
List: I
Status: A species discovered 1929. Since then no species was seen until 1968, when a small population was discovered in Kazakhstan. Now also known from several places in China.
Lit.: 79, 129, 133, 153, 167, 191, 311, 354, 410, 418

PLATE 39

1. CALOENAS NICOBARICA

Formerly known as: Columba nicobarica

English: Nicobar Pigeon, White-tailed Pigeon, Hackled Pigeon, Vulturine Pigeon; *French:* Nicobar à camail, Pigeon Nicobar, Pigeon à collerette; *German:* Kragentaube, Nikobartaube, Mähnentaube; *Italian:* Colomba dal pavera, Colomba delle Nicobare; *Spanish:* Paloma Nicobar

Range: Nicobar Islands to Luzon, New Guinea
Identification:
Length 34cm. **Male:** Head and neck black with purplish blue wash, neck with a ruff of long, lanceolate, blue-green feathers which hang over mantle and shoulders. Upperparts shining dark green with coppery red, feathers long and pointed. Wings bluish. Underparts dark iridescent green. Upper- and undertail-coverts green and blue. Short tail white.
Female: Slightly smaller, often more coppery red on mantle.
Young birds: Duller. Tail blackish green.
Bare parts: Bill with knob black. Legs purplish, juvenile blackish. Eyes brown, sometimes whitish.

Geographical variation:
Birds from Palau Islands are smaller and more bluish. Ruff shorter and some feathers are bifurcated at tips.
List: I
Lit.: 9, 37, 61, 66, 94, 129, 410

2. GALLICOLUMBA LUZONICA

Formerly known as: Columba luzonica

English: Luzon Bleeding Heart, Luzon Punalada, Luzon Blood-breasted Pigeon; *French:* Gallicolombe poignardée, Colombe poignardée; *German:* Dolchstichtaube; *Italian:* Colomba pugnalata; *Spanish:* Paloma apuñalada de Luzon

Range: Luzon, Polillo Islands, (Philippines)
Identification:
Length 30cm. Sexes similar. Forehead and crown grey. Dark band from bill through eyes and down neck. Hindneck and upperparts dark greyish with broad purple or metallic green fringes, according to light. Wings brown, wing-coverts blue-grey with three red-brown bars across closed wing. Side of face, front of neck and underparts white, breast centre with a large, longitudinal , blood-red spot, like a fresh wound. Belly pinkish, flanks and undertail-coverts buff. Tail grey with subterminal black bar, two central feathers brown.
Young birds: Undescribed.
Bare parts: Bill blackish, base grey. Legs red. Eyes greyish violet.
List: II
Lit.: 37, 61, 66, 94, 104, 129, 410, 426

3. GOURA SCHEEPMAKERI

English: Southern Crowned-Pigeon, Maroon-breasted Crowned Pigeon, Scheepmaker's Crowned Pigeon, Sclater's Crowned Pigeon; *French:* Goura de Sclater, Goura de Scheepmaker; *German:* Scheepmakers Krontaube, Sclaters Krontaube, Rotbrustkrontaube; *Italian:* Gura di Scheepmaker; *Spanish:* Paloma crestada de Scheepmaker

Range: S. E. New Guinea
Identification:
Length 75cm. Sexes similar. Large blue-grey crest with silvery tinge and laterally compressed. General coloration dark greyish-blue. Breast, belly and most wing-coverts dark purplish red. Large white-grey area on wings. Pale band on tip of tail.
Young birds: Purplish-red on wings replaced by chestnut. White-grey patch in wings with a creamy wash.
Bare parts: Bill bluish-grey. Legs purplish red. Eyes red.
Geographical variation:
Birds from S. New Guinea have white wing patch and grey-blue belly.
List: II
Lit.: 9, 37, 94, 129, 195, 410

4. GOURA CRISTATA

Formerly known as: Columba cristata; Goura cinerea; Goura coronata

English: Western crowned-Pigeon, Blue Crowned Pigeon, Grey Crowned Pigeon, Blue Goura, Grey Goura; *French:* Goura couronné; *German:* Krontauube, Blauschopfkrontaube; *Italian:* Gura coronata; *Spanish:* Paloma crestada azul

Range: W. Papuan Islands and N. W. New Guinea
Identification:
Length 66cm. Sexes similar. Like number 3, scheepmakeri but mantle and forewing dark purplish-red and breast and belly blue-grey. White area on wings. Irregular black patches are common, especially on head, upper tail-coverts and belly.
Young birds: Wing patches not white but cream and grey. Dark purplish-red on wings replaced by chestnut.
Bare parts: Bill black. Legs dark red. Eyes red.
List: II
Lit.: 9, 37, 94, 129, 410

5. GOURA VICTORIA

English: Victoria Crowned Pigeon, White-tipped Crowned Pigeon, Victoria Goura, White-tipped Goura; *French:* Goura de Victoria; *German:* Victoria-Krontaube, Fächertaube; *Italian:* Gura Victoria, Colomba coronata di Victoria, *Spanish:* Paloma crestada Victoria

Range: Yapen Island, Biak Island, (New Guinea), N. New Guinea
Identification:
Length 66cm. Sexes similar. Tip of crest feathers spatulate and with white tipping. Wing patch pale grey-blue edged dark purple. Otherwise like number 3, scheepmakeri.
Young birds: Duller. Breast more brownish.
Bare parts: Bill dark grey. Legs red. Eyes red.
List: II
Lit.: 9, 37, 94, 129, 410

6. DUCULA MINDORENSIS

English: Mindoro Imperial-Pigeon, Mindoro Zone-tailed Pigeon, Great Mindoro Fruit Pigeon, Pink-throated Imperial Pigeon; *French:* Carpophage de Mindoro; *German:* Mindorofruchttaube, Grosse Mindoro Fruchttaube; *Italian:* Colomba imperiale di Mindoro; *Spanish:* Paloma de Mindoro

Range: Mindoro Island (Philippines)
Identification:
Length 47cm. Sexes alike but female smaller and eye colour differs. Head and neck pale bluish grey. Side of face and throat pinkish white. Area around eye dusky. Upperparts iridescent green, purple, and bronzy-red, forming a purple-black V on mantle. Underparts grey, darker on belly. Tail blackish-green with a pale grey central band.
Young birds: Not described.
Bare parts: Bill black. Legs reddish-pink. Eyes, male: yellow with red outer ring, skin around eye dark red; female: eyes yellow with brown outer ring, skin around eye orange-yellow.
List: I
Lit.: 37, 61, 66, 94, 104, 129, 410

PLATE 40

1. COLUMBA LIVIA

Dutch: Rotsduit;*English:* Rock Dove, Rock Pigeon;*French:* Pigeon biset; *German:* Felsentaube;*Italian:* Piccione selvatico;*Spanish:* Paloma bravía; *Swedish:* Klippduva

Range: Europe to India and China, N. W. Africa, Canary Islands
Identification:
Length 31-34cm. **Male:** Mainly blue-grey with glossy green or deep purple (depending on angle of light) on neck, upper mantle and breast. Lower back and rump white or grey. Two distinct black wing bars and one black terminal tail bar. Underparts blue-grey.
Female: Slightly duller and less intense gloss on neck.
Young birds: Less glossy and with a brown tinge.
Bill lead-coloured with mealy white cere. Legs red. Eyes orange.

Geographical variation:
There are great differences in colour; birds from Senegal and Ghana are the darkest of all populations, and the eyes are yellow (**2.**)
Birds from Libya are palest and have dark grey wing bars, not black.
Birds from Asia Minor and Iraq have legs partly feathered.
Note: About 20-30% of all feral pigeons from most towns of the world are not distinguishable from Rock Doves.
List: III Ghana
Lit.: 51, 89, 94, 129, 191, 305, 410

3. COLUMBA GUINEA

English: Speckled Pigeon; *French:* Pigeon de Guinée; *German:* Guineataube

Range: Africa south of Sahara
Identification:
Length 38cm. Sexes similar but female slightly duller. Head and upper neck grey. Lower neck, upper breast, and shoulder feathers stiff and bifurcate, chestnut with silvery pink tips. Upperparts vinous chestnut, rump pale grey. Wings with white bands of triangular spots. Underparts light bluish-grey. Tail grey with a pale grey subterminal band.
Young birds: Brown where adults are grey. Wing spots ill-defined.
Bare parts: Bill blackish, cere whitish. Legs purplish pink. Eyes yellow to white or reddish pink, outer ring orange to purple. Bare skin around eyes wine-red, juvenile brown.

Geographical variation:
Birds from S. Africa are slightly smaller, darker grey and white spots on wings smaller.
Birds from Namibia are paler.
List: III Ghana
Lit.: 94, 129, 253, 305, 410

4. COLUMBA IRIDITORQUES

Formerly known as: Turturoena iriditorques; Columba malherbii

English: Western Bronze-naped Pigeon;*French:* Pigeon à nuque bronzée; *German:* Glanzkopftaube

Range: Sierra Leone to Angola, Zaire
Identification:
A. Length 28cm. **Male:** Head and throat dark blue-grey, hind crown, nape and lower hind neck glossed green. Upper part of mantle golden bronze glosséd pink. Rest of upperparts and wings slaty black with green to violet-blue fringes. Underparts dark mauve-pink. Tail slate tipped dark grey.
B. **Female:** Crown, nape and underparts rufous. Face and chin pale brown. Hindneck and mantle glossy green or violet.
Young birds: Less iridescent.
Bare parts: Bill blue-grey, tip whitish, cere dark red. Legs red. Eyes variously pink, red, pinkish grey or greenish yellow to blue.
List: III Ghana
Lit.: 7, 94, 129, 305, 410

5. COLUMBA UNICINCTA

English: Afep Pigeon, African Wood Pigeon; *French:* Pigeon gris; *German:* Kongotaube

Range: Liberia to Zaire, Uganda
Identification:
Length 40cm. **Male:** Head and upperparts grey, darker on back and rump, each feather edged pale grey, giving a scaly appearance. Underparts pale vinous pink. Flanks and sides of belly pale grey, belly and undertail-coverts white. Tail dark grey, broadly tipped black and with a broad pale grey central band.
Female: Like male but breast duller pink.
Young birds: Darker, breast brownish vinous.
Bare parts: Bill slate blue, tip paler. Legs grey. Eyes orange or red.
List: III Ghana
Lit.: 77, 94, 129, 173, 268, 305, 410

6. COLUMBA MAYERI

Formerly known as: Nesoenas mayeri; Columba maveri

English: Pink Pigeon, Mauritius Pigeon, Chestnut-tailed Pigeon;*French:* Pigeon rose, Pigeon de Maurice; *German:* Rosentaube, Rosataube, Mauritius-Rosentaube; *Italian:* Colomba delle Mauritius; *Spanish:* Paloma de Mauricio

Range: Mauritius
Identification:
Length 32cm. **Male:** Head, neck and underparts pink, forehead whitish. Mantle brownish pink to dusky pink. Rump pale bluish-grey to chestnut on uppertail-coverts and tail. Wings dark brown.
Female: Less bright colours and rump more brown.
Young birds: Undescribed.
Bare parts: Bill yellow, base red. Legs red. Eyes yellow.
List: III Mauritius
Status: About 18 birds in nature, and about 100 birds in zoos.
Lit.: 25, 37, 45, 94, 129, 196, 341, 410

7. OENA CAPENSIS

English: Namaqua Dove, Long-tailed Dove;*French:* Tourterelle à masque de fer; *German:* Kaptäubchen; *Spanish:* Tórtola de Cabo

Range: Senegal to Arabia and south to Cape Province, Madagascar
Identification:
A. Length 25cm. **Male:** Black mask, foreneck and frontal part of breast. Rest of head, sides of neck, breast and most wing-coverts bluish grey. Hindneck and upperparts light earth brown, upper rump crossed by two black and one brown band. Wings rufous with 3-5 large amethyst spots. Underparts white, under tail-coverts black. Long tail graduated blue-grey, central feathers dull silver grey, shading to black at tip.
B. **Female:** Mask, foreneck and breast greyish-brown.
Young birds: Like female but with a speckled buffish appearance.
Bare parts: Bill orange-yellow, base crimson, female reddish black. Legs purplish red. Eyes dark brown.

Geographical variation:
Birds from Madagascar are slightly darker and greyer.
List: III Ghana
Lit.: 94, 129, 253, 268, 305, 410

PLATE 41

1. STREPTOPELIA DECIPIENS

English: Mourning Collared Dove, Deceptive Turtle Dove, Angola Dove, Dongola Dove; *French:* Tourterelle pleureuse; *German:* Brillentaube

Range: Senegal to Ethiopia and southwards to Transvaal
Identification: Length 30cm. Sexes similar. Forehead, crown and cheeks ash-grey. Nape and neck mauve-pink, black patch on either side of lower neck. Upperparts and wings earth-brown washed slate on outer wing-coverts and rump. Under wing-coverts grey. Underparts vinous grey, chin, throat, and belly whitish to grey. Two central tail feathers earth brown, next two paler with whitish tip, outermost grey-brown with broad whitish tip, tail below black, tip white. Under tail-coverts grey tipped white.
Young birds: Like adult but browner.
Bare parts: Bill black. Legs wine-red. Eyes whitish to orange-red or red with narrow yellow inner ring. Eye-ring reddish.

Geographical variation:
Birds from Senegal to Nigeria are paler above and under tail-coverts grey.
Birds from Cameroon and N. Zaire are like last but have vinous grey breast.
Birds from Angola, S. Zaire and Zambia have darker grey under tail-coverts, boldly edged white. Eyes yellow.
Birds from Somalia to Malawi and Mozambique have breast more pinkish than lilac. Belly white extending to lower breast. Under tail-coverts with broader white edges.
List: III Ghana
Lit.: 88, 94, 129, 171, 253, 268, 305, 410

2. STREPTOPELIA SEMITORQUATA

English: Red-eyed Dove, Half-collared Dove, Black Dove; *French:* Tourterelle à collier; *German:* Halbmondtaube

Range: Africa south of Sahara and Arabia
Identification: Length 33-36cm. Sexes similar. Like number 1, Mourning Collared Dove, but larger and darker. Forehead whitish. Central tail feathers earth-brown, all others black with terminal quarter brown. Tail below like above, but tip dirty white.
Young birds: Duller and browner. Black patch on side of neck poorly developed.
Bare parts: Bill black. Legs wine red, juvenile dull grey. Eyes red, orange or yellow, juvenile dull grey. Dark red eye-ring very narrow, juvenile dull grey.
List: III Ghana
Lit.: 94, 129, 173, 174, 268, 305, 410
Note: 'Black Dove' is a rather misleading name.

3. STREPTOPELIA ROSEOGRISEA

English: African Collared Dove, Rose-grey Turtle Dove, Pink-headed Dove; *French:* Tourterelle rose-et-grise; *German:* Lachtaube

Range: Sahara, Senegal to Somalia and Arabia
Identification: Length 29cm. Sexes similar. Head, nape, neck, breast, and belly pale mauve-pink, shading to white on chin and belly. Black collar on hindneck edged above with white. Upperparts pale earth brown, mantle more isabelline. Wings pale earth brown, longest flight-feathers blackish brown. Outer wing-coverts pale blue-grey. Underwing-coverts white. Central tail feathers grey-brown, all others greyer with increasingly broad white tips, outermost with most white.
Young birds: Black collar on neck not well defined. Upperparts and wings with pale margins.
Bare parts: Bill black. Legs wine red, juvenile dull pink. Eyes dark red. Eye-ring white.

Geographical variation:
Birds from Mali, Nigeria and Chad are slightly darker and under wing-coverts white washed grey.
Birds from Ethiopia, Somalia and Arabia are darker. Under wing-coverts pale grey.
List: III Ghana
Lit.: 7, 94, 129, 305, 410

4. STREPTOPELIA VINACEA

English: Vinaceous Dove, Vinaceous Turtle Dove, Vinaceous Ring Dove; *French:* Tourterelle vineuse; *German:* Röteltaube
Range: Senegal to Sudan
Identification: Length 27cm. Sexes similar. Head, neck, and underparts pale pink to pinkish-vinous, darkest on nape and breast. A very narrow black line between eyes and bill. Collar on hindneck black with greyish upper edge. Upperparts and wings pale earth brown. Long flight-feathers blackish, edged whitish. Outer wing-coverts pale greyish blue. Underwings pale grey. Under tail-coverts white. Central tail feathers greyish-brown, rest brownish-black to black, terminal third greyish-white.
Young birds: Duller with pale edges on wings giving scaly appearance.
Bare parts: Bill black. Legs vinous red. Eyes dark brown.

Geographical variation:
Birds from Chad and Cameroon are paler and wing-coverts edged white.
Birds from Sierra Leone to Zaire are darker.
List: III Ghana
Lit.: 7, 94, 129, 305, 410

5. STREPTOPELIA TURTUR

Dutch: Tortelduit; *English:* European Turtle Dove, Common Turtle Dove, Isabelline Turtle Dove; *French:* Tourterelle des bois; *German:* Turteltaube; *Italian:* Tortora; *Spanish:* Tórtola común; *Swedish:* Turturduva

Range: Europe, Asia Minor to China, N. Africa, Madeira, Canary Islands, Cape Verde
Identification: A. Length 26-28cm. **Male:** Top of head, nape and hindneck grey-blue. Large black and white oval patch on side of neck. Upperparts light brown, feather centres usually black, upper mantle bluish-grey. Rump grey mottled light brown. Wings dark brown edged whitish, innermost wing-feathers and wing-coverts black with broad cinnamon-rufous fringes, outermost wing-coverts mostly blue-grey. Chin pale pinkish, throat pink, breast dark vinous pink, flanks blue-grey tinged pink. Belly and undertail-coverts white. Central tail feathers grey-brown, rest dark bluish-grey broadly tipped white, white extending up entire outside of outermost tail feathers.
Female: Duller breast, but not always distinguishable.
C. Young birds: Paler, lack black and white neck patch until October.
Bare parts: Bill black. Legs dark pink. Eyes orange-red to yellow or brown. Skin around eye reddish-purple.

Geographical variation:
N. Africa, S. W. Asia: Slightly smaller and paler.
Sahara: Crown buff. Back and wings more rufous.
Libya, Egypt: Crown and back pale isabelline brown. Wings and tail more rufous.
List: III Ghana
Lit.: 7, 51, 94, 129, 305, 410

6. STREPTOPELIA SENEGALENSIS

English: Laughing Dove, Senegal Dove, Palm Dove; *French:* Tourterelle maillée; *German:* Palmtaube; *Spanish:* Tórtola Senegalesa

Range: S. E. Europe to China, Africa
Identification: Length 24cm. Sexes alike. Head dark mauve-pink. Neck and upperparts light brown, each feather edged pale rufous; back and rump grey-blue. A collar of bifurcated feathers on side and front of lower neck with black bases and glossy golden-copper tips. Wings dark brown, longest feathers edged dull white, outermost wing-coverts slate blue. Underparts vinous pink shading to whitish belly and under tail-coverts. Central tail feathers grey-brown, rest patterned in dark grey-brown and white.
Young birds: Collar on foreneck absent. Otherwise duller.
Bare parts: Bill blackish. Legs purplish-red. Eyes dark brown or brown.

Geographical variation:
Morocco, Algeria, Tunisia: Larger and browner.
Dakhla Oasis, Libya: Paler above and more vinous below.
Nile Valley, Egypt: Darker, collar more rufous.
Sao Thomé Island: Smaller and head brighter pink.
Socotra Island: Smaller and very pale.
Iran, India: Smaller and much duller.
List: III Ghana
*Lit.:*7, 12, 94, 129, 305, 410

PLATE 42

1. TRERON C. CALVA

Formerly known as: Treron australis (in part); Vinago calva

English: African Green Pigeon, Green Fruit Pigeon, African Fruit Pigeon; *French:* Pigeon vert à front nu; *German:* Rotnasen-Grüntaube

Range: W. C. & N. E. Africa
Identification:
Length 28cm. **Male:** Head and neck yellowish green tinged yellow on neck. Upperparts olive-green, area between hindneck and upper mantle tinged grey. Wings black edged yellow also on greater wing-coverts; forewing olive-green with mauve spot on shoulder. Underparts olive-green tinged yellow, flanks darker. Tail bluish-grey with broad grey terminal band. Under tail-coverts pale chestnut.
Female: Like male but duller, and cere smaller.
Young birds: Green colour more greyish with yellow-green fringe to most feathers. Vent more yellow. Without mauve patch on shoulder.
Bare parts:Bill greyish, tip paler, cere and base of bill red. Legs yellow to orange-yellow. Eyes pale blue, outer ring brown to red or purple.

Subspecies:

2. TRERON CALVA DELALANDII

English: Delalande's Green Pigeon
Range: S. E. Africa
Larger. Bright yellowish green on upperparts. Tail dark yellow-green, terminal band whitish-green. Legs red.

Geographical variation:
Birds from E. Ethiopia and Kenya have a clear blue-grey area on hind neck and upper mantle. Cere smaller and legs red.
Some birds from E. Kenya and Tanzania are smaller and have green tail and red legs.
Birds from Zambia have reddish tail in males and longer upper tail-coverts dark olive-chestnut.
List: III Ghana
Lit.: 7, 94, 103, 112, 129, 171, 172, 173, 174, 222, 245, 253, 268, 305, 410

3. TRERON WAALIA

English: Yellow-bellied Green Pigeon, Bruce's Green Pigeon; *French:* Pigeon vert waalia; *German:* Waaliataube

Range: Senegal to Arabia
Identification:
Length 31cm. Sexes similar. Head, neck, and upper breast greyish green. Lower breast and belly bright yellow. Otherwise like above, African Green Pigeon.
Young birds: Duller, mauve patch on shoulder incomplete.
Bare parts: Bill whitish or pale bluish grey, cere dark purple, lilac to red. Legs orange. Eyes pale blue usually with red outer-ring.
List: III Ghana
Lit.: 94, 112, 126, 129, 268, 305, 410

4. TURTUR ABYSSINICUS

English: Black-billed Wood Dove, Abyssinian Wood Dove, Black-billed Blue-spotted Dove;*French:* Tourtelette d'Abyssinie;*German:*Erzflecktaube

Range: Senegal to Ethiopia
Identification:
Length 22cm. Sexes alike. Head blue-grey, forehead whitish. Hindneck brown with grey wash. Upperparts pale earth brown with two blackish bars on rump. Uppertail-coverts with two narrow black bands. Wings brown, rufous, and earth brown with some dark blue metallic spots. Underwing rufous. Foreneck and underparts pale greyish pink, chin, throat, and belly whitish, undertail-coverts black and grey. Tail grey-brown with blackish tip.
Young birds: Most feathers barred buff or rufous. Wing spots smaller.
Bare parts: Bill black, base dull carmine. Legs purple. Eyes dark brown.
List: III Ghana
Lit.: 94, 112, 129, 173, 305, 410

5. TURTUR AFER

English: Blue-spotted Wood Dove, Sapphire-spotted Dove, Red-billed Wood Dove, Red-billed Blue-spotted Wood Dove; *French:* Tourtelette améthystine; *German:* Stahlflecktaube

Range: Senegal to Ethiopia and south to Transvaal
Identification:
Length 22cm. Sexes similar. Like Black-billed Wood Dove but brown colour darker, not earth-brown. Chin pale vinous. Belly buffish.
Young birds: Browner than adult, upperparts with buff tips to feathers.
Bare parts: Bill yellow, base half red. Legs reddish. Eyes dark brown.
List: III Ghana
Lit.: 94, 112, 129, 173, 268, 305, 410

6. TURTUR BREHMERI

Formerly known as: Calopelia brehmeri

English: Blue-headed Wood Dove, Maiden Dove; Blue-headed Dove; *French:* Tourtelette demoiselle; *German:* Maidtaube

Range: Sierra Leone to Cameroon, Gabon, Zaire
Identification:
Length 25cm. Sexes similar. Head bright blue-grey. Upperparts dark chestnut with vinous wash, rump a little paler. Wings dark chestnut with 4-6 gold-copper patches. Underparts rufous. Tail rufous-brown.
Young birds: Like adults but upperparts with blackish barring.
Bare parts: Bill dark red, tip dull greenish. Legs dark red. Eyes dark brown.

Geographical variation:
Birds from Sierra Leone to Cameroon have metallic green wing spots.
List: III Ghana
Lit.: 94, 112, 129, 173, 268, 305, 410

7. TURTUR TYMPANISTRIA

Formerly known as: Tympanistria tympanistria

English: Tambourine Dove, White-breasted Wood Dove, Forest Dove, White-breasted Pigeon; *French:* Tourtelette tambourette; *German:* Tambourintaube

Range: Sierra Leone to Ethiopia and Tanzania, Zimbabwe, Natal to Cape Province
Identification:
A. Length 23cm. **Male:** Forehead and area behind eye white. Rest of head, hindneck, and upperparts uniformly dark brown. Wings uniformly dark brown, chestnut and with 5-6 green wing spots, less lustrous than in all other wood doves. Underparts white, flanks pale buffish to pale brown. Under tail-coverts and tail dark brown.
B. **Female:** White colour of the male more greyish. Wing spots black.
C. Young birds: Barred on most feathers with rufous and brown.
Bare parts: Bill deep purple, tip blackish. Legs purplish red. Eyes dark brown.
List: III Ghana
Lit.: 94, 112, 129, 172, 174, 253, 268, 305, 410

PLATE 43

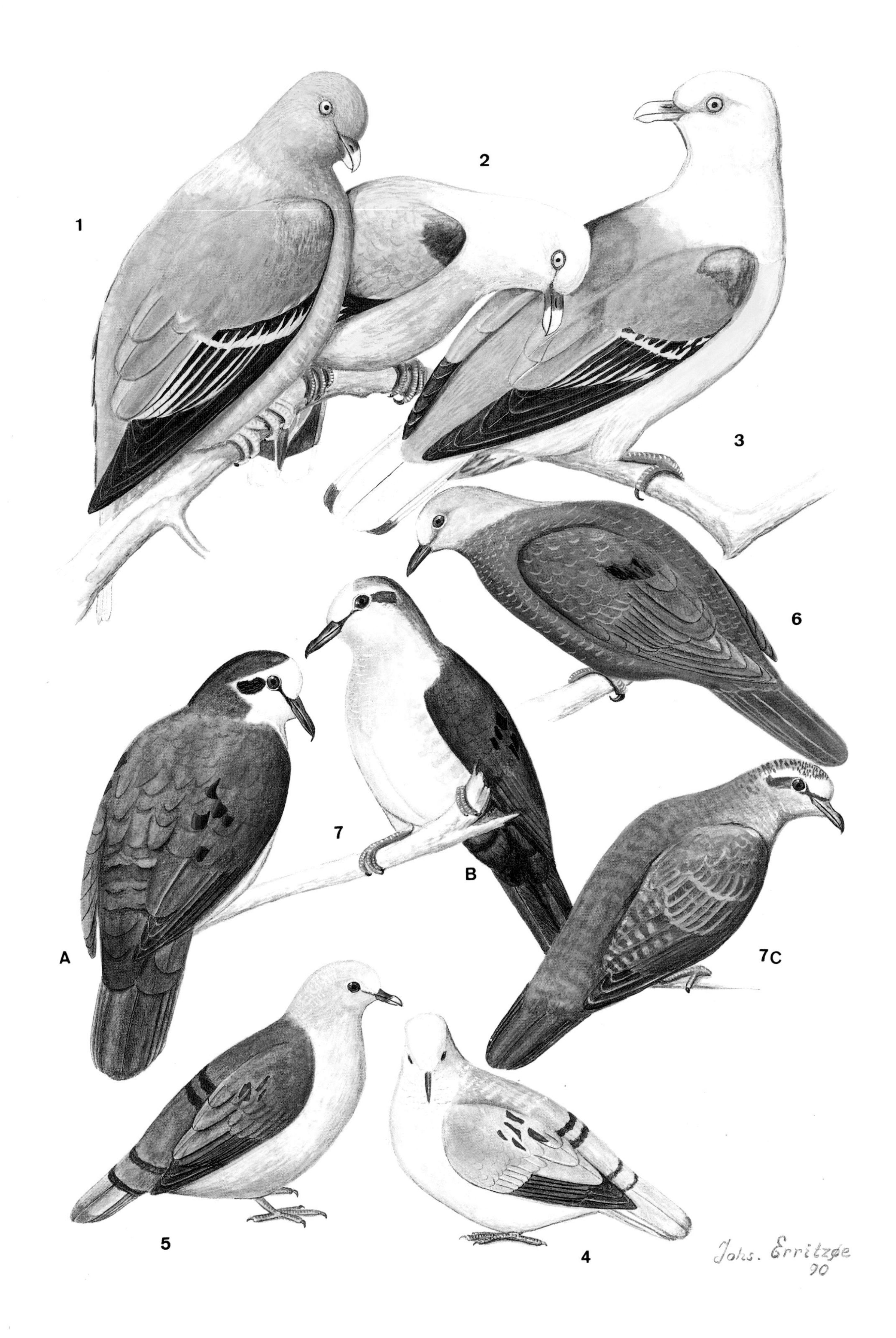

1. AMAZONA ARAUSIACA

English: Red-necked Amazon, Red-necked Parrot, Dominican Amazon; *French:* Amazone de Bouquet, Amazone à tête bleue; *German:* Blaukopfamazone; *Italian:* Amazzonia dal collo rosso; *Spanish:* Amazona cabenazul, Amazona cabeza azul

Range: Dominica Island
Identification:
Length 40cm. Sexes alike. Forepart of head violet-blue. Foreneck red. Rest of head, neck and upperparts green, edged black on nape, neck and mantle. Wings green with a red, yellow and blue area, long flight-feathers dark green becoming violet-blue toward tips. Underparts green. Tail green tipped yellowish-green, outermost feathers with red markings near base.
Young birds: Undescribed but said to be similar to adults.
Bare parts: Bill horn, tip grey. Legs grey. Eyes orange, juvenile brown. Bare skin around eyes whitish.
List: I
Status: 200 estimated in 1981.
Lit.: 37, 81, 129, 168, 361, 410

2. AMAZONA BARBADENSIS

English: Yellow-shouldered Amazon, Yellow-shouldered Parrot; *French:* Amazone à épaulettes jaunes, Amazone de la Barbade; *German:* Gelbflügelamazone; *Italian:* Amazzonia a spalla gialla; *Spanish:* Amazona de espalda amarilla

Range: Blanquilla, Margarita, (Venezuela), Bonaire
Identification:
Length: 33cm. Sexes similar. Forehead white, forecrown and round eyes yellow. Lower cheeks, throat and underparts green tinged blue. Hindneck and upperparts green edged black. Bend of wings and thighs yellow, rest of wings green with red patch and long flight-feathers becoming violet-blue near tips. Tail green and tip yellow-green with a blue wash on outer webs of outer tail feathers and near base some orange-red.
Young birds: Like adults, but below not tinged with blue.
Bare parts: Bill pale grey. Legs grey. Eyes orange. Skin round eye pale grey-blue.

Geographical variation:
Birds from Netherlands Antilles have less yellow area mixed with red on bend of wings.
List: I
Lit.: 37, 81, 129, 168, 192, 410

3. AMAZONA BRASILIENSIS

Brazilian: Papagaio-de-cara-roxa; *English:* Red-tailed Amazon, Blue-faced Amazon; *French:* Amazone à joues bleues, Amazone à joues vertes, Amazone à queue rouge, Amazone du Brésil; *German:* Rotschwanzamazone, Taubenhalsamazone; *Italian:* Amazzonia dalla coda rossa; *Spanish:* Papagayo de cara roja

Range: Brazil
Identification:
Length 37cm. Sexes similar. Forecrown and crown red. Chin, ear-coverts, throat and upper breast violet-blue, hindneck and upperparts green edged black on neck and mantle. Wings green, greater coverts edged pale yellow-green; front of wings red and blue area on lower wing. Underparts paler green, under tail-coverts yellow-green. Central tail feathers green tipped yellow, rest purple-blue at base, red in middle and tipped yellow.
Young birds: Undescribed.
Bare parts: Bill pale grey. Legs grey. Eyes brown. Bare skin round eye pale grey.
List: I
Lit. 37, 81, 129, 168, 410, 431

4. AMAZONA RHODOCORYTHA

Also known as: Amazona dufresniana rhodocorytha

Brazilian: Chauá; *English:* Red-crowned Amazon, Red-browed Amazon, Red-topped Amazon; *French:* Amazone à joues bleues; *German:* Goldmaskenamazone; *Italian:* Amazzonia a corona rossa; *Spanish:* Amazona crestirroja

Range: Brazil
Identification:
Length 35cm. Sexes alike. Forehead, forecrown, and area on wings red with some green or blue feathers on top of head. Nape dull reddish-purple. Between bill and eye, lower cheeks and throat orange. Ear-coverts and upper cheeks blue and green. Hind neck and body green, neck and mantle tipped black. Wings green, front yellow-green, central flight feathers green bordered purple-blue. Tail green tipped paler and with dark red spots except two central which are only green. Note: Colour of head variable.
Young birds: Crown green with a red wash. Less red in wings and tail.
Bare parts: Bill grey, reddish at base of upper bill. Legs grey. Eyes brown-orange.
List: I
Lit.: 37, 81, 129, 168, 410, 431

5. AMAZONA GUILDINGII

English: St. Vincent Amazon, St. Vincent Parrot, Guilding's Amazon; *French:* Amazone de Guilding, Amazone de St.-Vincent; *German:* Königsamazone; *Italian:* Amazzonia di Guilding, Amazzonia di San Vincente; *Spanish:* Amazona de San Vincente

Range: St. Vincent
Identification:
Length 40cm. Sexes alike but very variable. Yellow-brown phase: Forepart of head creamy-white, merging into orange and violet-blue behind eyes. Hindneck olive-green washed blue. Body bronze-brown, belly and under tail-coverts washed green, and upper tail coverts tipped green; feathers on hindneck and body edged black. Wings bronze-brown, orange, green, blue, and yellow. Tail dark violet-blue, broadly tipped yellow and orange at base.
D.+ Green phase: Like above but head without orange, wings with less orange areas, and upperparts predominantly dusky green.
Young birds: Just after leaving nest duller than adults. After first moult brighter than parents, particularly on heads.
Bare parts: Bill horn, washed olive-green and grey marks at base. Legs pale grey. Eyes orange, juvenile brown.
List: I
Lit.: 81, 129, 160, 168, 206, 410
+ Scale applies to all birds except No. 5D

6. AMAZONA IMPERIALIS

Formerly also known as: Amazona augusta

English: Imperial Amazon, Imperial Parrot; *French:* Amazone impériale; *German:* Kaiseramazone; *Italian:* Amazzonia imperiale; *Spanish:* Amazona imperial

Range: Dominica Island
Identification:
Length 45cm. Sexes alike. Head dark maroon-purple edged black with some green-blue, ear-coverts reddish-brown, cheeks brownish-maroon. Upperparts green edged black, wings green with small red egdes and a dark maroon area, long flight feathers violet-blue towards tips. Underparts purple edged black. Thighs, flanks and under tail-coverts greenish, tipped greenish-blue. Tail red-brown tipped greenish-blue, central feathers washed green.
Young birds: More green on heads.
Bare parts: Bill greyish-horn. Legs grey. Eyes yellow to orange-red, juvenile brown.
List: I
Status: Total population 1990 about 80 birds!
Lit.: 37, 81, 129, 168, 196, 361, 410

PLATE 44

1. AMAZONA LEUCOCEPHALA

English: Cuban Amazon; *French:* Amazone à face rouge, Amazone à tête blanche, Amazone de Cuba; *German:* Kubaamazone; *Italian:* Amazzonia dalla testa bianca; *Spanish:* Amazona de Cuba

Range: Cuba and nearby islands, Bahamas
Identification:
Length 32cm. Sexes alike. Forehead and area around eyes white. Hindcrown, nape and upperparts green edged black. Wings blue and green on outer webs of inner flight-feathers. Underparts green, often with rose-red feathers on breast, belly dull red. Cheeks and throat rose-red. Tail green tipped yellowish-green, outermost feathers with red markings at base and edged blue.
Young birds: Like adults but less black edging and belly with little dull red.
Bare parts: Bill horn. Legs pinkish. Eyes pale olive-green.
Bare skin round eyes white.

Geographical variation:
Birds from W. Cuba and Isle of Pines have throat deeper red and belly more purplish.
Birds from Abaco and Great Inagua (Bahama Islands) have little or no red on abdomen.
Birds from Grand Cayman Island have green colour more yellow-green.
Birds from Little Cayman Island and Cayman Brac Island are smaller and have abdomen more purple-red and throat red.
List: I
Status: World population of last mentioned about 130 birds in 1976.
Lit.: 37, 81, 129, 152, 168, 368, 410, 413, 416

2. AMAZONA PRETREI

Brazilian: Papagaio-da-serra; *English:* Red-spectacled Amazon, Prêtre's Amazon; *French:* Amazone de Prêtre; *German:* Prachtamazone; *Italian:* Amazzonia dalla faccia rossa; *Spanish:* Amazona cabecirroja, Loro de cabeza roja

Range: Brazil, Uruguay, Argentina
Identification:
Length 32cm. Sexes alike. Forehead, crown, between bill and eye, round eye, edges and bend of wings and thighs red. Wings green becoming blue at end. Rest of plumage green, distinctly edged black on body. Tail green tipped yellow, three outermost feathers with a small red area near base.
Young birds: Less red.
Bare parts: Bill and legs yellowish-brown. Eyes orange. Bare skin round eyes whitish.
List: I
Lit.: 37, 81, 129, 168, 410, 431

3. AMAZONA VERSICOLOR

English: St. Lucia Amazon, St. Lucia Parrot; *French:* Amazone versicolore, Amazone de Santa Lucia; *German:* Blaumaskenamazone; *Italian:* Amazzonia variopinta, Amazzonia di St. Lucia; *Spanish:* Amazona de Santa Lucia

Range: St. Lucia Island
Identification:
Length 43cm. Sexes alike. Forepart of head violet-blue gradually becoming green on hindcrown and nape. Upperparts green and like head edged black. Wings green with red area, tip violet-blue. Throat and upper breast red, lower breast and belly green, reddish-brown and tipped black. Thighs green, undertail-coverts yellowish-green. Tail green tipped paler, outermost feathers on basal half blue and near base hidden red spots.
Young birds: Said to be like adults.
Bare parts: Bill and legs grey. Eyes orange, juvenile brown. Bare skin round eyes pale grey.
List: I
Status: Wild population in 1981 about 100.
Lit.: 29, 37, 81, 129, 152, 168, 410

4. AMAZONA VINACEA

Brazilian: Papagaio-de-peito-roxo; *English:* Vinaceous Amazon, Vinaceous-breasted Parrot; *French:* Amazone vineuse, Amazone bourgogne; *German:* Taubenhalsamazone; *Italian:* Amazzonia vinacea; *Spanish:* Amazona pechivinosa, Loro pecho vinoso

Range: Brazil, Paraguay, Argentina
Identification:
Length 30cm. Sexes alike. Forehead and area between bill and eyes red. Hindneck and sides of neck green strongly washed blue with broad black edging. Upperparts green edged black. Wings green, front of wings and area in middle red, tip of long flight-feathers blue. Chin pink red, breast dull mauve-red edged dusky.Under tail- coverts yellow-green. Tail green tipped paler, outermost feathers barred red at base.
Young birds: Duller. Less red on head and greenish tinge on breast.
Bare parts: Bill dark pink-red, tip horn coloured, juvenile horn coloured, dull pinkish at base of upper bill. Legs pale grey. Eyes red, juvenile brown. Bare skin round eye blue-white.
List: I
Lit.: 37, 81, 129, 168, 410, 431

5. AMAZONA VITTATA

English: Puerto Rican Amazon, *French:* Amazone de Puerto Rico, Amazone à bandeauu rouge; *German:* Puertoricoamazone; *Italian:* Amazzonia di Porto Rico; *Spanish:* Amazona de Puerto Rico

Range: Puerto Rico
Identification:
Length 29cm. Sexes alike. Forehead and area between bill and eyes red. Rest of head and body green, edged black except on hind part of body. Wings green, long flight feathers and their coverts dark blue, short flight feathers blue edged dull green. Under tail-coverts yellow-green. Green belly sometimes slightly washed with dull red. Tail green, tipped yellowish-green, red and blue areas on base of lateral feathers.
Young birds: Like adults.
Bare Parts: Bill yellowish-horn. Legs yellowish-brown. Eyes brown. Bare skin round eyes white.
List: I.
Status: Wild population in 1986: 29 birds! In captivity also 29 birds.
Lit.: 37, 81, 129, 152, 168, 196, 410

PLATE 45

1. ANODORHYNCHUS HYACINTHINUS

Brazilian: Arara-azul-grande; *English:* Hyacinth Macaw; *French:* Ara hyacinthe; *German:* Hyazinthara; *Italian:* Cacatua bianco; *Spanish:* Guacamaya azul

Range: Brazil
Identification:
Length 100cm, largest of all parrots. Sexes alike. Plumage rich blue-violet, only underside of tail dark grey.
Young birds: Tail shorter.
Bare parts: Bill and legs dark grey. Eyes dark brown. Bare skin round eyes and at base of lower bill yellow.
List: I
*Lit.:*37, 81, 129, 168, 410, 431

2. ANODORHYNCHUS GLAUCUS

Brazilian: Arara-azul-pequena; *English:* Glaucous Macaw; *French:* Ara glauque, Ara bleu; *German:* Türkisara, Meerblauer Ara, Blauara; *Italian:* Ara blu; *Spanish:* Guacamayo violáceo, Ara azul

Range: Brazil, Paraguay, Uruguay, Argentina
Identification:
Length 72cm. Sexes alike. Plumage greenish-blue, more greenish below. Throat grey-brown and underside of tail dark grey.
Young birds: Undescribed.
Bare parts: Bill and legs dark grey. Eyes dark brown. Bare skin round eyes yellow and at base of lower bill pale yellow.
List: I
Status: Almost certainly now extinct. No records from this century.
Lit.: 37, 81, 129, 168, 410, 431

3. ANODORHYNCHUS LEARI

Brazilian: Arara-azul-de Lear; *English:* Indigo Macaw, Lear's Macaw; *French:* Ara de Lear, Ara cobalt; *German:* Lears Ara, Kleiner Hyazinthara; *Italian:* Ara di Lear; *Spanish:* Guacamayo cobalto

Range: Rasa da Caterina (Brazil)
Identification:
Length 74cm. Sexes alike. Head, neck, and underside greenish-blue feathers edged paler on underparts. Underside of tail dark grey, rest violet-blue.
Young birds: Undescribed
Bare parts: Bill and legs dark grey. Eyes dark brown. Bare skin round eyes yellow, at base of lower bill pale yellow.
List: I
Status: Less than 200 birds in wild.
Lit.: 77, 81, 129, 168, 275, 328, 410, 431

4. CYANOPSITTA SPIXII

Formerly known as: Ara spixii

Brazilian: Ararinha-azul; *English:* Spix's Macaw, Little Blue Macaw; *French:* Ara de Spix, Ara à face grise; *German:* Spixara, Spix Blauara; *Italian:* Ara di Spix; *Spanish:* Guacamayito azul

Range: Brazil
Identification:
Length 56cm. Sexes alike. Head grey tinged blue. Upperparts, wings and tail upperside dark blue, underparts paler blue, breast and belly with a little tinge of green. Tail underside dark grey.
Young birds: Darker and shorter tail.
Bare parts: Bill grey-black. Legs dark brownish-grey. Eyes yellow, juvenile brown. Bare skin round eyes and between bill and eyes dark grey, juvenile white.
List: I
Status. Very rare. In captivity about 40.
Lit.: 37, 81, 129, 168, 296, 384, 410, 431

5. ARA AMBIGUA

English: Great Green Macaw, Buffon's Macaw; *French:* Ara de Buffon; *German:* Bechsteinara, Grosser Soldatenara

Range: Nicaragua to Colombia and Ecuador
Identification:
Length 85cm. Sexes alike. Forehead and feathered lines between bill and eyes scarlet. Facial area with wide rows of black feathers. Rest of head green. Back and wings green tinged olive. Lower back, rump, upper and under tail-coverts and greater part of wings blue. Underparts green-olive. Tail brownish-red tipped pale blue; underside of tail and flight feathers olive-yellow.
Young birds: Duller. Indistinct yellow on shoulders, wings and central tail feathers.
Bare parts: Bill dark grey, tip paler. Legs dark grey. Eyes yellow, juvenile brown. Bare facial skin pinkish-white.

Geographical variation:
Birds from W. Ecuador have smaller and narrower bill. Underside of wings and tail more greenish.
List: I
Lit.: 81, 129, 168, 410

6. ARA GLAUCOGULARIS

Also known as: Ara caninde

English: Blue-throated Macaw, (Caninde Macaw, Wagler's Macaw - mis applied); *French:* Ara bleu et jaune, (Ara Caninde - mis-applied); *German:* (Kaninde - mis applied); *Italian:* Ara gialla e blu, Ara di Wagler, *Spanish:* Guacamayo amarilla

Range: Bolivia, Argentina, and (?) Paraguay
Identification:
Length 85cm. Sexes alike. Crown, nape, entire upperparts, wings, throat, under tail-coverts and tail blue. Facial area with blue-green lines of feathers. A narrow yellow line runs from ear to side of breast. Underparts orange-yellow. Tail olive-yellow below.
Young birds: Throat darker blue.
Bare parts: Bill and legs dark grey. Eyes yellow, juvenile brown.
List: I
Status: First found in nature few years ago and presumed rare.
Latest news: In August 1992 Charles Munn found four nests in Bolivia!
Lit.: 37, 81, 134, 168, 196, 410, 430

PLATE 46

1. ARA MACAO

Brazilian: Arara-canga; *English:* Scarlet Macaw; *French:* Ara rouge; *German:* Arakanga; *Spanish:* Guacamayo bandera

Range: Mexico to Peru and Brazil
Identification:
Length 85cm. Sexes alike. Facial skin with fine inconspicuous red lines. General colour scarlet. Lower back, rump, upper and under tail-coverts pale blue. Wings scarlet with yellow area tipped green and hind part blue.Tail above red narrow tipped blue, below red.
Young birds: Tail shorter.
Bare parts: Upper bill horn lower base black, lower bill blackish. Legs dark grey. Eyes yellow, juvenile brown. Bare facial skin creamy-white.
List: I
Lit.: 81, 129, 168, 410, 431

2. ARA MILITARIS

English: Military Macaw; *French:* Ara militaire; *German*: Soldatenara; *Spanish:* Guacamayo verde

Range: Mexico, Colombia, Ecuador, Peru, Bolivia, Argentina
Identification:
Length 70cm. Sexes alike. Forehead and feathered lines between bill and eyes red, transverse lines on facial skin of fine green-black feathers.Throat olive-brown.Wings green and blue. Rump, upper, and under tail-coverts blue. Tail red-brown tipped blue. Undertail and underside of wings olive-yellow. Rest green with olive tinge on back and wings. Note: Similar to Great Green Macaw number 5 of plate 46, but smaller, more green and narrower lines on facial skin and entire bill dark grey.
Young birds: Undescribed.
Bare parts: Bill and legs dark grey. Eyes yellow.

Geographical variation:
Birds from Mexico are larger.
Birds from Bolivia and Argentina have reddish-brown throat and ear-coverts reddish at base.
List: I
Lit.: 81, 129, 168, 410

3. ARA RUBROGENYS

English: Red-fronted Macaw, Red-cheeked Macaw; *French:* Ara de Lafresnaye; *German:* Rotohrara, *Italian:* Ara a fronte rossa; *Spanish:* Guacamayo frentirroja

Range: Bolivia
Identification:
Length 60cm. Sexes alike. Forehead, crown, area behind ear-coverts, and thighs red. Bare facial skin with fine blackish-brown feather lines. Front edges and upper part of wings orange-red, long flight-feathers blue. Tail olive-green tipped blue, underside olive-yellow. Rest of plumage olive-green to green.
Young birds: Little or no red colour.
Bare parts: Bill and legs dark grey. Eyes orange. Bare facial skin white.
List: I
Lit.: 37, 81, 129, 168, 208, 410

4. ARATINGA GUAROUBA

Formerly known as: Guaruba guarouba; Heliopsitta guarouba

Brazilian: Guaruba; *English:* Golden Parakeet, Golden Conure, Queen of Bavaria Conure; *French:* Perruche dorée, Conure doré, Guarouba jaune; *German:* Goldsittich; *Italian:* Conuro dorato, Parrocchetto dorato; *Spanish:* Cotorra amarilla

Range: Brazil
Identification:
Length 34cm. Sexes alike. Hind parts of wings dark green. Rest rich yellow.
Young birds: Much more greenish.
Bare parts: Bill horn coloured. Legs pinkish. Eyes brown.
List: I
Lit.: 37, 81, 129, 168, 410, 431

Cyanoramphus a. auriceps
English: Yellow-fronted Parakeet
Range: New Zealand
Identification:
Length 23cm. Sexes alike. Forehead and patch on each side of rump red. Crown golden yellow. Violet-blue on wings. Undertail brown. Rest of plumage green.
Young birds: Tail shorter.
Bare parts: Bill pale blue, darker greyish at tip. Legs grey-brown. Eyes orange-red, juvenile pale brown.
List: II

Subspecies:

5. CYANORAMPHUS AURICEPS FORBESI
English: Forbes' Yellow-fronted Parakeet, *French:* Perruche à tête d'or de Forbes; *German:* Forbes Springsittich; *Italian:* Kakariki a pileo giallo delle Chatam, Parrochetto dalla testa d'oro di Forbes; *Maori:* Kakariki; *Spanish:* Periquito frentiamarillo de Forbes
Range: Chatham Islands (New Zealand)
Identification:
Larger. Red on forehead does not reach eyes. Blue on wings more greenish-blue. Underparts more yellowish.
List: I
Status: 50 birds estimated in 1978.
Lit.: 37, 81, 168, 129

6. CYANORAMPHUS N. NOVAEZELANDIAE

English: Red fronted Parakeet; *French:* Perruche de Sparrman, Perruche de Nouvelle-Zélande à bandeau; *German:* Ziegensittich, Laufsittich; *Italian:* Kakariki a fronte rossa, Parrochetto della Nuova Zelanda; *Maori:* Kakariki; *Spanish:* Periquito cabecirrojo

Range: New Zealand and outlying islands.
Identification:
Length 27cm. Sexes similar. Forehead, crown, spot behind eye, and patch on each side of rump red. Flight-feathers violet-blue on outer webs. Rest green.
Young birds: Less red on head.
Bare parts: Bill bluish-grey, dark grey towards tip. Legs greyish-brown. Eyes red, juvenile pale brown.

Subspecies:

Cyanoramphus novaezelandiae saissetti
English: New Caledonian Parakeet
Range: New Caledonia
Head and underparts more yellowish. Red on crown brighter.

Geographical variation:
Birds from Kermadec Islands have more blue in wings. Tail blue-green.
Birds from Chatham Islands have head emerald green.
Birds from Antipodes Island are larger. Red on head orange-red. Otherwise more yellowish.
List: I
Lit.: 37, 81, 129, 152, 168, 410

PLATE 47

1. PEZOPORUS OCCIDENTALIS

Also known as: Geopsittacus occidentalis

English: Night Parrot; *French:* Perruche nocturne; *German:* Höhlensittich, Nachtsittich, Höhlenpapagei; *Italian:* Parrocchetto notturno; *Spanish:* Loro nocturno

Range: Australia
Identification:
Length 23cm. Sexes similar. Plumage yellow-green mottled with dark brown, black and yellow. Belly yellow. Tail short.
Young birds: Said to be duller with yellow on throat and neck.
Bare parts: Bill horn. Legs brown. Eyes black.
List: I
Status: Not seen in 67 years but in 1979 three-four birds were seen.
Lit.: 37, 81, 129, 152, 168, 196, 410

2. PEZOPORUS WALLICUS

English: Ground Parrot; *French:* Perruche terrestre, Perruche de terre; *German:* Erdsittich; *Italian:* Parrocchetto terragnolo; *Spanish:* Periquito de tierra

Range: S.W. and S. E. Australia, Tasmania
Identification:
Length 30cm. Sexes alike. Forehead with narrow orange-red band. Rest green strongly mottled and barred black and yellow. Belly greenish-yellow barred black. Tail long.
Young birds: Without red on forehead.
Bare parts: Bill brownish. Legs brown. Eyes dull yellow.

Geographical variation:
Birds from S. W. Australia have bright yellow belly with broken black barring.
List: I
Status: Population from S. W. Australia endangered.
Lit.: 37, 81, 129, 152, 168, 196, 410

3. NEOPHEMA CHRYSOGASTER

English: Orange-bellied Parrot, Orange-bellied Grass Parakeet; *French:* Perruche à ventre orange, Perruche à lunettes vertes; *German:* Goldbauchsittich; *Italian:* Parrocchetto dal ventre arancione; *Spanish:* Loro ventrinaranja

Range: S. E. Australia, Tasmania
Identification:
A. Length 20cm. **Male:** Forehead blue, upper border pale blue. Rest of head and upperparts bright grass-green. Throat and breast green-yellow. Rest of underparts yellow with a big orange area on belly. Wings blue and green. Tail green with a blue tinge, underside yellow.
B. **Female:** Duller. Forehead paler blue without paler passage to crown.
Young birds: Duller than female, orange patch smaller.
Bare parts: Bill greyish-brown. Legs greyish. Eyes brown.
List: I
Status: Total population estimated at less than 200 birds.
Lit.: 21, 37, 81, 129, 168, 410

4. OGNORHYNCHUS ICTEROTIS

English: Yellow-eared Parrot, Yellow-eared Conure, Yellow-eared Macaw; *French:* Perruche à joues d'or, Perruche aux oreilles jaunes; *German:* Gelbohrsittich; *Italian:* Conuro a orecchie gialle; *Spanish:* Periquito orejiamarillo

Range: Colombia, Ecuador
Identification:
Length 42cm. Sexes similar. Forehead, area between bill and eye, upper cheeks and ear-coverts yellow. Rest of head, upperparts and wings dark green. Underparts yellowish-green, darker green on cheeks, thighs and under tail-coverts. Tail above green, below dusky orange-red.
Young birds: No data available.
Bare parts: Bill dark grey. Legs grey. Eyes orange.
List: I
Lit.: 37, 81, 129, 168, 410

Cyclopsitta d. diophthalma
Also known as: Opopsitta diophthalma

English: Double-eyed Fig Parrot
Range: New Guinea & islands
Identification:
Length 14cm. **Male:** Forehead, crown and cheeks scarlet, becoming orange-yellow on hindcrown. Above eye blue. Wings blue and green, innermost wing-coverts red. Flanks yellow. Rest green, darker on upperparts.
Female: Like male but lower cheeks buffish-brown.
Young birds: Like female.
Bare parts: Bill pale grey, tip black. Legs greenish-grey. Eyes dark brown.
List: II (also six other subspecies)

Subspecies:

5. CYCLOPSITTA DIOPHTHALMA COXENI

Also known as: Opopsitta diophthalma coxeni

English: Coxen's Double-eyed Fig Parrot, *French:* Psittacule à double-oeil de Coxen, Perroquet masqué de Coxen; *German:* Rotwangen-Zwergpapagei, Coxen's Zwergpapagei; *Italian:* Pappagallo dei finchi di Coxen, Parrocchetto mascherato di Coxen; *Spanish:* Lorito de Coxen, Lorito enmascarado de Coxen
Range: S. Queensland and N. New South Wales (Australia)
Identification:
A. Length 15cm. **Male:** Centre of forehead blue. Band on side of cheeks mauve-blue. Between bill and eyes, below eyes, ear-coverts some red feathers and innermost wing-coverts red. Long flight-feathers deep blue. Flanks yellow-green. Rest green, darker on upperparts.
B. **Female:** Like male, but red and blue on head less extensive.
Young birds: Like female.
Bare parts: Bill pale grey, tip black. Legs greenish-grey. Eyes dark brown.
List: I
Lit.: 37, 81, 129, 168, 410

6. PIONOPSITTA PILEATA

Brazilian: Cuiú-cuiú; *English:* Pileated Parrot, Red-capped Parrot; *French:* Caïque mitré, Perroquet à oreilles; *German*: Scharlachkopfpapagei; *Italian:* Parrocchetto a cappuccio rosso, Parrocchetto con le orecchie; *Spanish:* Lorito carirrojo, Cuiu cuiu

Range: Brazil, Paraguay, Argentina
Identification:
A. Length 22cm. **Male:** Forehead, crown, between bill and naked skin around eye and upper part of ear-coverts red. Lower part of ear-coverts brownish-purple. Rest green with some blue in wings and tail.
B. **Female:** Without red on head. Forehead green washed with blue.
Young birds: Males have a narrow red band on forehead followed by a orange-yellow patch. Female like adult females but blue in wings paler and more greenish.
Bare parts: Bill greyish-green, tip horn. Legs grey. Eyes dark brown.
List: I
Lit.: 37, 81, 129, 168, 410, 431

PLATE 48

1. PROBOSCIGER ATERRIMUS

English: Palm Cockatoo; *French:* Cacatoès noir; *German:* Arakakadu

Range: Aru Islands, Misbol Island, Papuan Islands New Guinea, Cape York, Queensland (N. E. Australia)
Identification:
Length 60cm. **Male:** Entire plumage greyish-black to black. Large crest conspicuous.
Female: Like male but upper bill smaller.
Young birds: Underparts edged pale yellow.
Bare parts: Bill grey-black. Legs grey. Eyes dark brown.

Geographical variation:
Birds from Yapen Island are larger and have much narrower crest feathers.
*List:*I
Lit.: 81, 129, 168, 410

2. PSEPHOTUS CHRYSOPTERYGIUS

English: Golden-shouldered Parrot, Golden-shouldered Parakeet; *French:* Perruche à ailes d'or, Perruche à épaules dorées; *German:* Goldschultersittich, Gelbschultersittich; *Italian:* Parrocchetto dalle ali gialle, Parrocchetto dalle ali dorate; *Spanish:* Loro hombroamarillo, Periquito de espalda dorada

Range: Cape York Peninsula, (N. Australia)
Identification:
A. Length 26cm. **Male:** Forehead pale yellow, crown and nape black. Upperparts brown, rump and underparts turquoise-blue, belly, thighs and under tail-coverts scarlet tipped white. Wings blue, black, and greenish-brown with a big yellow patch on forewing. Tail greenish-blue.
B. **Female:** Crown and nape pale bronze-brown. Rest of head and body dull yellowish-green. Belly pale blue with red and white bands.
Young birds: Like female, but young males brighter on sides of head and under tail-coverts.
Bare parts: Bill greyish. Legs greyish-brown. Eyes dark brown.
List: I
Lit.: 37, 80, 81, 168, 218, 410

3. PSEPHOTUS DISSIMILIS

English: Hooded Parrot, Hooded Parakeet; *French:* Perruche à capuchon noir; Perruche à capuchon; *German:* Hooded Sittich, Schwarzkappensittich; *Spanish:* Periquito encapuchado

Range: Northern Territory (N. Australia)
Identification:
Length 28cm. **Male:** Like number 2 but top of head and nape black. Dark brown mantle. Rump and underparts turquoise-blue. Under tail-coverts salmon-pink bordered white. Yellow wing patch bigger and brighter than Golden-shouldered Parrot.
Female: Crown and face pale greyish-green. Under tail-coverts salmon-pink, otherwise like Golden-shouldered Parrot female.
Young birds: Like female but duller.
Bare parts: Bill greyish. Legs greyish-brown. Eyes dark brown.
List: I
Lit.: 37, 80, 81, 129, 168, 218, 410

4. PSEPHOTUS PULCHERRIMUS

English: Paradise Parrot, Paradise Parakeet; *French:* Perruche de paradis, Perruche magnifique; *German:* Paradiessittich; *Italian:* Parrocchetto del paradiso; *Spanish:* Loro del paraíso

Range: Queensland, New South Wales, (Australia)
Identification:
A. Length 27cm. **Male:** Forehead red, crown and nape brownish-black. Mantle brown, rump turquoise-blue. Below turquoise-blue, often washed with green on breast. Belly, thighs, and under tail-coverts scarlet. Wings blackish to brown with a big red patch on forewing. Tail bronze-green washed blue.
B. **Female:** Forehead pale yellow-green tipped with red.Crown and nape dark brown. Side of head and breast buff-yellow with spots of brownish-orange. Belly pale blue, some feathers edged red. Wings like male but red patches smaller.
Young birds: Like female, young males have darker crown and a little blue or green on face.
Bare parts: Bill greyish. Legs greyish-brown. Eyes brown.
List: I
Status: Extremely rare, if not already extinct.
Lit.: 37, 74, 80, 81, 129, 168, 218, 410

5. PSITTACULA ECHO

Also known as: Psittacula eques; Psittacula krameri echo

English: Mauritius Parakeet, Echo Parakeet; *French:* Perruche de Maurice, Perruche à collier de Maurice; *German*: Mauritiussittich; *Italian:* Parrocchetto dal collare dell'Isola di Mauritius; *Spanish:* Cotorra de Mauricio

Range: Mauritius
Identification:
A. Length 42cm. **Male:** Plumage mainly green, darker above. Between bill and eye a narrow black stripe and a broad black stripe from chin to sides of neck. On side of neck to nape a rose-pink patch. Hind neck and nape washed with blue. Tail green above, brown below.
B. **Female:** Like male but no blue on hind neck and nape. Rose-pink patch of male yellow-green. Dark green stripe on lower cheek. Central tail feathers tinged blue.
Young birds: Like female.
Bare parts: Bill: upper dark red, lower black, female and juvenile both black, very young birds both red. Legs grey. Eyes pale yellow.
List: I
Status: Total population estimated in 1984 at less than ten!
Lit.:37, 81, 129, 152, 168, 196, 410

Psittacus e. erithacus
English: Grey Parrot
Range: Ivory Coast to Kenya, Tanzania & Angola
Identification:
Length 33cm. Sexes alike. Plumage pale grey, on head and neck edged greyish-white, on belly edged dark grey. Rump very pale grey. Tail and tail-coverts red. Note: some individuals are so dark that they are impossible to distinguish from princeps.
Young birds: Tail darker red towards tip.Under tail-coverts washed grey.
Bare parts: Bill black. Legs dark grey. Eyes pale yellow, juvenile grey. Bare skin around eyes whitish.

Subspecies:

Psittacus erithacus timneh
Range: Guinea to Ivory Coast
Smaller. Plumage darker. Tail dark maroon edged brownish. Upper bill reddish tipped black.
List: II

6. PSITTACUS ERITHACUS PRINCEPS

English: Principe Grey Parrot; *French:* Perroquet jaco de Fernando Poo, Jacquot de Fernando Poo; *German:* Fernando-Poo Graupapagei; *Italian*: Pappagallo cennerino dell'Isola di Fernando Poo; *Spanish:* Koo, Loro gris de Fernando Poo
Range: Principe Island, Bioko
Plumage darker, otherwise like Grey Parrot.
List I
Lit.: 37, 81, 129 168

PLATE 49

1. PSITTACULA KRAMERI

English: Rose-ringed Parakeet, Ring-necked Parakeet; *French:* Perruche à collier; *German:* Halsbandsittich; *Italian:* Parrocchetto dal collare; *Spanish:* Periquito de collar Africano, Cotorra de Kramer

Range: Senegal to Sudan, Uganda, Somalia, S. Asia, Burma
Identification:
A. Length 40cm. **Male:** Like Mauritius Parakeet (plate 49, No. 5) but green colour paler, more yellowish. Central tail feathers bluish washed with greenish, all other tail feathers green.
B. **Female:** Like male but without black line, blue, and rose-pink on head. Tail shorter.
Young birds: Similar to female.
Bare parts: Bill dark red, tip black, lower bill black with some red spots at base, juvenile coral-pink with pale tip. Legs greenish-grey, juvenile grey. Eyes pale yellow, juvenile greyish-white.

Geographical variation:
Birds from Sudan and Somalia are more green on head. Bill smaller and upper bill more red.
Birds from Pakistan, N. India and Burma are larger with a blue tinge on sides of head. Underparts greyish-green. Whole bill coral red, lower often with black spots.
Birds from S. India and Sri Lanka are larger and lower bill black.
List: III Ghana
Lit.: 37, 81, 129, 410, 424

2. PYRRHURA CRUENTATA

Brazilian: Fura-mato; *English:* Blue-throated Conure, Red-rumped Conure, Red-eared Conure, Ochre-marked Parakeet; *French:* Perruche tiriba, Conure à gorge bleue; *German:* Blaulatzsittich; *Italian:* Conuro a gola azzurra, Conuro dalla gola blu; *Spanish:* Perico grande

Range: Brazil
Identification:
Length 30cm. Sexes similar. Top of head dark brown edged with narrow buff, on nape more prominent. Between bill and eye, below eye, and ear-coverts rufous-red. Cheeks and mantle green. Rump, a big area on centre of belly, and forewing red. Throat, upper breast, and around hindneck blue. Rest of underparts green. Wings green and blue. Tail olive, below brownish-red.
Young birds: Duller and less red on wings.
Bare parts: Bill greyish-brown. Legs grey. Eyes yellow-orange.
List: I
Lit.: 37, 81, 129, 168, 410, 431

3. RHYNCHOPSITTA PACHYRHYNCHA

Formerly also known as: Rhynchopsitta terrisi

English: Thick-billed Parrot, Maroon-fronted Parrot; *French:* Perruche à gros bec, Perruche ara; *German:* Kiefernsittich, Arasittich; *Italian:* Parrocchetto dal grosso becco, Pappagallo a fronte castana; *Spanish:* Periquito de pico grueso, Loro piquigordo

Range: Mexico
Identification:
Length 38cm. Sexes alike. Forehead, forecrown, stripe over and behind eye, bend of wings and thighs red. Underside of wings bright yellow and grey. Underside of tail grey. Rest bright green, with a yellowish wash on cheeks and ear-coverts.
Young birds: Duller and less red on wings.
Bare parts: Bill greyish-brown. Legs grey. Eyes yellow-orange.
Bare skin around eyes pale pinkish.
List: I
Lit.: 37, 81, 129, 168, 410

4. RHYNCHOPSITTA TERRISI

Also known as: Rhynchopsitta pachyrhyncha terrisi

English: Maroon-fronted Parrot; *French:* Perruche à front brun; *German:* Maronenstirnsittich

Range: Nuevo Leon, (Mexico)
Identification:
Length 40cm. Sexes alike. Like Thick-billed Parrot, but larger and darker. Red on head maroon-brown. Underside of wings grey with a little wash of olive. Bend of wings and thighs brownish-red.
Young birds: Like number 3.
Bare parts: Like number 3.
List: I
Lit.: 37, 81, 129, 164, 168, 410

5. STRIGOPS HABROPTILUS

English: Kakapo, Owl Parrot; *French:* Strigops kakapo, Kakapo, Perroquet nocturne, Perroquet hibou; *German:* Kakapo, Eulenpapagei; *Italian:* Strigope, Kakapo; *Spanish:* Cacapo

Range: New Zealand
Identification:
Length 64cm. Sexes similar. Forehead, face, and ear-coverts yellowish-brown. Stripe over eye yellow. Crown, nape, and upperparts bright green barred and streaked with brown, yellow, and black. Underparts greenish-yellow barred like upperparts.
Tail dull green with brown and buffish-yellow barring.
Young birds: Duller, less yellowish on head.
Bare parts: Bill grey, lower bill horn. Legs pinkish to bluish-grey. Eyes dark brown.
List: I
Status: On the brink of extinction. In 1981 estimated to be about 108 birds.
Lit.: 37, 69, 81, 129, 164, 168, 196, 410

PLATE 50

ORDER
PSITTACIFORMES*

Family: PSITTACIDAE

English: Parrots; *German:* Papageien

Identification:
Length 8-100cm. Sexes usually alike. Colour often very bright; green predominant. Bill short and hooked. Legs very short and with two toes facing forward, two backward.

All other parrots:

	Chalcopsitta atra	Black Lory
	Chalcopsitta duivenbodei	Brown Lory
	Chalcopsitta sintillata	Yellow-streaked Lory
	Chalcopsitta cardinalis	Cardinal Lory
	Eos histrio	Red-and Blue Lory
	Eos squamata	Violet-necked Lory
	Eos bornea	Red Lory
	Eos reticulata	Blue-streaked Lory
	Eos cyanogenia	Black-winged Lory
	Eos semilarvata	Blue-eared Lory
	Pseudeos fuscata	Dusky Lory
	Trichoglossus ornatus	Ornate Lorikeet
	Trichoglossus haematodus	Rainbow Lorikeet
	Trichoglossus rubritorquis	Red-collared Lorikeet
	Trichoglossus euteles	Olive-headed Lorikeet
	Trichoglossus flavoviridis	Yellow-and-Green Lorikeet
	Trichoglossus johnstoniae	Mindanao Lorikeet
	Trichoglossus rubiginosus	Pohnpei Lorikeet
	Trichoglossus chlorolepidotus	Scaly-breasted Lorikeet
	Psitteuteles versicolor	Varied Lorikeet
	Psitteuteles iris	Iris Lorikeet
	Psitteuteles goldiei	Goldie's Lorikeet
	Lorius garrulus	Chattering Lory
	Lorius domicella	Purple-naped Lory
1.	Lorius lory	Black-capped Lory
	Lorius hypoinochrous	Purple-bellied Lory
	Lorius albidinuchus	White-naped Lory
	Lorius chlorocercus	Yellow-bibbed Lory
	Phigys solitarius	Collared Lory
	Vini australis	Blue-crowned Lorikeet
	Vini kuhlii	Kuhl's Lorikeet
	Vini stepheni	Stephens's Lorikeet
	Vini peruviana	Tahitian Lorikeet
	Vini ultramarina	Ultramarine Lorikeet
	Glossopsitta concinna	Musk Lorikeet
	Glossopsitta pusilla	Little Lorikeet
	Glossopsitta porphyrocephala	Purple-crowned Lorikeet
	Charmosyna palmarum	Palm Lorikeet
	Charmosyna rubrigularis	Red-chinned Lorikeet
	Charmosyna meeki	Meek's Lorikeet
	Charmosyna toxopei	Blue-fronted Lorikeet
	Charmosyna multistriata	Striated Lorikeet
	Charmosyna wilhelminae	Pygmy Lorikeet
	Charmosyna rubronotata	Red-fronted Lorikeet
	Charmosyna placentis	Red-flanked Lorikeet
	Charmosyna diadema	New Caledonian Lorikeet
	Charmosyna amabilis	Red-throated Lorikeet
	Charmosyna margarethae	Duchess Lorikeet
	Charmosyna pulchella	Fairy Lorikeet
	Charmosyna josefinae	Josephine's Lorikeet
	Charmosyna papou	Papuan Lorikeet
	Oreopsittacus arfaki	Plum-faced Lorikeet
	Neopsittacus musschenbroekii	Yellow-billed Lorikeet
	Neopsittacus pullicauda	Orange-billed Lorikeet
	Calyptorhynchus baudinii	White-tailed Black-Cockatoo
	Calyptorhynchus latirostris	Slender-billed Black-Cockatoo
	Calyptorhynchus funereus	Yellow-tailed Black-Cockatoo
2.	Calyptorhynchus banksii	Red.tailed Black-Cockatoo
	Calyptorhynchus lathami	Glossy Black-Cockatoo
	Callocephalonn fimbriatum	Gang-Gang Cockatoo
	Eolophus roseicapillus	Galah
	Cacatua leadbeateri	Pink Cockatoo
	Cacatua sulphurea	Yellow-crested Cockatoo
	Cacatua galerita	Sulphur-crested Cockatoo
	Cacatua ophthalmica	Blue-eyed Cockatoo
3.	Cacatua alba	White Cockatoo
	Cacatua sanguinea	Little Corella
	Cacatua pastinator	Western Corella
	Cacatua tenuirostris	Long-billed Corella
	Cacatua ducorpsii	Ducorps's Cockatoo
4.	Nestor notabilis	Kea
	Nestor meridionalis	Common Kaka
	Micropsitta keiensis	Yellow-capped Pygmy-Parrot
	Micropsitta geelvinkiana	Geelvink Pygmy-Parrot
5.	Micropsitta pusio	Buff-faced Pygmy-Parrot
	Micropsitta meeki	Meek's Pygmy-Parrot
	Micropsitta finschii	Finsch's Pygmy-Parrot
	Micropsitta bruijnii	Red-breasted Pygmy-Parrot
	Cyclopsitta gulielmitertii	Orange-breasted Fig-Parrot
	Cyclopsitta diophthalma (- coxeni)	Double-eyed Fig-Parrot
	Psittaculirostris desmarestii	Large Fig-Parrot
	Psittaculirostris edwardsii	Edwards's Fig-Parrot
	Psittaculirostris salvadorii	Salvadori's Fig-Parrot
	Bolbopsittacus lunulatus	Guaiabero
	Psittinus cyanurus	Blue-rumped Parrot
	Psittacella brehmii	Brehm's Tiger-Parrot
	Psittacella picta	Painted Tiger-Parrot
	Psittacella modesta	Modest Tiger-Parrot
	Psittacella madaraszi	Madarasz's Tiger-Parrot
	Geoffroyus geoffroyi	Red-cheeked Parrot
	Geoffroyus simplex	Blue-collared Parrot
	Geoffroyus heteroclitus	Singing Parrot
6.	Prioniturus montanus	Luzon Racquet-Tail
	Prioniturus waterstradti	Mindanao Racquet-Tail
	Prioniturus platenae	Blue-headed Racquet-Tail
	Prioniturus luconensis	Green Racquet-Tail
	Prioniturus discurus	Blue-crowned Racquet-Tail
	Prioniturus verticalis	Blue-winged Racquet-Tail
	Prioniturus flavicans	Yellowish-breasted Racquet-Tail
	Prioniturus platurus	Golden-mantled Racquet-Tail
	Prioniturus mada	Buru Racquet-Tail

PLATE 51

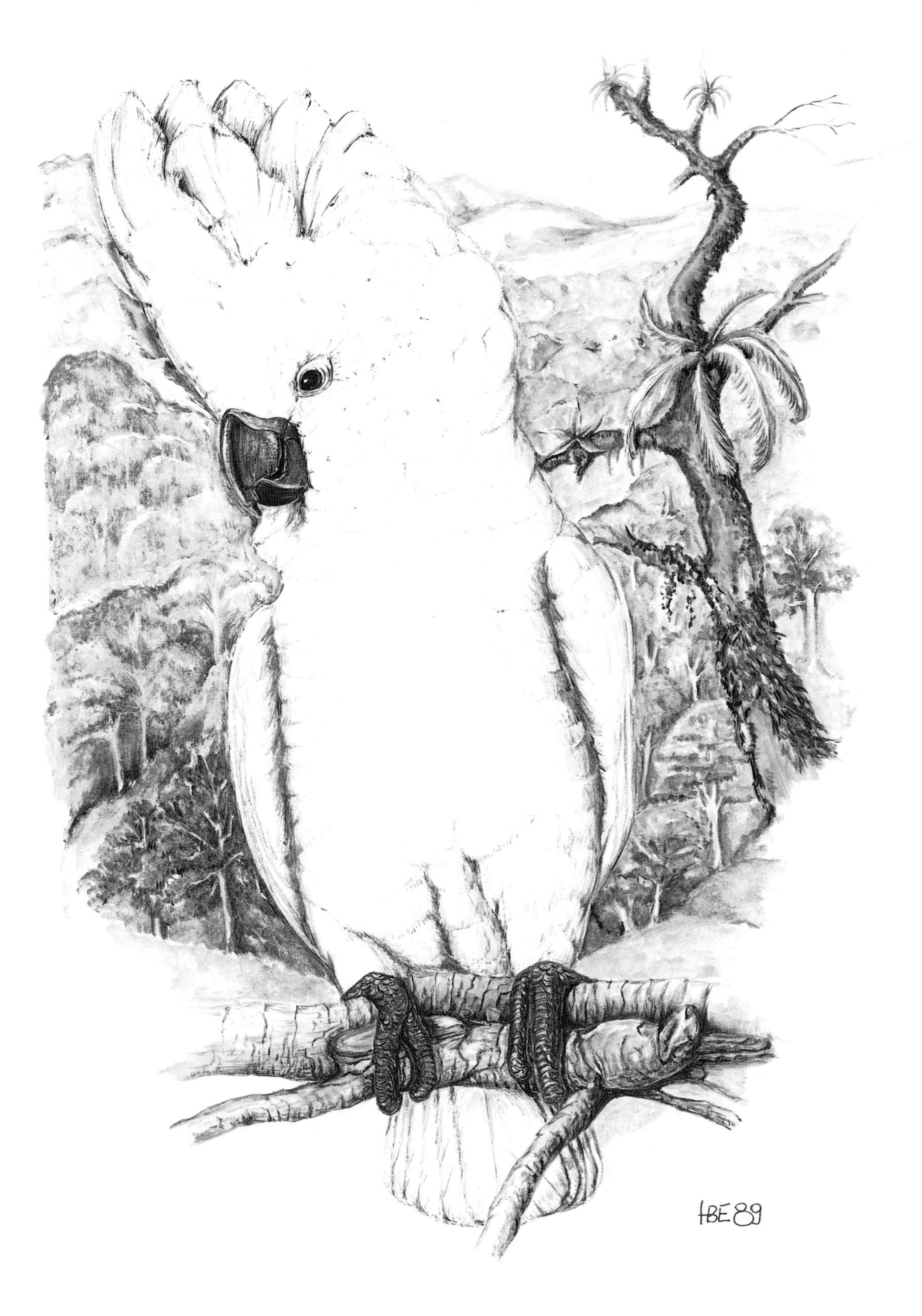

WHITE COCKATOO 3

	Scientific name	Common name
7.	Tanygnathus megalorynchos	Great-billed Parrot
	Tanygnathus lucionensis	Blue-naped Parrot
	Tanygnathus sumatranus	Blue-backed Parrot
	Tanygnathus gramineus	Black-lored Parrot
	Eclectus roratus	Eclectus Parrot
8.	Psittrichas fulgidus	Pesquet's Parrot
	Prosopeia splendens	Crimson Shining-Parrot
	Prosopeia personata	Masked Shining-Parrot
	Prosopeia tabuensis	Red Shining-Parrot
	Alisterus scapularis	Australian King-Parrot
	Alisterus amboinensis	Moluccan King-Parrot
	Alisterus chloropterus	Papuan King-Parrot
	Aprosmictus jonquillaceus	Olive-shouldered Parrot
	Aprosmictus erythropterus	Red-winged Parrot
	Polytelis swainsonii	Superb Parrot
	Polytelis anthopeplus	Regent Parrot
	Polytelis alexandrae	Alexandra's Parrot
	Purpureicephalus spurius	Red-capped Parrot
	Platycercus zonarius	Port Lincoln Ringneck
	Platycercus barnardi	Green Rosella
9.	Platycercus elegans	Crimson Rosella
	Platycercus flaveolus	Yellow Rosella
	Platycercus venustus	Northern Rosella
	Platycercus adscitus	Pale-headed Rosella
	Platycercus eximius	Eastern Rosella
	Platycercus icterotis	Western Rosella
10.	Northiella haematogaster	Bluebonnnet
	Psephotus haematonotus	Red-rumped Parrot
	Psephotus varius	Mulga Parrot
	Cyanoramphus unicolor	Antipodes Parakeet
	Cyanoramphus cookii	Norfolk Parakeet
	Cyanoramphus auriceps (- forbesi)	Yellow-fronted Parakeet
	Eunymphicus cornutus	Horned Parakeet
	Neopsephotus bourkii	Bourke's Parrot
	Neophema chrysostoma	Blue-winged Parrot
	Neophema elegans	Elegant Parrot
	Neophema petrophila	Rock Parrot
	Neophema pulchella	Turquoise Parrot
	Neophema splendida	Scarlet-chested Parrot
11.	Lathamus discolor	Swift Parrot
12.	Coracopsis vasa	Vasa Parrot
	Coracopsis nigra	Black Parrot
	Psittacus erithacus (-princeps)	Grey Parrot
	Poicephalus robustus	Brown-necked Parrot
	Poicephalus gulielmi	Red-fronted Parrot
	Poicephalus senegalus	Senegal Parrot
	Poicephalus crassus	Niam-Niam Parrot
	Poicephalus meyeri	Meyer's Parrot
	Poicephalus flavifrons	Yellow-fronted Parrot
	Poicephalus rufiventris	Red-bellied Parrot
	Poicephalus cryptoxanthus	Brown-headed Parrot
	Poicephalus rueppellii	Rüppell's Parrot
	Agapornis canus	Grey-headed Lovebird
	Agapornis pullarius	Red-headed Lovebird
	Agapornis taranta	Black-winged Lovebird
	Agapornis swindernianus	Black-collared Lovebird
	Agapornis roseicollis	Rosy-faced Lovebird
	Agapornis fischeri	Fischer's Lovebird
13.	Agapornis personatus	Yellow-collared Lovebird
	Agapornis lilianae	Lilian's Lovebird
	Agapornis nigrigenis	Black-cheeked Lovebird
	Loriculus vernalis	Vernal Hanging-Parrot
	Loriculus beryllinus	Ceylon Hanging-Parrot
	Loriculus philippensis	Colasisi
14.	Loriculus galgulus	Blue-crowned Hanging-Parrot
	Loriculus stigmatus	Sulawesi Hanging-Parrot
	Loriculus amabilis	Moluccan Hanging-Parrot
	Loriculus catamene	Sangihe Hanging-Parrot
	Loriculus aurantiifrons	Orange-fronted Hanging-Parrot
	Loriculus tener	Green-fronted Hanging-Parrot
	Loriculus exilis	Red-billed Hanging-Parrot
	Loriculus pusillus	Yellow-throated Hanging-Parrot
	Loriculus flosculus	Wallace's Hanging-Parrot
	Psittacula eupatria	Alexandrine Parakeet
	Psittacula himalayana	Slaty-headed Parakeet
	Psittacula finschii	Grey-headed Parakeet
	Psittacula intermedia	Intermediate Parakeet
	Psittacula cyanocephala	Plum-headed Parakeet
	Psittacula roseata	Blossom-headed Parakeet
	Psittacula columboides	Malabar Parakeet
	Psittacula calthropae	Layard's Parakeet
	Psittacula derbiana	Derbyan Parakeet
	Psittacula alexandri	Red-breasted Parakeet
	Psittacula caniceps	Nicobar Parakeet
	Psittacula longicauda	Long-tailed Parakeet
	Ara ararauna	Blue-and-Yellow Macaw
	Ara chloropterus	Red-and-Green Macaw
	Ara servera	Chestnut-fronted Macaw
	Ara manilata	Red-bellied Macaw
	Ara couloni	Blue-headed Macaw
	Ara auricollis	Yellow-collared Macaw
	Ara nobilis	Red-shouldered Macaw
	Aratinga acuticaudata	Blue-crowned Parakeet
	Aratinga holochlora	Green Parakeet
	Aratinga strenua	Pacific Parakeet
	Aratinga wagleri	Scarlet-fronted Parakeet
	Aratinga mitrata	Mitred Parakeet
	Aratinga erythrogenys	Red-masked Parakeet
	Aratinga finschi	Crimson-fronted Parakeet
	Aratinga leucophthalmus	White-eyed Parakeet
	Aratinga euops	Cuban Parakeet
	Aratinga chloroptera	Hispaniolan Parakeet
	Aratinga solstitialis	Sun Parakeet
	Aratinga jandaya	Jandaya Parakeet
	Aratinga auricapilla	Golden-capped Parakeet
	Aratinga weddellii	Dusky-headed Parakeet
	Aratinga nana	Olive-throated Parakeet
	Aratinga canicularis	Orange-fronted Parakeet
	Aratinga aurea	Peach-fronted Parakeet
	Aratinga pertinax	Brown-throated Parakeet
	Aratinga cactorum	Cactus Parakeet
	Nandayus nenday	Nanday Parakeet
	Leptosittaca branickii	Golden-plumed Parakeet
	Cyanoliseus patagonus	Burrowing Parakeet

PLATE 51A

Pyrrhura devillei	Blaze-winged Parakeet
Pyrrhura frontalis	Maroon-bellied Parakeet
Pyrrhura perlata	Pearly Parakeet
Pyrrhura rhodogaster	Crimson-bellied Parakeet
Pyrrhura molinae	Green-cheeked Parakeet
15. Pyrrhura pica	Painted Parakeet
Pyrrhura leucotis	White-eared Parakeet
Pyrrhura viridicata	Santa Marta Parakeet
Pyrrhura egregia	Fiery-shouldered Parakeet
Pyrrhura melanura	Maroon-tailed Parakeet
Pyrrhura orcesi	El Oro Parakeet
Pyrrhura rupicola	Black-capped Parakeet
Pyrrhura albipectus	White-necked Parakeet
Pyrrhura calliptera	Brown-breasted Parakeet
Pyrrhura hoematotis	Red-eared Parakeet
Pyrrhura rhodocephala	Rose-headed Parakeet
Pyrrhura hoffmanni	Sulphur-winged Parakeet
Enicognathus ferrugineus	Austral Parakeet
Enicognathus leptorhynchus	Slender-billed Parakeet
Myiopsitta monachus	Monk Parakeet
Bolborhynchus aymara	Grey-hooded Parakeet
Bolborhynchus aurifrons	Mountain Parakeet
Bolborhynchus lineola	Barred Parakeet
Bolborhynchus orbygnesius	Andean Parakeet
Bolborhynchus ferrugineifrons	Rufous-fronted Parakeet
Forpus cyanopygius	Mexican Parrotlet
Forpus passerinus	Green-rumped Parrotlet
Forpus xanthopterygius	Blue-winged Parrotlet
Forpus conspicillatus	Spectacled Parrotlet
Forpus sclateri	Dusky-billed Parrotlet
Forpus coelestis	Pacific Parrotlet
Forpus xanthops	Yellow-faced Parrotlet
Brotogeris tirica	Plain Parakeet
Brotogeris versicolurus	Canary-winged Parakeet
Brotogeris chiriri	Yellow-chevroned Parakeet
Brotogeris pyrrhopterus	Grey-cheeked Parakeet
Brotogeris jugularis	Orange-chinned Parakeet
Brotogeris cyanoptera	Cobalt-winged Parakeet
Brotogeris chrysopterus	Golden-winged Parakeet
Brotogeris sanctithomae	Tui Parakeet
Nannopsittaca panychlora	Tepui Parakeet
Touit batavica	Lilac-tailed Parrotlet
Touit huetii	Scarlet-shouldered Parrotlet
Touit costaricensis	Red-fronted Parrotlet
Touit dilectissima	Blue-fronted Parrotlet
Touit purpurata	Sapphire-rumped Parrotlet
Touit melanonotus	Brown-backed Parrotlet
Touit surda	Golden-tailed Parrotlet
Touit stictoptera	Spot-winged Parrotlet
Pionites melanocephala	Black-headed Parrot
Pionites leucogaster	White-bellied Parrot
Pionopsitta haematotis	Brown-hooded Parrot
Pionopsitta pulchra	Rose-faced Parrot
Pionopsitta barrabandi	Orange-cheeked Parrot
Pionopsitta pyrilia	Saffron-headed Parrot
Pionopsitta caica	Caica Parrot
Gypopsitta vulturina	Vulturine Parrot
Hapalopsittaca melanotis	Black-winged Parrot
Hapalopsittaca amazonina	Rusty-faced Parrot
Hapalopsittaca fuertesi	Indigo-winged Parrot
Hapalopsittaca pyrrhops	Red-faced Parrot
Graydidascalus brachyurus	Short-tailed Parrot
Pionus menstruus	Blue-headed Parrot
Pionus sordidus	Red-billed Parrot
Pionus maximiliani	Scaly-headed Parrot
Pionus tumultuosus	Speckle-faced Parrot
Pionus senilis	White-crowned Parrot
16. Pionus chalcopterus	Bronze-winged Parrot
Pionus fuscus	Dusky Parrot
Amazona collaria	Yellow-billed Parrot
Amazona ventralis	Hispaniolan Parrot
Amazona albifrons	White-fronted Parrot
Amazona xantholora	Yellow-lored Parrot
Amazona agilis	Black-billed Parrot
Amazona viridigenalis	Red-crowned Parrot
Amazona finschi	Lilac-crowned Parrot
Amazona autumnalis	Red-lored Parrot
Amazona dufresniana	Blue-cheeked Parrot
Amazona festiva	Festive Parrot
Amazona xanthops	Yellow-faced Parrot
Amazona aestiva	Blue-fronted Parrot
Amazona oratrix	Yellow-headed Parrot
Amazona auropalliata	Yellow-naped Parrot
Amazona ocrocephala	Yellow-crowned Parrot
Amazona amazonica	Orange-winged Parrot
Amazona mercenaria	Scaly-naped Parrot
Amazona farinosa	Mealy Parrot
Deroptyus accipitrinus	Red-fan Parrot
Triclaria malachitacea	Blue-bellied Parrot
New species:	
Nannopsittaca dachilleae	Amazonian Parrotlet

List: II

Lit.: 2, 9, 37, 43, 61, 77, 81, 114, 125, 129, 168, 218, 222, 253, 307, 368, 397, 410

Reading List:

Forshaw, J. M. & W. T. Cooper 1989: *Parrots of the World*, Blandford Press, London.

Two parrots are not listed:

Budgerigar	Melopsittacus undulatus
Cockatiel	Nymphicus hollandicus

PLATE 52

1. TAURACO CORYTHAIX

Formerly known as: Tauraco persa; Tauraco corythaix persa; Tauraco corythaix schuettii; Tauraco corythaix fischeri; Tauraco corythaix schalowi Tauraco corythaix livingstonii

English: Knysna Turaco; *French:* Touraco vert; *German:* Helmturako

Range: Natal, Transvaal, Cape Province
Identification: Length 41cm. Sexes similar. Head, neck, mantle, and breast green. Crest 48-56 mm long with white tips. Two white stripes from gape to top of eyes and from gape below eyes to ear-coverts, between them a black line. Rest of upperparts dark iridescent green-blue, forewing and innermost wing feathers violet blue and green. Outermost 16 flight-feathers brilliant red with black outer edgings, only visible in flight. Underparts blackish, tinged green. Tail blackish green with violet-blue gloss.
Young birds: Duller with no white on crest. Red on wings only present on 8 outermost flight-feathers.
Bare parts: Bill orange-red, juvenile brownish. Legs black. Eyes brown, eye-ring red, juvenile dusky.

Geographical variation:
Birds from Transvaal have back and wings tinged more steel-blue.
List: II
Lit.: 7, 83, 129, 171, 172, 173, 174, 195, 253, 410

TAURACO PERSA (not depicted)
English: Guinea Turaco
Range: Gambia to Cameroon, Central African Republic, S. W. Zaire, N.W. Angola
Like No. 1 Knysna Turaco, but short crest (45 mm) not tipped white. Back and rump dull black.Wings and tail blackish with violet gloss.
List: II

TAURACO SCHUETTII (not depicted)
English: Black-billed Turaco
Range: Congo to N. Angola, Zaire, Sudan, Uganda, W. Kenya
Like No. 1 Knysna Turaco, but bill black.
List: II

TAURACO SCHALOWI (not depicted)
English: Schalow's Turaco
Range: S.W. Kenya to N. E. Botswana
Like No. 1 Knysna Turaco but crest longer 70-100 mm. Tail glossed violet-purple.
List: II

TAURACO FISCHERI (not depicted)
English: Fischer's Turaco
Range: S. Somalia, E. Kenya and N. E. Tanzania
Like No. 1 Knysna Turaco but greater part of crest and nape dull crimson with white tips.
List: II

TAURACO LIVINGSTONII (not depicted)
English: Livingstone's Turaco
Range: Burundi to Mozambique, Zimbabwe and Natal
Like No. 1 Knysna Turaco but crest a little longer, 62-72 mm and pointed. Tail blackish with strong greenish gloss.
List: II
Lit.: 7, 83, 129, 171, 172, 173, 174, 253, 325, 410

2. MUSOPHAGA PORPHYREOLOPHUS

Also known as: Tauraco porphyreolophus; Gallirex porphyreolophus

English: Purple crested Turaco, Violet-crested Turaco; *French:* Touraco à huppe splendide; *German:* Glanzhaubenturako

Range: Kenya to Mozambique, Zimbabwe, Transvaal, Natal
Identification: Length 47cm. Sexes similar. Rounded, iridescent violet crest, nape and chin distinctive. Sides of head iridescent green. Neck, breast, and mantle pale green, tinged pink on breast and mantle. Rump, wings, and tail glossy dark bluish-purple. Crimson outermost feathers not visible in folded wings.
Young birds: Like adult except red in wings less extensive and duller.
Bare parts: Bill and legs black. Eyes dark brown. Eye-ring scarlet.
Geographical variation:
Birds from Kenya to Mozambique lacks pink tinge on mantle and breast.
List: II
Lit.: 7, 83, 129, 172, 174, 253, 325, 410

3. TAURACO MACRORHYNCHUS

English: Yellow-billed Turaco, Crested Turaco; *French:* Touraco à gros bec; *German:* Blaurückenturako

Range: Sierra Leone to Ivory Coast, Nigeria to Zaire
Identification: Length 43cm. Sexes alike. Crest green with black tips and subterminal white lines. Otherwise resembles number 1, Knysna Turaco, but has no white line above eye, and wings and tail more dark blue-violet.
Young birds: Duller and crest lacks black and white tips.
Bare parts: Bill yellow, base red. Legs black. Eyes brown. Eye-ring red.

Geographical variation:
Birds from Nigeria to Zaire have crest with crimson tips.
List: III Ghana
Lit.: 7, 83, 129, 173, 195, 410

4. CORYTHAEOLA CRISTATA

English: Great Blue Turaco, Blue Plantain-eater; *French:* Touraco géant; *German:* Riesenturako

Range: Kenya, Guinea-Bissau to Angola
Identification: Length 76cm. Sexes similar. Crest blue-black. Rest of head, neck, upper breast, upperparts, wings, and central tail feathers greyish-blue. Lower breast green, rest of underparts yellow and chestnut. Outer tail feathers dull yellow with blue base. Whole tail with a broad black bar and blue tips.
Young birds: Duller with small crest.
Bare parts: Bill yellow, tip orange-red. Legs blackish, soles yellow. Eyes dark brown.
List: III Ghana
Lit.: 7, 83, 129, 172, 173, 325, 410

5. CRINIFER PISCATOR

English: Western Grey Plantain-eater, Grey Platain-eater; *French:* Touraco gris; *German:* Schwarzschwanz-Lärmvogel

Range: Senegal to Zaire
Identification: Length 50cm. Sexes alike. Crest dark brown, edged white. General colour dark grey-brown to silvery grey streaked blackish. Lower breast, abdomen, flanks, thighs and under tail-coverts white with dark brown shaft-streaks. Tail dark brown, two central tail feathers grey gradually becoming dark brown at tip.
Young birds: Crest smaller.
Bare parts: Bill yellow, base greenish. Legs brownish black. Eyes dark brown.
List: III Ghana
Lit.: 7, 83, 129, 173, 410

6. MUSOPHAGA VIOLACEA

English: Violet Turaco, Violet Plantain-eater; *French:* Touraco violet; *German:* Schildturako

Range: Senegal to Cameroon, Central African Republic, Chad
Identification:
Length 48cm. Sexes similar.Crown and nape feathers short, velvety and crimson. White line below and behind face. General colour otherwise glossy violet-blue on upperparts, mantle and wing-coverts purple. Long flight-feathers crimson tipped brown.
Underparts blackish, breast tinged green. Tail violet-blue with a faint tinge of green.
Young birds: Without red on head.
Bare parts: Bill red, base yellow with yellow shield extending to back to eyes, juvenile blackish without shield. Legs black to greenish-black. Eyes dark brown. Bare skin around eyes red.
List: III Ghana
Lit.: 83, 173, 268

PLATE 53

1. TYTO SOUMAGNEI

Formerly known as: Heliodilus soumagneii

English: Soumagne's Owl, Madagascar Red Owl, Madagascar Grass Owl; *French:* Effraie jaune, Effraie de Madagascar; *German:* Malegasseneule; *Italian:* Barbagianni del Madagascar; *Spanish:* Lechuza de Madagascar

Range: Madagascar
Identification:
Length 30cm. Sexes alike. Whole bird reddish with a few dusky spots. Facial disc edged paler reddish. Wing tip 2cm shorter than tail tip. Tail faintly barred.
Young birds: Undescribed.
Bare parts: Bill pale yellow. Legs feathered. Eyes black.
List: I
Status: May be already extinct. Last collected in 1930.
Lit.: 30, 37, 56, 101, 129, 152, 162, 410

2. ATHENE BLEWITTI

English: Forest Owlet, Forest Spotted Owlet, Forest Little Owl; *French:* Chevêche forestière, Chouette des forêts; *German:* Blewittkauz, Bänder-Steinkauz; *Italian:* Civetta macchiata delle foreste; *Spanish:* Mochuelo forestal

Range: India
Identification:
Length 21-23cm. Sexes alike. Top of head and neck dark earthy-brown without spots. Upperparts and wings same colour with a few white spots, on wings more white wing bars. Throat with a strongly marked dark brown band, below this a white area. Rest of underparts whitish with dark barring on flanks. Tail white with dark barring.
Young birds: Undescribed.
Bare parts: Bill not described, but photo yellow. Legs feathered. Eyes bright yellow.
List: I
Status: Last collected 1914, last seen 1968. Certainly very rare if not extinct.
Lit.: 2, 30, 37, 101, 129, 152, 410

Ninox n. novaeseelandiae
English: Morepork, Boobook Owl
Range: South Island (New Zealand)
Identification:
Length 29cm. Sexes alike. White face with black shafts. Feathers around bill black. Crown, nape, and upperparts brownish grey to dusky brown speckled with buff or white spots or streaks, but very variable, so crown may lack white spots or have pale stripes or spots may form a collar on hindneck. Underparts dark brown to dark rufous with white to buffish spots or stripes. Tail brownish-grey with pale bars.
Young birds: Undescribed.
Bare parts: Bill varies from dark brown to yellow with dark cutting edges. Legs feathered. Eyes golden yellow.
List: II

Subspecies:

3. NINOX NOVAESEELANDIAE UNDULATA

Also known as: Ninox novaeseelandiae royana
English: Norfolk Island Boobook Owl, Norfolk Boobook; *French:* Chouette-coucou de l'île Norfolk; *German:* Norfolkinsel-Kuckuckskauz; *Italian:* Ulula Australiana; *Spanish:* Lechuza gavilana de Norfolk
Range: Norfolk Island (New Zealand)
Identification:
Not possible for a layman to distinguish from other Moreporks.
List: I
There are 13 other races of Moreporks, most of them from small islands in Pacific Ocean, rest from Australia (4), New Guinea (1), and Tasmania (1). Impossible for a layman to distinguish from each other.
List: II
Note: Sibley & Monroe have only two races: undulata and albaria, latter from Lord Howe Island possibly extinct.
Lit.: 30, 37, 69, 101, 124, 129, 181, 182, 222, 232, 249, 302, 410

NINOX S. SQUAMIPILA
English: Moluccan Hawk-Owl
Range: Wallacea
Identification:
Length 25-36cm. Sexes alike. Feathers around bill whitish with black shafts. Rest of head, upperparts, and tail mahogany red with a few white spots particularly on shoulders. Underparts white with rufous (2mm) bands, each feather 4 bands. Tail with 8 diffuse dusky bars.
Young birds: Undescribed.
Bare parts: Bill bluish, tip yellow. Legs feathered pale yellow. Eyes dark brown.
List: II

Subspecies:

4. NINOX SQUAMIPILA NATALIS
English: Christmas Island Owl, Christmas Hawk-Owl; *French:* Chouette épervière de l'île Christmas; *German:* Weihnachtinsel-Buschkauz; *Italian:* Ulula delle Isole Christmas; *Spanish:* Lechuza gavilana (de las Molucas)
Range: Christmas Island (Australia)
Identification:
Paler reddish-brown, and more white spots, particularly on back and wings.
List: I
Status: Less than 100 birds.
Lit.: 30, 37, 101, 129, 152, 166, 410

5. MIMIZUKU GURNEYI

Also known as: Otus gurneyi

English: Giant Scops-Owl; *French:* Petit-Duc de Gurney; *German:* Rotohreule

Range: Mindanao, Dinagat, Siargao, (Philippines)
Identification:
Length 31cm. Sexes alike. Large ear-tufts and face rufous and white. Crown and upperparts rufous with dark stripes. Nape pure rufous. Throat whitish. Rest of underside pale buffish with big dark brown spots.
Young birds: Not described.
Bare parts: Bill yellowish. Legs feathered, toes yellow. Eyes yellow.
List: I
Lit.: 30, 61, 66, 129, 410

PLATE 54

ORDER

STRIGIFORMES

Family: TYTONIDAE

English: Barn Owls; *French:* Effraies; *German:* Schleiereulen

Identification:
Length 30-53cm. Sexes alike but female sometimes larger. Colour brown, spotted white, buffish, orange-buff,, grey-brown or white. Pattern spotted, barred or vermiculated. Often two colour phases. Small eyes directed forward, surrounded by heart-shaped, white or buffish disc. Wings long and rounded. Tail short. Bill hooked, cere at base. Legs long and feathered except sparsely feathered toes, outer toe reversible, claws sharp and hooked.

All other barn owls:

Species	English name
Tyto multipunctata	Lesser Sooty-Owl
Tyto tenebricosa	Greater Sooty-Owl
Tyto inexspectata	Minahassa Masked-Owl
Tyto nigrobrunnea	Taliabu Masked-Owl
Tyto sororcula	Lesser Masked-Owl
Tyto manusi	Manus Masked-Owl
Tyto aurantia	Bismarck Masked-Owl
Tyto novaehollandiae	Australian Masked-Owl
Tyto castanops	Tasmanian Masked-Owl
Tyto rosenbergii	Sulawesi Owl
Tyto alba	Barn Owl
Tyto glaucops	Ashy-faced Owl
Tyto capensis	African Grass-Owl
Tyto longimembris	Eastern Grass-Owl
Phodilus prigoginei	Congo Bay-Owl
Phodilus badius	Oriental Bay-Owl

List: II

Family: STRIGIDAE

English: Owls; *German:* Eulen

Identification:
Length 13-69cm. Sexes alike but female larger. Colour brown, grey to white. Often two colour phases. Pattern spotted, barred or streaked. Large eyes directed forward, surrounded by feathered round disc. Many with ear-like tufts. Wings broad and rounded. Tail short to medium. Bill hooked and short, cere at base. Legs short to medium, feathered in most species. Outer toe reversible, claws sharp and hooked.

All other owls:

	Species	English name
	Otus sagittatus	White-fronted Scops-Owl
	Otus rufescens	Reddish Scops-Owl
	Otus icterorhynchus	Sandy Scops-Owl
	Otus ireneae	Sokoke Scops-Owl
	Otus balli	Andaman Scops-Owl
	Otus spilocephalus	Mountain Scops-Owl
	Otus mirus	Mindanao Scops-Owl
	Otus brucei	Pallid Scops-Owl
	Otus flammeolus	Flammulated Owl
	Otus scops	Eurasian Scops-Owl
	Otus senegalensis	African Scops-Owl
	Otus sunia	Oriental Scops-Owl
	Otus manadensis	Sulawesi Scops-Owl
	Otus longicornis	Luzon Scops-Owl
	Otus mindorensis	Mindoro Scops-Owl
	Otus alfredi	Flores Scops-Owl
	Otus angelinae	Javan Scops-Owl
	Otus umbra	Simeulue Scops-Owl
	Otus hartlaubi	Sao Tome Scops-Owl
	Otus elegans	Elegant Scops-Owl
	Otus mantananensis	Mantanani Scops-Owl
	Otus magicus	Moluccan Scops-Owl
	Otus enganensis	Enggano Scops-Owl
	Otus rutilus	Madagascar Scops-Owl
	Otus brookii	Rajah Scops-Owl
	Otus bakkamoena	Indian Scops-Owl
	Otus lempiji	Collared Scops-Owl
	Otus mentawi	Mentawai Scops-Owl
	Otus megalotis	Philippine Scops-Owl
	Otus fuliginosus	Palawan Scops-Owl
	Otus silvicola	Wallace's Scops-Owl
	Otus leucotis	White-faced Scops-Owl
	Otus kennicottii	Western Screech-Owl
	Otus asio	Eastern Screech-Owl
	Otus seductus	Balsas Screech-Owl
	Otus cooperi	Pacific Screech-Owl
	Otus trichopsis	Whiskered Screech-Owl
	Otus choliba	Tropical Screech-Owl
	Otus koepckeae	Koepcke's Screech-Owl
	Otus roboratus	West Peruvian Screech-Owl
	Otus barbarus	Bridled Screech-Owl
	Otus vermiculatus	Vermiculated Screech-Owl
	Otus atricapillus	Variable Screech-Owl
	Otus watsonii	Tawny-bellied Screech-Owl
	Otus usta	Austral Screech-Owl
	Otus petersoni	Cinnamon Screech-Owl
	Otus marshalli	Cloud-Forest Screech-Owl
	Otus ingens	Rufescent Screech-Owl
	Otus clarkii	Bare-shanked Screech-Owl
	Otus nudipes	Puerto Rican Screech-Owl
7.	Otus lawrencii	Bare-legged Owl
	Otus podarginus	Palau Scops-Owl
	Otus albogularis	White-throated Screech-Owl
	Otus hoyi (new species)	Montane Forest Screech-Owl
	Bubo virginianus	Great Horned Owl
1.	Bubo bubo	Eurasian Eagle-Owl
	Bubo bengalensis	Rock Eagle-Owl
	Bubo ascalaphus	Pharaoh Eagle-Owl
	Bubo capensis	Cape Eagle-Owl
	Bubo africanus	Spotted Eagle-Owl
	Bubo poensis	Fraser's Eagle-Owl
	Bubo nipalensis	Spot-bellied Eagle-Owl
	Bubo sumatranus	Barred Eagle-Owl
	Bubo shelleyi	Shelley's Eagle-Owl
	Bubo lacteus	Verreaux's Eagle-Owl
	Bubo coromandus	Dusky Eagle-Owl
	Bubo leucostictus	Akun Eagle-Owl
	Bubo philippensis	Philippine Eagle-Owl
	Ketupa blakistoni	Blakiston's Fish-Owl
	Ketupa zeylonensis	Brown Fish-Owl
	Ketupa flavipes	Tawny Fish-Owl
	Ketupa ketupu	Buffy Fish-Owl
	Scotopelia peli	Pel's Fishing-Owl
	Scotopelia ussheri	Rufous Fishing-Owl
2.	Scotopelia bouvieri	Vermiculated Fishing-Owl
4.	Nyctea scandiaca	Snowy Owl
	Strix seloputo	Spotted Wood-Owl
	Strix ocellata	Mottled Wood-Owl
	Strix leptogrammica	Brown Wood-Owl
	Strix aluco	Tawny Owl
	Strix butleri	Hume's Owl
	Strix occidentalis	Spotted Owl
	Strix varia	Barred Owl
	Strix fulvescens	Fulvous Owl
	Strix hylophila	Rusty-barred Owl
	Strix rufipes	Rufous-legged Owl
	Strix uralensis	Ural Owl
	Strix davidi	Sichuan Wood-Owl

PLATE 55

LONG-EARED OWL

9.	Strix nebulosa	Great Grey Owl
	Strix virgata	Mottled Owl
	Strix nigrolineata	Black-and-White Owl
8.	Strix huhula	Black-banded Owl
	Strix albitarsus	Rufous-banded Owl
	Strix woodfordii	African Wood-Owl
	Jubula lettii	Maned Owl
	Lophostrix cristata	Crested Owl
3.	Pulsatrix perspicillata	Spectacled Owl
	Pulsatrix melanota	Band-bellied Owl
	Pulsatrix koeniswaldiana	Tawny-browed Owl
5.	Surnia ulula	Northern Hawk Owl
	Glaucidium passerinum	Eurasian Pygmy-Owl
	Glaucidium brodiei	Collared Owlet
	Glaucidium perlatum	Pearl-spotted Owlet
	Glaucidium californicum	Northern Pygmy-Owl
	Glaucidium gnoma	Mountain Pygmy-Owl
6.	Glaucidium minutissimum	Least Pygmy-Owl
	Glaucidium brasilianum	Ferruginous Pygmy-Owl
	Glaucidium jardinii	Andean Pygmy-Owl
	Glaucidium siju	Cuban Pygmy-Owl
	Glaucidium tephronotum	Red-chested Owlet
	Glaucidium sjostedti	Sjostedt's Owlet
	Glaucidium cuculoides	Asian Barred Owlet
	Glaucidium castanopterum	Javan Owlet
	Glaucidium radiatum	Jungle Owlet
	Glaucidium castanonotus	Chestnut-backed Owlet
	Glaucidium capense	African Barred Owlet
	Glaucidium castaneum	Chestnut Owlet
	Glaucidium ngamiense	Ngami Owlet
	Glaucidium schefferi	Scheffler's Owlet
	Glaucidium albertinum	Albertine Owlet
	Xenoglaux loweryi	Long-whiskered Owlet (Front cover)
	Micrathene whitneyi	Elf Owl
	Athene noctua	Little Owl
	Athene brama	Spotted Owlet
	Speotyto cunicularia	Burrowing Owl
	Aegolius funereus	Boreal Owl
11.	Aegolius acadicus	Northern Saw-Whet Owl
	Aegolius ridgwayi	Unspotted Saw-Whet Owl
	Aegolius harrisii	Buff-fronted Owl
	Ninox rufa	Rufous Owl
	Ninox strenua	Powerful Owl
	Ninox connivens	Barking Owl
	Ninox rudolphi	Sumba Boobook
	Ninox boobook	Southern Boobook
	Ninox novaeseelandiae (- undulata)	Morepork
	Ninox scutulata	Brown Hawk-Owl
	Ninox affinis	Andaman Hawk-Owl
	Ninox superciliaris	Madagascar Hawk-Owl
	Ninox philippensis	Philippine Hawk-Owl
	Ninox ochracea	Ochre-bellied Hawk-Owl
	Ninox squamipila (- natalis)	Moluccan Hawk-Owl
	Ninox theomacha	Jungle Hawk-Owl
	Ninox meeki	Manus Hawk-Owl
	Ninox punctulata	Speckled Hawk-Owl
	Ninox variegata	Bismarck Hawk-Owl
	Ninox odiosa	Russet Hawk-Owl
	Ninox jacquinoti	Solomon Hawk-Owl
	Uroglaux dimorpha	Papuan Hawk-Owl
	Sceloglaux albifacies	Laughing Owl
	Pseudoscops grammicus	Jamaican Owl
	Asio strygius	Stygian Owl
10.	Asio otus	Long-eared Owl
	Asio abyssinicus	Abyssinian Owl
	Asio madagascariensis	Madagascar Owl
	Asio clamator	Striped Owl
	Asio flammeus	Short-eared Owl
	Asio capensis	Marsh Owl
	Nesasio solomonensis	Fearful Owl

List: II

Lit.: 30, 101, 193, 307, 316, 336, 375, 386, 410

Reading List:

Burton, J. A. (ed.) 1984: *Owls of the World.* Collins, Glasgow.

Hume, R. & T. Boyer 1991: *Owls of the World.* Dragonsworld, Limpsfield.

PLATE 56

ORDER

TROCHILIFORMES *

FAMILY: TROCHILIDAE

English: Hummingbirds; *French:* Colibris; *German:* Kolibris

Identification:
Length 6 to 22cm. Most very small, weight down to 2g. Smallest bird of the world is No. 27, Bee Hummingbird. Biggest hummingbird is No. 17, Giant Hummingbird. Colour mainly green often with iridescent blue, green, red, purple or gold areas on head, throat or back. Wings adapted for hovering both forwards and backwards, outer flight-feathers long and narrow, innermost few and short. Tail varied. Legs and feet very small and weak. Bill slender, straight or curved. In most species females have drab colours.

All other hummingbirds:

No.	Scientific name	English name
	Glaucis aenea	Bronzy ermit
2.	Glaucis hirsuta	Rufous-breasted Hermit
	Threnetes niger	Sooty Barbthroat
	Threnetes leucurus	Pale-tailed Barbthroat
	Threnetes ruckeri	Band-tailed Barbthroat
	Phaethornis yaruqui	White-whiskered Hermit
	Phaethornis guy	Green Hermit
	Phaethornis syrmatophorus	Tawny-bellied Hermit
	Phaethornis superciliosus	Long-tailed Hermit
	Phaethornis malaris	Great-billed Hermit
	Phaethornis eurynome	Scale-throated Hermit
	Phaethornis hispidus	White-bearded Hermit
	Phaethornis anthophilus	Pale-bellied Hermit
	Phaethornis bourcieri	Straight-billed Hermit
	Phaethornis koepckeae	Koepcke's Hermit
	Phaethornis philippii	Needle-billed Hermit
	Phaethornis squalidus	Dusky-throated Hermit
3.	Phaethornis augusti	Sooty-capped Hermit
	Phaethornis pretrei	Planalto Hermit
	Phaethornis subochraceus	Buff-bellied Hermit
	Phaethornis nattereri	Cinnamon-throated Hermit
	Phaethornis gounellei	Broad-tipped Hermit
	Phaethornis ruber	Reddish Hermit
	Phaethornis stuarti	White-browed Hermit
	Phaethornis griseogularis	Grey-chinned Hermit
	Phaethornis longuemareus	Little Hermit
	Phaethornis idaliae	Minute Hermit
4.	Eutoxeres aquila	White-tipped Sicklebill
	Eutoxeres condamini	Buff-tailed Sicklebill
1.	Androdon aequatorialis	Tooth-billed Hummingbird
	Ramphodon naevius	Saw-billed Hermit
	Doryfera johannae	Blue-fronted Lancebill
	Doryfera ludovicae	Green-fronted Lancebill
	Phaeochroa cuvierii	Scaly-breasted Hummingbird
	Campylopterus curvipennis	Wedge-tailed Sabrewing
	Campylopterus excellens	Long-tailed Sabrewing
	Campylopterus largipennis	Grey-breasted Sabrewing
	Campylopterus rufus	Rufous Sabrewing
	Campylopterus hyperythrus	Rufous-breasted Sabrewing
	Campylopterus duidae	Buff-breasted Sabrewing
	Campylopterus hemileucurus	Violet Sabrewing
	Campylopterus ensipennis	White-tailed Sabrewing
	Campylopterus falcatus	Lazuline Sabrewing
	Campylopterus phainopeplus	Santa Marta Sabrewing
	Campylopterus villaviscensio	Napo Sabrewing
	Eupetomena macroura	Swallow-tailed Hummingbird
5.	Florisuga mellivora	White-necked Jacobin
6.	Melanotrochilus fuscus	Black Jacobin
	Colibri delphinae	Brown Violet-Ear
	Colibri thalassinus	Green Violet-Ear
	Colibri coruscans	Sparkling Violet-Ear
	Colibri serrirostris	White-vented Violet-Ear
	Anthracothorax viridigula	Green-throated Mango
	Anthracothorax prevostii	Green-breasted Mango
	Anthracothorax nigricollis	Black-throated Mango
	Anthracothorax mango	Jamaican Mango
	Anthracothorax dominicus	Antillean Mango
	Anthracothorax viridis	Green Mango
	Avocettula recurvirostris	Fiery-tailed Awlbill
	Eulampis jugularis	Purple-throated Carib
	Eulampis holosericeus	Green-throated Carib
	Chrysolampis mosquitus	Ruby-Topaz Hummingbird
	Orthorhyncus cristatus	Antillean Crested Hummingbird
	Klais guimeti	Violet-headed Hummingbird
	Abeillia abeillei	Emerald-chinned Hummingbird
	Stephanoxis lalandi	Plovercrest
	Lophornis ornatus	Tufted Coquette
	Lophornis gouldii	Dot-eared Coquette
8.	Lophornis magnificus	Frilled Coquette
7.	Lophornis delattrei	Rufous-crested Coquette
	Lophornis stictolophus	Spangled Coquette
	Lophornis chalybeus	Festive Coquette
	Lophornis pavoninus	Peacock Coquette
	Lophornis helenae	Black-crested Coquette
	Lophornis adorabilis	White-crested Coquette
9.	Popelairia popelairii	Wire-crested Thorntail
	Popelairia langsdorffi	Black-bellied Thorntail
	Popelairia letitiae	Coppery Thorntail
	Popelairia conversii	Green Thorntail
10.	Discosura longicauda	Racket-tailed Coquette
	Chlorestes notatus	Blue-chinned Sapphire
	Chlorostilbon canivetii	Fork-tailed Emerald
	Chlorostilbon assimilis	Garden Emerald
11.	Chlorostilbon mellisugus	Blue-tailed Emerald
	Chlorostilbon aureoventris	Glittering-bellied Emerald
	Chlorostilbon ricordii	Cuban Emerald
	Chlorostilbon bracei	Brace's Emerald
	Chlorostilbon swainsonii	Hispaniolan Emerald
	Chlorostilbon maugaeus	Puerto Rican Emerald
	Chlorostilbon gibsoni	Red-billed Emerald
	Chlorostilbon russatus	Coppery Emerald
	Chlorostilbon stenura	Narrow-tailed Emerald
	Chlorostilbon alice	Green-tailed Emerald
	Chlorostilbon poortmani	Short-tailed Emerald
	Cynanthus sordidus	Dusky Hummingbird
	Cynanthus latirostris	Broad-billed Hummingbird
	Cyanophaia bicolor	Blue-headed Hummingbird

PLATE 57

Thalurania colombica	Crowned Woodnymph
Thalurania furcata	Fork-tailed Woodnymph
Thalurania watertonii	Long-tailed Woodnymph
Thalurania glaucopis	Violet-capped Woodnymph
Panterpe insignis	Fiery-throated Hummingbird
12. Damophila julie	Violet-bellied Hummingbird
Lepidopyga coeruleogularis	Sapphire-throated Hummingbird
Lepidopyga lilliae	Sapphire-bellied Hummingbird
Lepidopyga goudoti	Shining-green Hummingbird
Hylocharis xantusii	Xantus's Hummingbird
Hylocharis leucotis	White-eared Hummingbird
Hylocharis eliciae	Blue-throated Goldentail
Hylocharis sapphirina	Rufous-throated Sapphire
Hylocharis cyanus	White-chinned Sapphire
Hylocharis pyropygia	Flame-rumped Sapphire
Hylocharis chrysura	Gilded Hummingbird
Hylocharis grayi	Blue-headed Sapphire
Chrysuronia oenone	Golden-tailed Sapphire
Goldmania violiceps	Violet-capped Hummingbird
Goethalsia bella	Rufous-cheeked Hummingbird
Trochilus polytmus	Streamertail
Leucochloris albicollis	White-throated Hummingbird
Polytmus guainumbi	White-tailed Goldenthroat
Polytmus theresiae	Green-tailed Goldenthroat
Leucippus fallax	Buffy Hummingbird
Leucippus baeri	Tumbes Hummingbird
Leucippus taczanowskii	Spot-throated Hummingbird
Leucippus chlorocercus	Olive-spotted Hummingbird
Taphrospilus hypostictus	Many-spotted Hummingbird
Amazilia viridicauda	Green-and-White Hummingbird
Amazilia chionogaster	White-bellied Hummingbird
Amazilia candida	White-bellied Emerald
Amazilia chionopectus	White-chested Emerald
Amazilia versicolor	Versicolored Emerald
Amazilia luciae	Honduran Emerald
Amazilia fimbriata	Glittering-throated Emerald
Amazilia distans	Tachira Emerald
Amazilia lactea	Sapphire-spangled Emerald
Amazilia amabilis	Blue-chested Hummingbird
Amazilia decora	Charming Hummingbird
Amazilia rosenbergi	Purple-chested Hummingbird
Amazilia boucardi	Mangrove Hummingbird
Amazilia franciae	Andean Emerald
14. Amazilia leucogaster	Plain-bellied Emerald
Amazilia cyanocephala	Azure-crowned Hummingbird
Amazilia cyanifrons	Indigo-capped Hummingbird
Amazilia beryllina	Berylline Hummingbird
Amazilia cyanura	Blue-tailed Hummingbird
Amazilia saucerrottei	Steely-vented Hummingbird
Amazilia tobaci	Copper-rumped Hummingbird
Amazilia viridigaster	Green-bellied Hummingbird
Amazilia edward	Snowy-breasted Hummingbird
Amazilia rutila	Cinnamon Hummingbird
Amazilia yucatanensis	Buff-bellied Hummingbird
Amazilia tzacati	Rufous-tailed Hummingbird
Amazilia castaneiventris	Chestnut-bellied Hummingbird
Amazilia amazilia	Amazilia Hummingbird

Amazilia viridifrons	Green-fronted Hummingbird
Amazilia violiceps	Violet-crowned Hummingbird
Eupherusa poliocerca	White-tailed Hummingbird
Eupherusa eximia	Stripe-tailed Hummingbird
Eupherusa cyanophrys	Blue-capped Hummingbird
Eupherusa nigriventris	Black-bellied Hummingbird
Elvira chionura	White-tailed Emerald
Elvira cupreiceps	Coppery-headed Emerald
15. Microchera albocoronata	Snowcap
Chalybura buffonii	White-vented Plumeleteer
Chalybura urochrysia	Bronze-tailed Plumeleteer
Aphanotochroa cirrochloris	Sombre Hummingbird
Lampornis clemenciae	Blue-throated Hummingbird
Lampornis amethystinus	Amethyst-throated Hummingbird
Lampornis viridipallens	Green-throated Mountain-Gem
Lampornis sybillae	Green-breasted Mountain-Gem
Lampornis hemileucus	White-bellied Mountain-Gem
Lampornis castaneoventris	Variable Mountain-Gem
Lamprolaima rhami	Garnet-throated Hummingbird
Adelomyia melanogenys	Speckled Hummingbird
Anthocephala floriceps	Blossomcrown
Phlogophilus hemileucurus	Ecuadorian Piedtail
Phlogophilus harterti	Peruvian Piedtail
Clytolaema rubricauda	Brazilian Ruby
Heliodoxa imperatrix	Empress Brilliant
Heliodoxa xanthogonys	Velvet-browed Brilliant
Heliodoxa gularis	Pink-throated Brilliant
Heliodoxa branickii	Rufous-webbed Brilliant
Heliodoxa schreibersii	Black-throated Brilliant
Heliodoxa aurescens	Gould's Jewelfront
Heliodoxa rubinoides	Fawn-breasted Brilliant
Heliodoxa jacula	Green-crowned Brilliant
Heliodoxa leadbeateri	Violet-fronted Brilliant
13. Eugenes fulgens	Magnificent Hummingbird
Hylonympha macrocerca	Scissor-tailed Hummingbird
Sternoclyta cyanopectus	Violet-crested Hummingbird
Topaza pyra	Fiery Topaz
16. Topaza pella	Crimson Topaz
Oreotrochilus chimborazo	Ecuadorian Hillstar
Oreotrochilus estella	Andean Hillstar
Oreotrochilus leucopleurus	White-sided Hillstar
Oreotrochilus melanogaster	Black-breasted Hillstar
Oreotrochilus adela	Wedge-tailed Hillstar
Urochroa bougueri	White-tailed Hillstar
17. Patagona gigas	Giant Hummmingbird
Aglaeactis cupripennis	Shining Sunbeam
Aglaeactis castelnaudii	White-tufted Sunbeam
Aglaeactis aliciae	Purple-backed Sunbeam
Aglaeactis pamela	Black-hooded Sunbeam
Lafresnaya lafresnayi	Mountain Velvetbreast

	Scientific name	Common name
	Pterophanes cyanopterus	Great Sapphirewing
	Coeligena coeligena	Bronzy Inca
	Coeligena wilsoni	Brown Inca
	Coeligena prunellei	Black Inca
18.	Coeligena torquata	Collared Inca
	Coeligena phalerata	White-tailed Starfrontlet
	Coeligena orina	Dusky Starfrontlet
	Coeligena bonapartei	Golden-bellied Starfrontlet
	Coeligena helianthea	Blue-throated Starfrontlet
	Coeligena lutetiae	Winged Starfrontlet
	Coeligena violifer	Violet-throated Starfrontlet
	Coeligena iris	Rainbow Starfrontlet
19.	Ensifera ensifera	Sword-billed Hummingbird
	Sephanoides sephaniodes	Green-backed Firecrown
	Sephanoides fernandensis	Juan Fernandez Firecrown
	Boissonneaua flavescens	Buff-tailed Coronet
	Boissonneaua matthewsii	Chestnut-breasted Coronet
	Boissonneaua jardini	Velvet-purple Coronet
	Heliangelus mavors	Orange-throated Sunangel
	Heliangelus spencei	Merida Sunangel
	Heliangelus amethysticollis	Amethyst-throated Sunangel
	Heliangelus strophianus	Gorgetted Sunangel
	Heliangelus exortis	Tourmaline Sunangel
	Heliangelus viola	Purple-throated Sunangel
	Heliangelus regalis	Royal Sunangel
	Eriocnemis nigrivestis	Black-breasted Puffleg
	Eriocnemis vestitus	Glowing Puffleg
20.	Eriocnemis godini	Turquoise-throated Puffleg
	Eriocnemis luciani	Sapphire-vented Puffleg
	Eriocnemis cupreoventris	Coppery-bellied Puffleg
	Eriocnemis mosquera	Golden-breasted Puffleg
	Eriocnemis glaucopoides	Blue-capped Puffleg
	Eriocnemis mirabilis	Colorful Puffleg
	Eriocnemis alinae	Emerald-bellied Puffleg
	Eriocnemis derbyi	Black-thighed Puffleg
	Haplophaedia aureliae	Greenish Puffleg
	Haplophaedia lugens	Hoary Puffleg
	Urosticte benjamini	Purple-bibbed Whitetip
	Urosticte ruficrissa	Rufous-vented Whitetip
21.	Ocreatus underwoodii	Booted Racket-tail
22.	Lesbia victoriae	Black-tailed Trainbearer
	Lesbia nuna	Green-tailed Trainbearer
	Sappho sparganura	Red-tailed Comet
	Polyonomus caroli	Bronze-tailed Comet
	Ramphomicron microrhynchum	Purple-backed Thornbill
	Ramphomicron dorsale	Black-backed Thornbill
	Metallura williami	Viridian Metaltail
	Metallura baroni	Violet-throated Metaltail
	Metallura odomae	Neblina Metaltail
	Metallura theresiae	Coppery Metaltail
	Metallura aupogon	Fire-throated Metaltail
	Metallura aeneocauda	Scaled Metaltail
	Metallura phoebe	Black Metaltail
	Metallura tyrianthina	Tyrian Metaltail
	Metallura iracunda	Perija Metaltail
	Chalcostigma ruficeps	Rufous-capped Thornbill
	Chalcostigma olivaceum	Olivaceous Thornbill
	Chalcostigma stanleyi	Blue-mantled Thornbill
	Chalcostigma heteropogon	Bronze-tailed Thornbill
	Challcostigma herrani	Rainbow-bearded Thornbill
23.	Oxypogon guerinii	Bearded Helmetcrest
	Opisthoprora euryptera	Mountain Avocetbill
	Taphrolesbia griseiventris	Grey-bellied Comet
24.	Aglaiocercus kingi	Long-tailed Sylph
	Aglaiocercus coelestis	Violet-tailed Sylph
	Oreonympha nobilis	Bearded Mountaineer
	Augastes lumachellus	Hooded Visorbearer
	Augastes scutatus	Hyacinth Visorbearer
	Augastes geoffroyi	Wedge-billed Hummingbird
25.	Heliothryx barroti	Purple-crowned Fairy
	Heliothryx aurita	Black-eared Fairy
	Heliactin cornuta	Horned Sungem
26.	Loddigesia mirabilis	Marvelous Spatuletail
	Heliomaster constantii	Plain-capped Starthroat
	Heliomaster longirostris	Long-billed Starthroat
	Heliomaster squamosus	Stripe-breasted Starthroat
	Heliomaster furcifer	Blue-tufted Starthroat
	Rhodopis vesper	Oasis Hummingbird
	Thaumastura cora	Peruvian Sheartail
	Philodice bryantae	Magenta-throated Woodstar
	Philodice mitchellii	Purple-throated Woodstar
	Doricha enicura	Slender Sheartail
	Doricha eliza	Mexican Sheartail
	Tilmatura dupontii	Sparkling-tailed Hummingbird
	Microstilbon burmeisteri	Slender-tailed Woodstar
	Calothorax lucifer	Lucifer Hummingbird
	Calothorax pulcher	Beautiful Hummingbird
	Archilochus colubris	Ruby-throated Hummingbird
	Archilochus alexandri	Black-chinned Hummingbird
	Calypte anna	Anna's Hummingbird
	Calypte costae	Costa's Hummingbird
	Calliphlox evelynae	Bahama Woodstar
	Calliphlox amethystina	Amethyst Woodstar
27.	Mellisuga helenae	Bee Hummingbird
	Mellisuga minima	Vervain Hummingbird
	Stellula calliope	Calliope Hummingbird
	Atthis heloisa	Bumblebee Hummingbird
	Atthis ellioti	Wine-throated Hummingbird
	Myrtis fanny	Purple-collared Woodstar
	Eulidia yarrellii	Chilean Woodstar
	Myrmia micrura	Short-tailed Woodstar
	Acestrura mulsant	White-bellied Woodstar
	Acestrura bombus	Little Woodstar

Acestrura heliodor	Gorgeted Woodstar
Acestrura astreans	Santa Marta Woodstar
Acestrura berlepschi	Esmeralda's Woodstar
28. Chaetocercus jourdanii	Rufous-shafted Woodstar
Selasphorus platycercus	Broad-tailed Hummingbird
Selasphorus rufus	Rufous Hummingbird
Selasphorus sasin	Allen's Hummingbird
Selasphorus flammula	Volcano Hummingbird
Selasphorus scintilla	Scintillant Hummingbird
Selasphorus ardens	Glow-throated Hummingbird

List: II
Lit.: 3, 10, 77, 125, 129, 189, 190, 192, 197, 238, 263. 264, 265, 266, 267, 295, 307, 371, 372, 373, 405 410

Reading List:
Fjeldså, J. & Krabbe, N. 1990: *Birds of the High Andes.* Zoologisk Museum & Apollo Books, Copenhagen & Svendborg.

Hilty, S. L. & Brown, W. L. 1986: *A Guide to the Birds of Colombia.* Princeton University Press, Princeton.

Meyer de Schauensee, R. & Phelps, W. H. 1978: *A Guide to the Birds of Venezuela.* Princeton University Press, Princeton.

National Geographic Soc. 1983: *Field Guide to the Birds of North America.* National Geographic Soc., Washington D. C.

Ridgely, R. S. 1981: *A Guide to the Birds of Panama.* Princeton University Press, Princeton.

PLATE 58

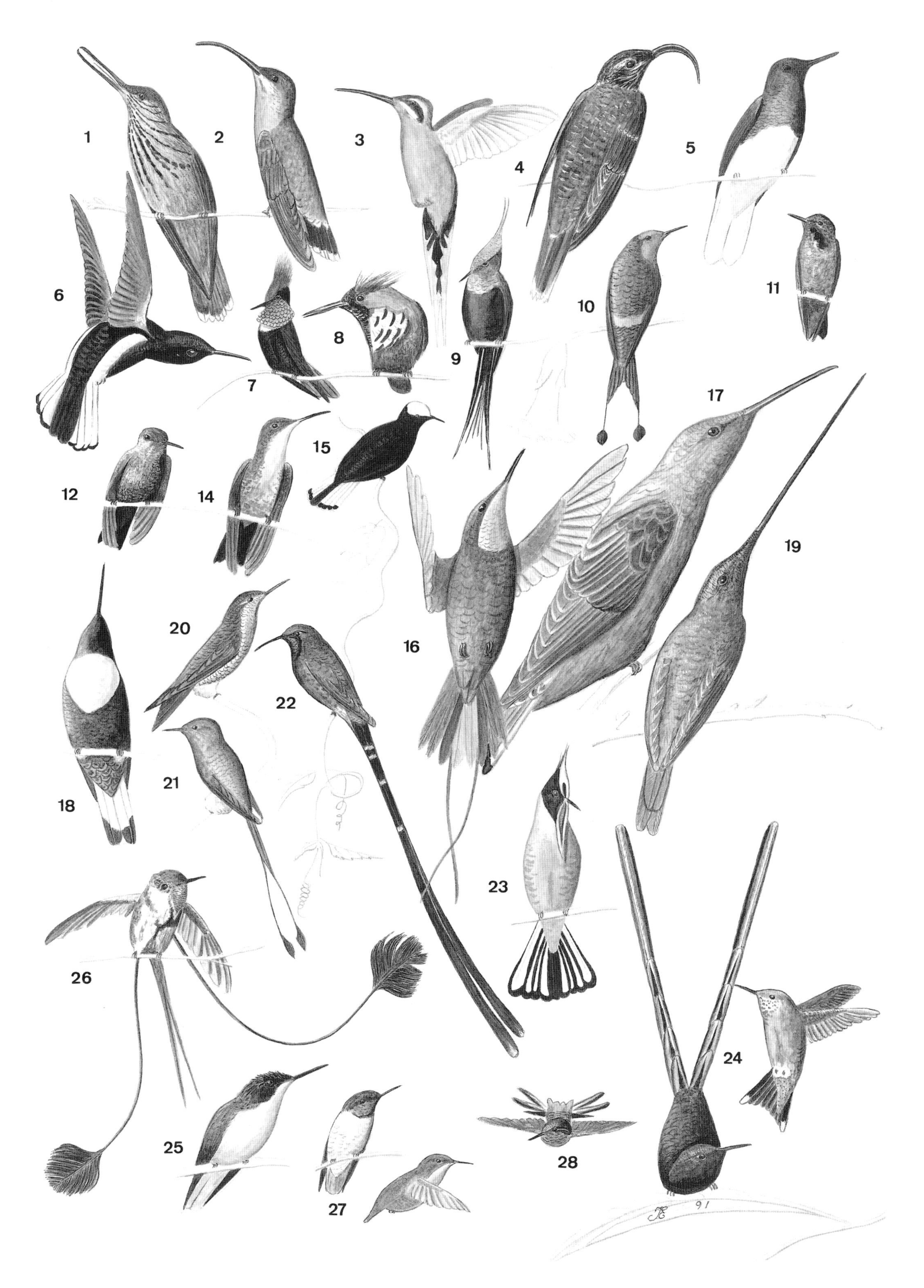

1. RAMPHODON DOHRNII
Also known as: Glaucis dohrnii

Brazilian: Balança-rabo-canela; *English:* Hook-billed Hermit; *French:* Ermite de Dohrn; *German:* Bronzeschwanz-Eremit

Range: Brazil
Identification:
Length 12cm. **Male:** Upperhead bronze-green. Stripe behind eye white. Upperparts reddish-bronze-green. Underparts cinnamon. Tail reddish-bronze-green with white tip.
Female: Like male but tail without white tip.
Young birds: Undescribed.
Bare parts: Bill heavy, straight, hooked at tip, upper black, lower yellowish. Legs yellow. Eyes undescribed.
List: I
Lit.: 10, 129, 190, 254, 410, 431

2. PHAROMACHRUS MOCINNO
Also known as: Pharomachrus paradiseus; Pharomachrus resplendens; Trogon resplandes

English: Resplendent Quetzal, Magnificent Quetzal; *French:* Quetzal resplendissant, Quetzal magnifique; *German:* Quetzal; *Italian:* Quetzal, Trogone splendido; *Local:* Kucú (maya), Kughug, Gugú (quiché), Cucui (quetchi), Quetzaltoltoti (azteca), Quetzalli (nahoa), Quetzal (nahualti); *Spanish:* Quetzal

Range: Mexico to Panama
Identification:
A. **Male:** Length 76-99cm. Above glittering golden green and extremely lengthened feathers of upper tail-coverts (38-76cm) and laterally flattened crest make this species unmistakable. Neck and forebreast like above. Rest scarlet. Underside of tail white.
B. **Female:** Length 36-38cm. Without crest and elongated upper tail-coverts. Above duller green, head more bronzy. Breast green, rest of underparts grey except for crimson vent and under tail-coverts. Tail slaty barred white.
Young birds: Undescribed.
Bare parts: Bill yellow, female black. Legs yellowish, two toes directed forward and two backward. Eyes not described.

Geographical variation:
Birds from Costa Rica to Panama have a blue tinge on upperparts. Only 4 (not 6) elongated upper tail-coverts.
List: I
Lit.: 32, 37, 129, 223, 238, 410

3. RAMPHASTOS SULFURATUS

English: Keel-billed Toucan; *French:* Toucan à carène; *German:* Fischertukan; *Spanish:* Piapoco pico verde

Range: Mexico to Venezuela and N. Colombia
Identification:
Length 48-50cm. Sexes alike. Top of head and above black, rump white. Below yellow on cheeks and breast. Narrow breast band and under tail-coverts scarlet. Wings, belly and tail black.
Young birds: Undescribed.
Bare parts: Bill brightly green, blue, red and orange, base black. Legs blue. Eyes brown. Bare skin around eyes greenish.

Geographical variation:
Birds from Guatemala to Colombia and Venezuela have shorter bills, 14cm (against 21cm) and red breast band broader.
List: II
Lit.: 125, 129, 192, 223, 254, 370, 410

4. CAMPEPHILUS IMPERIALIS

English: Imperial Woodpecker; *French:* Pic impérial; *German:* Kaiserspecht

Range: Mexico
Identification:
A. The world's largest woodpecker 50-56cm! **Male:** Curled long crest and hindneck red. Big white patch on hind wings and two narrow white stripes on upper back forming a V. Rest black.
B. **Female:** Like male but curled crest black.
Young birds: Browner.
Bare parts: Bill ivory-white. Legs greyish. Eyes yellow.
List: I
Status: Very close to extinction if still extant.
Lit.: 129, 152, 223, 274, 410

5. DRYOCOPUS JAVENSIS RICHARDSI

English: Korea White-bellied Woodpecker; *French:* Pic à ventre blanc; *German:* Weissbauchspecht; *Indonesian:* Caladi (platuk) besar hitam; *Japanese:* Kitataki

Range: Korea
Identification:
A. Length 40cm. **Male:** Forehead, crest and malar orange-red. Hind back, rump, wing tips, and belly white. Thighs and hind flanks barred black and white. Rest black.
B. **Female:** Like male but lacks orange-red, having entire head black.
Young birds: Paler and duller.
Bare parts: Bill grey-black, lower bill paler. Legs dark grey. Eyes yellowish., juvenile grey.
List: I
Status: About 40 pairs in 1969
Lit.: 129, 152, 195, 274, 299, 410
(Sibley and Monroe have no subspecies).
Note: There are 15 other races of White-bellied Woodpecker, ALL NOT LISTED

PLATE 59

1. ACEROS NARCONDAMI

Also known as: Rhyticeros narcondami

English: Narcondam Hornbill; *French:* Calao de Narcondam; *German:* Narcondam-Jahrvogel; *Italian:* Calao di Narcondam, Bucero di Narcondam; *Spanish:* Cálao de Narcondam

Range: Andaman Island, Narcondam Island (?)
Identification:
A. **Male:** Length 64cm. Head, crest and neck rufous. Upperparts black with a greenish wash. Underparts dull black. Tail white with yellow shafts.
B. **Female:** Length 50cm. Head and neck with long, narrow, and shining black feathers. On belly a little brownish spot. Otherwise like male.
Young birds: Undescribed.
Bare parts: Bill pale yellow, red-brown at base; small casque with furrows, which vary with age. Legs blackish, soles yellow. Eyes red with a fine yellow circle of pale yellow next to pupil, female olive-brown with yellow ring. Bare skin round eye blue.
List: II
Lit.: 2, 37, 129, 147, 258, 410

2. BUCEROS BICORNIS

English: Great Hornbill, Great Indian Hornbill; *French:* Calao bicorne; *German:* Doppelhornvogel; *Malay:* Enggang papan

Range: W. India, Himalayas to Indochina, Malaya, Sumatra
Identification:
Length 130cm. Sexes alike but female slightly smaller. Forehead and upperparts black. Wings black with two white bars. Back of head, neck, tail and tail-coverts white, tail with a broad black band near middle. Belly brownish.
Young birds: Without casque till about half a year old.
Bare parts: Bill and large casque yellowish with black at both ends of casque in male. Females have smaller casque with some red at back end. Casque in both sexes U-shaped when seen from front. Legs black with greenish wash. Eyes red, female white.

Geographical variation
3. Birds from S. W. India are larger.
List: I
Lit.: 2, 37, 129, 147, 258, 410

4. BUCEROS HYDROCORAX

English: Rufous Hornbill; *French:* Calao à casque plat; *German:* Feuerhornvogel

Range: Philippine Islands
Identification:
Length 80-94cm. Sexes alike but female smaller. Forehead black. Rest of head and neck rufous. Upper throat white. Upperparts and wings brown, inner flight-feathers with buff edges. Breast and upper belly blackish. Lower belly, thighs, and short undertail feathers pale rufous. Tail white, but soon becoming dirty yellow or buffish.
Young birds: Rufous areas, breast and belly with yellowish-white.
Bare parts: Bill and large flat casque bright red, juvenile black, base of lower bill and tip red. Legs reddish. Eyes red. Bare skin around eyes yellowish.
List: II
Lit.: 37, 61, 104, 129, 147, 410

5. BUCEROS RHINOCEROS

English: Rhinoceros Hornbill; *French:* Calao rhinocéros; *German:* Rhinozerosvogel; *Indonesia:* Rangkong; *Italian:* Bucero rinoceronte; *Malay:* Burong enggang; *Spanish:* Cálao rinoceronte

Range: Malaya, Thailand, Sumatra, Borneo, Java
Identification:
Length 120cm. Sexes alike but female smaller (92cm). General plumage black with a little glossy lilac tinge of upperparts, except white rump, belly and under tail-coverts. Tail also white with a broad subterminal black band.
Young birds: Undescribed.
Bare parts: Bill and upturned casque yellowish to ivory, casque with some reddish behind. Legs greenish-yellow. Eyes brown, female white. Bare skin around eyes yellowish, reddish posteriorly.

Geographical variation:
Birds from Java have horn not curved upwards at tip. Black tail band broader.
Birds from Borneo are smaller. Horn shorter, broader and tip upward-curved.
List: II
Lit.: 37, 129, 147, 150, 258, 281, 306, 367, 410

6. BUCEROS VIGIL

Also known as: Rhinoplax vigil

English: Helmeted Hornbill; *French:* Calao à casque rond; *German:* Schildschnabel; *Iban:* Tajai; *Italian:* Calao dal casco, Bucero dal casco; *Malay:* Burong torak; *Spanish:* Cálao de casco

Range: Malaya, Sumatra, Borneo
Identification:
Length 130cm. Sexes alike but female smaller. Crown and crest black. Upperparts and wings dark brown, flight-feathers with white tips. Breast blackish-brown, belly white. Tail white with a broad black band in middle; two central tail feathers longer, grey-brown, tip black and outermost white.
Young birds: Tail band brown.
Bare parts: Bill and high and rounded casque dark red, distal part yellowish. Legs reddish-brown. Eyes red-brown, female paler. Bare skin on head and neck rufous to dark crimson, juvenile shading from dark blue to pale green.
List: I
Lit.: 34, 37, 129, 147, 150, 258, 281, 410

PLATE 60

1. COTINGA MACULATA

Brazilian: Crejoá; *English:* Banded Cotinga; *French:* Cotinga cordon-bleu; *German:* Prachtkotinga

Range: Brazil
Identification:
A. Length 20cm. **Male:** Throat, breast, and abdomen reddish-purple. Across breast a blue band. Hind wings and tail blackish. Rest deep blue.
B. **Female:** Brown with white edges to feathers on upperparts. Underparts white with brown stripes. Impossible to distinguish from other female Cotingas.
Young birds: Not described.
Bare parts: Bill and legs blackish. Eyes dark brown.
List: I
Lit.: 129, 284, 410, 431

2. XIPHOLENA ATROPURPUREA

Brazilian: Anambé-de-asa-branca; *English:* White-winged Cotinga; *French:* Cotinga porphyrion; *German:* Weissflügelkotinga

Range: Brazil
Identification:
A. Length 20cm. **Male:** Wings white. Rest black with purple tinge.
B. **Female:** Drab grey and white. Impossible to tell from the other Xipholenas.
Young birds: Not described.
Bare parts: Bill and legs brown. Eyes pale yellow, juvenile brown.
List: I
Lit.: 129, 284, 410, 431

3. RUPICOLA RUPICOLA

Brazilian: Galo-da-serra; *English:* Guianan Cock-of-the-Rock; *French:* Coq-de-roche orange; *German:* Cayenneklippenvogel, Felsenhahn; *Spain:* Gallito de las rocas

Range: Colombia, Guianas, Brazil
Identification:
A. Length 32cm. **Male:** Crest with maroon subterminal band. Wing blackish edged white. Tail black with orange bars. Rest orange.
B. **Female:** Olive-brown with faint, paler bars on tail.
Young birds: Like female, but more whitish.
Bare parts: Bill yellow, base orange, female black, tip and ridge of culmen yellow. Legs yellowish-orange, female dark brownish. Eyes orange, female pale brown.
List: II
Lit.: 129, 192, 284, 410, 431

RUPICOLA PERUVIANA

English: Andean Cock-of-the-Rock; *French:* Coq-de-roche péruvien; *German:* Andenklippenvogel; *Spanish:* Gallito de las Sierras

Range: Colombia, Venezuela, Ecuador, Peru, Bolivia
Identification:
Length 32cm. **Male:** Crest without band. Wings and tail black, three long and wide innerwing-feathers pearly grey. Rest pale orange.
Female: Wings dark brown. Tail a little paler. Rest dark reddish-brown.
Young birds: Some birds washed with olive on body feathers, otherwise like female.
Bare parts: Bill yellow, female blackish, tip pale yellow. Legs yellow or orange-yellow, female dark brown. Eyes yellow, females' vary from brown to pinkish-grey.

Geographical variation:
4. Birds from W. Colombia and Ecuador have:
A. Male blood-red colour, not pale orange, and eyes red.
B. Female light red to pinkish-grey-brown.
Birds from Venezuela, Colombia, E. Ecuador and N. Peru have bright orange males. Grey of three inner wing-feathers restricted so that black basal area is exposed. Eyes: Male bright yellow, female white to brown. Birds from S. E. Peru and Bolivia are male deep orange-red. Eyes: Male white to pale blue, female pale blue, sometimes brown.
List: II
Lit.: 129, 284, 410

Note:
Cephalopterus ornatus — Amazonian Umbrella bird
Cephalopterus penduliger — Long-wattled Umbrella bird
listed 1989 on List III. See plate 79.

PLATE 61

1. PITTA KOCHI

English: Whiskered Pitta, Koch's Pitta; *French:* Brève de Koch; *German:* Luzonpitta

Range: Luzon
Identification:
Length 22cm. Sexes alike. Head red-brown and dark brown with a wide lilac-white malar stripe. Upperside olive-brown. Wings olive-brown and grey-blue with white spot visible when wings are spread. Throat dark brown with paler margins. Upper breast bluish, lower breast and abdomen red. Flanks grey-brown. Tail grey-blue.
Young birds: White spots on throat and breast. Rest darker and duller.
Bare parts: Bill dark brown. Legs brownish. Eyes dark brown.
List: I
Lit.: 23, 61, 66, 104, 129, 353, 410

2. PITTA NYMPHA
Also known as: Pitta brachyura nympha

English: Fairy Pitta, Blue-winged Pitta; *French:* Brève du Bengale; *German:* Nympfenpitta

Range: Japan, China, Indochina, Borneo
Identification:
Length 18cm. Sexes alike. Wide eye stripe black, over eye a buffy white stripe, a rusty brown stripe, and on middle of crown a black stripe. Back and wings green, wings with a blue and a white patch, long flight-feathers black. Upper tail-coverts shining sky-blue. Throat white. Breast and flanks dull buff. Centre of belly and under tail-coverts red. Tail black, blue at tip.
Young birds: Colours less bright and crown narrowly margined with black.
Bare parts: Bill brown with orange tinge. Legs pinkish flesh. Eyes dark brown.
List: II
Lit.: 129, 187, 191, 281, 410
Note: Sibley and Monroe have no races.

3. PITTA GUAJANA

English: Banded Pitta, Blue-tailed Pitta; *French:* Brève azurine; *German:* Blauschwanzpitta; *Indonesian:* Burung paok

Range: Thailand, Malaya, Sumatra, Borneo, Java, Bali
Identification:
A. Length 20cm. **Male:** Top of crown and sides of head black. Sides of crown bright orange to yellow towards bill. Upperparts dark hazel, with white spots in wings. Throat whitish. Broad (about 10mm) black breastband. Rest of underparts yellow heavily barred black. Tail blue.
B. **Female:** Like male, but crown brown, and sides of crown buff. Underparts barred brownish black and pale buff.
Young birds: Like female but paler and more speckled brownish.
Bare parts: Bill black. Legs grey-brown. Eyes dark grey to brown.

Geographical variation:
4. Birds from Malaysia and Sumatra are flame-orange on sides of hind crown, yellow towards eyes. Underparts dark purplish blue, margined with pale orange on sides, breast narrowly barred with black.
Birds from Thailand have golden orange area not restricted to nape but usually extends to eyes. Mantle paler, less rich purplish blue in centre of breast and abdomen and orange bars on sides of breast.
Birds from Banka Island are buffish on sides of crown behind eye. Centre of crown brown. Underparts narrowly barred with a yellowish ground colour.
Birds from Java have narrower black breast band (about 6mm in males and 3-6mm in females). Underparts yellow-orange heavily barred blue.
Birds from Borneo have golden yellow sides of crown behind eye. Small blue patch on belly. Rest of underparts yellow narrowly barred with black.
List: II
Lit.: 57, 112, 127, 129, 154, 188, 246, 281, 298, 410, 419

5. PITTA GURNEYI

English: Gurney's Pitta; *French:* Brève de Gurney; *German:* Goldkehlpitta

Range: Burma, Thailand
Identification:
A. Length 20cm. **Male:** Crown, nape, and tail bright blue. Face black. Upperparts pale chestnut-brown. Upper breast and a broad band on flanks bright yellow, flanks barred black. Rest of underparts black.
B. **Female:** Top of head and nape yellowish-chestnut. Underparts pale buff barred black.
Young birds: Speckled brownish on top of head, nape and underparts.
Bare parts: Bill black. Legs whitish or brownish flesh. Eyes brown.
List: I
Lit.: 46, 99, 129, 246, 252, 410, 419

PLATE 62

1. ATRICHORNIS CLAMOSUS

English: Noisy Scrub-Bird, Western Scrubbird; *German:* Braunbauch-Dickicksvogel

Range: S. W. Australia
Identification:
A. Length 20-23cm. **Male:** Above brown with fine dark brown bars. Wings short and rounded. Throat white, centre of throat to upperbreast black outlining an inverted white V. Lower breast dull white. Flanks, abdomen, and under tail-coverts buffish-brown to rufous. Tail long, slightly down-curved, and dark brown.
B. **Female:** Smaller. No black on throat and upper breast.
C. Young birds: Like female but without barring; rump, flanks, and belly grey.
Bare parts: Bill red-brown, lower pale pinkish. Legs pinkish-brown. Eyes dark brown.
List: I
Status: Rediscovered 1961 near Two People Bay. In 1981 about 60 breeding pairs; 1989 about 500 birds.
Lit.: 36, 53, 54, 129, 225, 233, 278, 279, 410

2. PSEUDOCHELIDON SIRINTARAE

English: White-eyed River Martin, *German:* Sirintaraschwalbe

Range: Thailand
Identification:
A. Length 18cm. Sexes similar. Head black with blue gloss. Upperparts black with green gloss and silvery- white rump. Wings blackish but pale brown along inner webs. Underparts black glossed blue-green. Tail black with green gloss, tips pale brown, two central feathers have long, narrow wire-like streamers.
C. Young birds: Head and underparts brownish, lacks long tail streamers.
Bare parts: Bill green-yellow, tip black. Legs flesh. Eyes white.
List: I
Status: First discovered 1968 at Lake Boraphet, breeding area still unknown. Population size seems to be very small.
Lit.: 129, 150, 196, 286, 303, 410, 419

3. BEBRORNIS RODERICANUS

Also known as: Acrocephalus rodericanus

English: Rodriguez Brush Warbler; *German:* Rodriguezrohrsänger, *Local:* Zoiseau long-bec

Range: Rodriguez Island (E. Mascarene Islands)
Identification:
Length 15cm. Sexes alike. Yellow stripe from bill to eye and round eye. Ear-coverts yellowish-green. Above olive-green. Wings and tail brown with broad grey-green edges. Underparts pale yellow.
Young birds: Undescribed.
Bare parts: Bill dark horn, lower pale yellowish. Legs pale brown. Eyes brown.
List: III Mauritius
Status: 29-48 birds were estimated to exist in 1983.
Lit.: 45, 129, 150, 204, 221, 287, 410

4. DASYORNIS BROADBENTI LITORALIS

English: Rufous Bristlebird, *German:* Rotkopf-Lackvogel

Range: S. W. Western Australia
Identification:
Length 23cm. Sexes similar. Head rich rufous with narrow whitish eye-ring. Nape pale chestnut, each feather edged black. Upperparts slaty-brown mottled with ashy-grey margins to feathers. Rump and wings chestnut-brown. Underparts grey and white, throat and breast with dark scaly pattern, flanks dark slate-brown and abdomen white with some scaly pattern, under tail-coverts reddish-brown. Tail dark brown, shaft of tail-feathers black.
Young birds: Undescribed.
Bare parts: Bill dark brown, lower paler at base. Legs dark brown. Eyes red.
List: I
Status: May be extinct.
Lit.: 36, 129, 194, 195, 225, 233, 278, 279, 410
(Sibley and Monroe have no races)

Note: Dasyornis b. broadbenti from South Australia, Victoria and Dasyornis broadbenti whitei from South Australia are both **not listed**. Both are larger, 23-27cm, and have a triangular pale grey area between bill and eye.

5. DASYORNIS LONGIROSTRIS

English: Western Bristlebird; *German:* Langschnabel-Lackvogel

Range: S. W. Western Australia
Identification:
Length 18-20cm. Sexes alike. Pale eye-brow. Head and upperparts brown, crown, nape, and mantle mottled grey. Rump and wings rufous-brown. Underparts grey-brown, throat and chest with scaly pattern; flanks dark brown. Tail dark brown and shorter than previous species.
Young birds: Above olive-brown without spotting, below grey.
Bare parts: Bill larger than last and dusky. Legs grey-brown. Eyes red
List: I
Status: Total population in 1975 estimated at about 50 pairs.
Lit.: 129, 150, 196, 225, 233, 278, 279, 410

6. CYORNIS RUCKII

Also known as: Niltava ruecki; Muscicapa ruecki

English: Rueck's Blue Flycatcher, Rueck's Niltava; *German:* Sumatrablauschnäpper

Range: Sumatra, Malaya (?)
Identification:
A. Length 18cm. **Male:** Plumage glossy indigo-blue, between bill and eye, and ear-coverts blackish-blue. Wings and tail dusky edged blue, lightest on tail. Throat pale indigo-blue, breast dark sapphire-blue, belly and undertail-coverts whitish-grey slightly tinged blue, flanks bluish-grey.
B. **Female:** Upperparts chestnut-brown, rump, upper tail-coverts and tail rufous-chestnut. Underparts rufous-buff, paler on throat and grading to whitish on belly.
Young birds: Forehead, eye-ring, throat and breast bright chestnut-buff. Rest of upperparts brown with buffish spots. Middle of belly whitish.
Bare parts: Bill black. Legs plumbeous. Eyes dark.
List: II
Status: Only known from 4 museum-skins without exact locality statements.
Lit.: 129, 150, 247, 248, 309, 410

PLATE 63

1. PICATHARTES GYMNOCEPHALUS

English: White-necked Rockfowl, White-necked Bald Crow; *German:* Gelbkopf-Felschüpfer

Range: Sierra Leone to Ghana
Identification:
Length 37cm. Sexes similar. Naked head except for some short bristles and a little erectile crest (3-4 feathers about 6mm long) along midline of crown. Upperparts bluish-grey. Wings dark brown and bluish-grey. Neck and underparts silky-white. Tail brown.
Young birds: Like adults.
Bare parts: Bill black. Legs pale blue to dusky grey. Eyes dark brown. Bare skin on head bright yellow, and a black patch on hind crown at either side.
List: I
Lit.: 7, 129, 173, 268, 304, 410

2. PICATHARTES OREAS

English: Grey-necked Rockfowl; Grey-necked Bald Crow, Grey-necked Picathartes; *German:* Buntkopf-Felshüpfer

Range: Cameroon
Identification:
Length 35cm. Sexes similar. Head naked except for some short bristles and a little transverse row of up to 9 erectile feathers (7mm long), on middle of crown at rear of blue skin.
Neck and upperparts slaty-grey with a bluish wash. Wings black, dark brown and grey. Chin and throat grey becoming greyish-white on breast and white on rest of underparts. Flanks grey. Tail slaty-grey with brown shafts.
Young birds: Undescribed.
Bare parts: Bill black. Legs grey. Eyes dark brown. Bare skin on head blue on forecrown, red on hindcrown and nape, sides of head black.
List: I
Lit.: 7, 129, 173, 268, 304, 410

3. TERPSIPHONE BOURBONNENSIS
Also known as: Tchitrea bourbonnensis

English: Mascarene Paradise Flycatcher; *German:* Maskarenen-paradiesschnäpper; *Local:* Coq des bois, Zoiseau La Vierze

Range: Mauritius, Reunion
Identification:
A. Length 16cm. **Male:** Head and nape deep glossy blue-black. Eye-ring blue. Hindneck grey with a blue wash. Upperparts, wings and tail cinnamon-rufous, flight-feathers dark brown with cinnamon-rufous edges. Chin, throat, and underpart grey, becoming whitish on belly and under tail-coverts.
B. **Female:** Like male, but head greyish with a faint blue wash.
Young birds: Undescribed.
Bare parts: Bill black. Legs brown. Eyes dark brown.
List: III Mauritius
Lit.: 129, 287, 410

4. GUBERNATRIX CRISTATA

Brazilian: Cardeal-amarelo; *English:* Yellow Cardinal; *French:* Cardinal vert; *German*: Grünkardinal; *Italian:* Cardinale verde; *Spanish:* Cardenal amarillo

Range: Brazil, Uruguay, Argentina
Identification:
A. Length 20cm. **Male:** Forehead, crown and crest black. Yellow eye and malar stripe. Rest of head, neck and upperparts olive with dusky streaks. Wings dark brown, flight-feathers edged pale yellow. Chin and throat black, rest of underparts olive-yellow. Tail dark olive to dark brown, outermost feathers yellow.
B. **Female:** Like male but slightly smaller and greyer.
Young birds: Like female but duller.
Bare parts: Bill blackish horn, lower paler. Legs brown. Eyes dark brown.
List: II
Lit.: 37, 129, 236, 240, 410, 431

5. PAROARIA CAPITATA

Brazilian: Cavalaria; *English:* Yellow-billed Cardinal; *German:* Mantel-kardinal

Range: Bolivia, Brazil, Paraguay, Argentina
Identification:
Length 16.5cm. Sexes similar. Head crimson.White collar almost closed on nape. Upperparts, wings, and tail glossy blue-black. Chin and throat black. Underparts white, flanks greyish.
Young birds: Head buffish and throat brown.
Bare parts: Bill pinkish yellow. Legs pale red-brown. Eyes orange.

Geographical variation:
Birds from Bolivia have pinkish-orange bill and legs brownish flesh.
List: II
Lit.: 14, 129, 236, 240, 410, 431

6. PAROARIA CORONATA

Brazilian: Cardeal; *English:* Red-crested Cardinal; *German:* Graukardinal

Range: Bolivia, Paraguay, Brazil, Uruguay, Argentina
Identification:
Length 19cm. Sexes similar. Head, crest, and long pointed area on throat bright scarlet. Upperparts grey. Wings black edged pale grey. Underparts and collar white, flanks grey. Tail black.
Young birds: Duller with greyish-brown head and small crest.
Bare parts: Bill whitish. Legs brown. Eyes dark brown.
List: II
Lit.: 129, 236, 240, 410, 431

PLATE 64

1. AGELAIUS FLAVUS
Also known as: Xanthopsar flavus; Oriolus flavus

Brazilian: Pássaro-preto-de-veste amarela; *English:* Saffron-cowled Blackbird; *French:* Ictéride à tête jaune; *German:* Gilbstärling, Gelbköpfiger Stärling; *Spanish:* Dragón, Tordo de cabeza amarilla

Range: Paraguay, Brazil, Uruguay, Argentina
Identification:
A. Length 20cm. **Male:** Between bill and eye black. Rest of head, rump, and entire underparts bright saffron yellow. Upperparts and tail glossy black.
B. **Female:** Eyebrow, rump and underparts yellow. Upperparts olive-brown streaked dusky, darkest on crown.
Young birds: Undescribed.
Bare parts: Bill black. Legs black. Eyes dark brown.
List: III Uruguay
Lit.: 37, 129, 190, 240, 369, 410, 431

2. ZOSTEROPS ALBOGULARIS

English: White-chested White-eye; *German:* Norfolkbrillenvogel

Range: Norfolk Island (Australia)
Identification:
Length 14.5cm. Sexes alike. Forehead, crown, and hindneck yellowish olive-green. Between bill and eye and area below eye black. Around eye white. Ear-coverts and side of neck greyish. Upperside olive-green, rump yellowish olive-green. Forewing green, rest of wing brown edged green. Underparts whitish, chin with faint buff wash, flanks very conspicuously brown. Under tail-coverts pale yellow. Tail brown with green edges.
Young birds: Undescribed.
Bare parts: Bill black. Legs brown. Eyes undescribed.
List: I
Lit.: 129, 249, 410

3. CARDUELIS CUCULLATA
Formerly known as: Spinus cucullatus

English: Red Siskin; *French:* Chardonneret rouge, Serin à tête noire; *German:* Kapuzenzeisig, Feuerzeisig, Kappenzeisig; *Italian:* Cardellino rosso; *Spanish:* Cardenalito, Lúgano encapuchado, Lúgano cardenalito

Range: Colombia, Venezuela, Trinidad
Identification:
A. Length 10cm. **Male:** Head and most of neck black. Body bright orange-vermilion, back darker. Wings and tail black, pale orange-vermilion patches in wings and inner flight-feathers edged white. Centre of belly white.
B. **Female:** Head, upperparts and throat greyish-brown and mouse brown tinged vermilion. Wing patch, breast, flanks and rump salmon-vermilion. Belly white.
Young birds: Like female, but more brown and duller.
Bare parts: Bill and legs brown. Eyes blackish-brown.
List: I
Lit.: 37, 44, 125, 129, 189, 190, 192, 240, 340, 410, 417

4. CARDUELIS YARRELLII
Also known as: Spinus yarrellii

Brazilian: Coroinha, Pintasilgo-do-nordeste; *English:* Yellow-faced Siskin, Yarrell's Siskin; *French:* Chardonneret de Yarrell; *German:* Yarrellzeisig; *Italian:* Cardellino di Yarrell; *Spanish:* Jilguero de cara amarilla, Jilguero de Yarrell, Lúgano de Yarrell

Range: Venezuela, Brazil
Identification:
A. Length 10cm. **Male:** Cap black. Face, neck, and underparts bright yellow. Back olive-green. Rump bright yellow. Wings black with large yellow patch. Tail black, basal third bright yellow.
B. **Female:** Lacks black cap. Above dull olive-grey, rump bright olive-yellow. Below grey. Breast and flanks olivaceous.
Young birds: Paler and less contrasting.
Bare parts: Bill and legs greyish-blue. Eyes dark brown.
List: II
Lit.: 37, 129, 190, 192, 240, 410, 417, 431

5. SERINUS CANICAPILLUS
Also known as: Serinus gularis; Crithagra canicapilla

English: West African Seedeater, West African Streaky-headed Seedeater; *German:* Braunwangengirlitz

Range: Ghana to Sudan, Zaire, Uganda, Kenya
Identification:
Length 15cm. Sexes alike. Heavy ashy-brown streaking on top of head. Stripe over eye white. Big patch on side of head and upperparts ashy-brown, back streaked with dusky. Chin and throat white with a black moustachial stripe. Rest of underparts pale earth-brown. Tail ashy-brown.
Young birds: More streaked on underside.
Bare parts: Bill horn. Legs pale brown. Eyes hazel.
List: III Ghana
Lit.: 40, 171, 172, 173, 174, 195, 222, 321, 410, 417

6. SERINUS LEUCOPYGIUS
Also known as: Ochrospiza leucopygia

English: White-rumped Seedeater, Grey Canary; *French:* Serin à croupion blanc, Chanteur d'Afrique; *German:* Weissbürzelgirlitz; *Spanish:* Canario culiblanco

Range: Africa south of Sahara
Identification:
Length 11.5cm. Sexes alike. Head greyish.Upperparts dull grey-brown, mottled dark brown. Rump white. Underparts creamy-white spotted with grey-brown. Tail dark brown.
Young birds: More streaked above.
Bare parts: Bill brown. Legs and eyes: No information available.

Geographical variation:
Birds from Senegal to Chad and Central African Republic have throat and breast whitish with dusky spotting.
Birds from Aïr to Nigeria are paler on upperparts and not so heavily streaked.
List: III Ghana
Lit.: 35, 129, 171, 195, 268, 320, 325, 410, 417

7. SERINUS MOZAMBICUS
Also known as: Ochrospiza mozambica

English: Yellow-fronted Canary; *French:* Serin à front jaune, Serin du Mozambique; *German:* Mocambiquegirlitz; *Spanish:* Canario de Mozambique

Range: Africa from Senegal to Cape Province
Identification:
Length 12cm. **Male:** Forehead, streak over eye, chin, throat and rest of underparts and rump yellow. Black moustachial stripe on either side of throat. Above dull green streaked dusky. Wings and tail dusky brown edged yellowish-green, tail tipped white.
Female: Slightly browner on head and back.
Young birds: Like female but duller, with streakings on sides of breast and flanks.
Bare parts: Bill brown, lower paler. Legs flesh brown. Eyes brown.

Geographical variation:
There are some minor variations.
List: III Ghana
Lit.: 39, 129, 171, 172, 173, 174, 268, 320, 325, 410, 417

PLATE 65

1. ESTRILDA ASTRILD

English: Common Waxbill; *French:* Astrild ondulé, Bec-de-corail, Bec de corail ondulé; *German:* Wellenastrild; *Spanish:* Astrilda común

Range: Sierra Leone to Cameroon and Ethiopia and S. Africa
Identification:
Length 10cm. **Male:** Head grey-brown with a conspicuous crimson stripe from bill through eye to top of ear-coverts. Upperparts grey-brown with fine dusky transverse barring. Wings uniformly dark brown. Underparts light brown with rosy-pink wash, barred brown, especially on flanks. Lower breast and belly have red central patch. Under tail-coverts black and tail dark brown.
Female: Less red and black on belly.
Young birds: Crown dark brown. Indistinctly barred below.
Bare parts: Bill red, juvenile black. Legs and eyes dark brown.

Geographical variation:
Birds from Ethiopia have darker upperparts with rosy wash but without red patch on belly.
Birds from Uganda, N. W. Tanzania and S. Zaire have no red patch on belly.
2. Birds from Kenya and N. Tanzania are darker above and less pinkish below.
Birds from W. Angola have very pale upperparts, barring poorly marked. Underparts also pale.
Birds from Gabon are heavily marked red on belly. Upper tail-coverts tinged carmine.
Birds from South Africa are darker and strongly washed olivaceous-brown on upperparts.
List: III Ghana
Lit.: 7, 38, 95, 129, 171, 172, 173, 174, 203, 205, 210, 222, 244, 268, 320, 325, 410

3. ESTRILDA TROGLODYTES

English: Black-rumped Waxbill; *French:* Astrild cendré, Bec de corail cendré; *German:* Grauastrild; *Spanish:* Astrilda culinegra

Range: Senegal to Ethiopia, Ghana, Zaire and Uganda
Identification:
Length: 9.5cm. Sexes similar. Same red eye stripe as last number 1. Upperparts including crown grey-brown, very finely barred, only visible when closely examined. Rump black. Cheeks white to buffish. Underparts greyish with pink tinge, middle of belly pinkish-red. Under tail-coverts white. Tail black, outer tail-feathers edged white.
Young birds: Lack red eye stripe.
Bare parts: Bill red, juvenile black. Legs blackish-brown. Eyes reddish-brown.
List: III Ghana
Lit.: 7, 95, 129, 173, 268, 325, 410

4. ESTRILDA MELPODA

English: Orange-cheeked Waxbill; *French:* Astrild à jous orangées, Joues-oranges; *German:* Orangebäckchen; *Spanish:* Astrilda carirroja

Range: Senegal to Cameroon and Chad, Angola, Zambia
Identification:
Length 11cm. Sexes similar. Upper head grey. Large orange patch on side of head. Upperparts and wings brown, rump crimson. Underparts pale grey, abdomen whitish to buffish yellow, sometimes with a rose wash. Tail blackish.
Young birds: Like adult but crown greyish-brown, and orange and crimson colours paler.
Bare parts: Bill red to orange-red. Legs brown. Eyes reddish-brown.

Geographical variation:
Birds from Cameroon and Chad are paler.
List: III Ghana
Lit.: 7, 95, 129, 173, 268, 410

5. ESTRILDA CAERULESCENS

Formerly known as: Lagonosticta caerulescens

English: Lavender Waxbill, Lavender Fire-Finch; *French:* Bengali gris bleu, Queue de vinaigre; *German:* (Rotschwanz-) Schönbürzel: *Spanish:* Astrilda ceniza Senegalesa

Range: Senegal to Central African Republic and Chad
Identification:
Length 10cm. **Male:** Entire plumage bluish-grey, paler below, except for carmine rump, upper- and under tail-coverts and central tail feathers. Black stripe between bill and eye. Flight feathers dark grey. Belly blackish in middle and lower flanks black spotted white.
Female: Like male but not so blackish on belly.
Young birds: Slightly paler.
Bare parts: Bill blue-grey and/or dark reddish tipped black. Legs dark brown. Eyes dark brown.
List: III Ghana
Lit.: 7, 95, 129, 173, 268, 410

6. LAGONOSTICTA V. VINACEA

Formerly known as: Estrilda larvata; Lagonosticta larvata; Lagonosticta nigricollis

English: Vinaceous Firefinch, Black-faced Firefinch; *French:* Amarante masqué; *German*: Schwarzkehlamarant; *Spanish*: Bengali carinegro

Range: Senegal, Guinea
Identification:
A. Length 10cm. **Male:** Upper head grey gradually becoming vinous-pink on sides of neck and upper mantle, rest of back and wings darker rosy-crimson. Rump and upper tail-coverts crimson-pink. Sides of head, chin, and throat black. Rest of underparts vinous-pink with numerous fine white dots on sides of breast and flanks. Lower abdomen and under tail-coverts black. Tail black with broad outer edging dull crimson.
B. **Female:** More brownish and without black mask.
Young birds: Duller than adult.
Bare parts: Bill said to be dark olive-green, lower paler. Legs said to be pale olive-green. Eyes: Dark red to red-brown.

Subspecies:

Lagonosticta vinacea nigricollis
Range: Upper Volga to Cameroon, Chad, Sudan, Uganda
Upperparts dark grey, only rump and tail vinous-pink. Underparts greyer.
List: III Ghana
Lit.: 7, 35, 95, 129, 173, 268, 325, 410

PLATE 66A

7. LAGONOSTICTA RARA

English: Black-bellied Firefinch; *French:* Sénégali à ventre noir; *German:* Schwarzbauchamarant

Range: Sierra Leone to Sudan, Uganda, Kenya, Zaire
Identification:
A. Length 10cm. **Male:** General colour pinkish-crimson to vinous-pink, brighter on upper tail-coverts. Wings brown. Middle of abdomen, under tail-coverts and tail black, tail edged crimson on central feathers. No white dots on breast and flanks.
B. **Female:** Head brown with grey sides of face and a crimson spot in front of eye. Upperparts dull crimson. Underparts buffy washed with crimson and with black area as in male, (but duller). Note: both sexes without white spots on breast and flanks.
Young birds: Sooty brown. Upper tail-coverts dull crimson. Young males said to have a crimson wash overall.
Bare parts: Bill black, sides of lower whitish to crimson. Legs grey to brown. Eyes dark brown to blackish.

Geographical variation:
Birds from Sierra Leone to Nigeria are brighter red. Females have grey throat tinged red and belly blacker.
List: III Ghana
Lit.: 7, 35, 95, 129, 171, 320, 325, 410

8. LAGONOSTICTA RUBRICATA

English: African Firefinch, Dark Firefinch; *French:* Sénégali à bec bleu, Amarante flambé; *German:* Dunkelamarant; *Spanish:* Bengali de Lichtenstein

Range: Guinea to Ethiopia and south to Cape Province
Identification:
A. Length 11cm. **Male:** Like No. 7, but top of head, nape, and hind neck dull grey. Few white dots on sides of breast and flanks. Belly dark grey-brown. First long flight feather notched.
B. **Female:** Brighter earth-brown above, paler below. Strongly pinkish on throat, breast, and flanks.
Young birds: Sides of face and underparts buffish-brown.
Bare parts: Bill blackish, juvenile paler. Legs and eyes brown.

Geographical variation:
Birds from Cameroon to N. Tanzania have lower belly black.
Birds from Tanzania and Mozambique have a pinkish wash on crown.
List: III Ghana
Lit.: 7, 88, 95, 129, 171, 172, 173, 174, 205, 268, 320, 410

9. LAGONOSTICTA SENEGALA

English: Red-billed Firefinch, Senegal Fire-Finch; *French:* Sénégali rouge, Amaranthe, Amarante commun; *German:* Senegalamarant; *Spanish:* Bengali Senegalés

Range: Mauritania and Senegal to Ethiopia, south to Cape Province
Identification:
A. Length 10cm. **Male:** General coloration crimson. Crown, nape, mantle, and back brown tinged crimson. Wings brown, wing-coverts washed crimson. First long flight-feathers notched. On sides of crimson breast a variable number of very fine white dots. Belly and under tail-coverts pale brown. Tail black outermost edged red. Note: birds in worn plumage browner.
Female: Sandy grey-brown above and buffish below. Crimson rump and spot in front of eye. Larger white spots on breast and flanks.
Young birds: Like female but lacking red spot on head and white spots below.
Bare parts: Bill red with stripe on back of upper bill, juvenile blackish. Legs brown. Eyes brown. Eye-ring yellow.

Geographical variation:
Birds from Coast of Guinea and Sierra Leone are darker red on crown, nape, mantle, and breast. Females have tawny-olive underparts instead of buff.
Birds from Uganda, W. Tanzania, S. E. Zaire and Zambia have more reddish on upperparts and more purplish-rose underparts.
Birds from S. Tanzania, S. E. Zaire and South Africa are more lilac-red below with more dots on breast.
List: III Ghana
Lit.: 7, 88, 95, 106, 129, 173, 174, 202, 205, 222, 268, 320, 410

10. LAGONOSTICTA RUFOPICTA

English: Bar-breasted Firefinch; *French:* Sénégali à poitrine barrée, Amarante pointé; *German:* Pünktchenamarant; *Spanish:* Bengali de Fraser

Range: Senegal to Central African Republic, Zaire and Uganda
Identification:
Length 10.5cm. Sexes in most cases alike. Similar to number 9 but more greyish earth brown on crown, nape and upperparts. Forehead dark pinkish red. Face rose-pink.
Numerous tiny white dots forming more or less broken bars on rose-pink breast. First long flight feather not notched.
Young birds: Dull brown except for dull crimson upper tail-coverts and a tinge of crimson on breast.
Bare parts: Bill red with black stripe on back of upper bill, juvenile dusky bill. Legs pinkish-brown, juvenile purplish to greyish. Eyes brown, eye-ring grey to whitish.
List: III Ghana
Lit.: 7, 95, 129, 171, 173, 268, 410

PLATE 66B

1. LONCHURA B. BICOLOR

Formerly known as: Spermestes bicolor

English: Black and White Munia, Black and White Mannikin, Blue-billed Mannikin; *French:* Spermète à bec bleu; *German:* Glanzelsterchen

Range: Guinea Bissau to Cameroon
Identification:
Length 9-10cm. Sexes similar. Head, neck, upperparts and breast black with a green tinge. Lower breast, abdomen and under tail-feathers white. Black flank with conspicuous white markings.
Young birds: Sooty brown above and more buffish-grey below.
Bare parts: Bill bluish-grey. Legs dark olive to blackish. Eyes dark brown.

Subspecies:

Lonchura bicolor poensis
Range: From S. Cameroon to Ethiopia, Kenya, Angola
Faint to dense trace of white barring on rump, upper tail-coverts and wings.
List: III Ghana
Lit.: 7, 95, 129, 173, 195, 234, 268, 320, 410, 417

2. LONCHURA NIGRICEPS
Also known as: Lonchura bicolor nigriceps; Spermestes bicolor nigriceps
Range: S. Africa from Angola to Somalia and south to Cape Province
English: Brown-backed Munia
Like last, number 1, but mantle and inner-wing chestnut to dark chocolate-brown.
NO MORE LISTED

3. LONCHURA CUCULLATA

Also known as: Spermestes cucullata

English: Bronze Munia, Bronze Mannikin; *French:* Tisserin des villages, Gendarme, Spermète nonnette; *German*: Kleinelsterchen; *Spanish:* Negrita bronceada

Range: Senegal to Ethiopia and south to Cape Province
Identification:
Length 9cm. Sexes alike. Top and sides of head, throat to upper breast, a narrow row on shoulders, and area on sides of breast blackish glossed bottle green and/or purplish on cheeks and breast.
Upperparts and wings dull brown. Rump, upper- and under tail-coverts, and hind flanks barred black and white. Tail black.
Young birds: Paler brown, buffish on belly and flanks.
Bare parts: Bill black, lower grey-blue, juvenile black. Legs dark grey. Eyes brown.

Geographical variation:
Birds from Ethiopia to Angola and Cape Province usually lacks glossy green area on side of breast.
List: III Ghana
Lit.: 7, 95, 129, 173, 268, 320, 410, 417

4. LONCHURA FRINGILLOIDES

Also known as: Spermestes fringilloides

English: Magpie Munia, Magpie Mannikin; *French:* Spermète-pie; *German:* Riesenelsterchen; *Spanish:* Negrita picaza

Range: Senegal to Somalia and south to Natal
Identification:
Length 12cm. Sexes alike. Resembles a large edition of last species. Head, neck, rump, and large patch on each side of breast glossy blue-black. Upperparts and wings slightly reddish earth-brown, fine white stripes on back. Underparts white to creamy or buffish white. Flanks rufous, black and white. Tail black.
Young birds: Duller
Bare parts: Bill blackish, lower bluish-grey. Legs blackish. Eyes brown to reddish-brown.
List: III Ghana
Lit.: 7, 95, 129, 173, 268, 320, 410, 417

5. LONCHURA CANTANS

Also known as: Eudice cantans; Lonchura malabarica

English: African Silverbill; *French:* Bec d'argent, Gros-bec chanteur; *German:* Silberschnäbelchen

Range: Senegal to Ethiopia, south to Tanzania
Identification:
Length 11cm. Sexes similar. Head, neck, and upperparts ashy-brown with a reddish tinge, most feathers edged paler. Innermost 4 flight-feathers same colour with darker transverse barring. Rest of wings and tail blackish, tail narrow and graduated. Underside whitish, flanks buffish, throat darkest ashy red-brown, edged pale buffish.
Young birds: Faint buff edges to feathers on upperparts and tail.
Bare parts: Bill silver blue-grey. Legs grey-brown. Eyes dark brown to blackish.

Geographical variation:
Birds from Somalia, Ethiopia to Tanzania and Yemen are browner and more distinctly barred on upperside.
List: III Ghana
Lit.: 7, 95, 111, 126, 129, 173, 320, 410, 417

6. MANDINGOA NITIDULA

Formerly known as: Hypargos nitidulus

English: Green-backed Twinspot; *French:* Bengali vert tacheté; *German:* Grüner Tropfenastrild

Range: Sierra Leone to Ethiopia, south to Cape Province
Identification:
A. Length 10cm. **Male:** Face tomato red. Rest of head, upperparts and breast dark olive-green. Wings blackish with olive-green borders. Lower underparts black with white spots. Tail black edged olive-green.
B. **Female:** Paler. Face pale orange. Lower underparts grey, spotted white.
Young birds: Greyish-olive, round eyes buff. Underside grey without white spots.
Bare parts: Bill black tipped reddish. Legs flesh. Eyes dark brown.

Geographical variation:
Birds from Sierra Leone to Zaire and Angola have throat and chest of male orange-red, female golden olive.
Birds from Bioko (Fernando Po Island) have orange tinge on back and rump. Whole bill reddish.
Birds from Ethiopia, Sudan, Kenya and Tanzania have throat and chest tinged golden olive or rarely orange-red. Bill smaller.
List: III Ghana
Lit.: 7, 95, 129, 173, 205, 320, 325, 410

7. NESOCHARIS CAPISTRATA

English: Grey-headed Oliveback, White-cheeked Olive Weaver; *French:* Bengali vert à joues blanches, Sénégali vert à joues blanches; *German:* Weisswangenastrild; *Spanish:* Olivino cariblanco

Range: Gambia to Sudan, Uganda
Identification:
Length 12cm. Sexes alike. Top of head blue-grey, face whitish. Black band from throat to behind ear-coverts. Upperparts, wings and tail bright yellowish olive-green. Underparts blue-grey, flanks yellowish.
Young birds: Duller and darker.
Bare parts: Bill black, juvenile whitish. Legs dark grey. Eyes dark red.
List: III Ghana
Lit.: 7, 95, 129, 173, 268, 320, 410

PLATE 67

1. NIGRITA BICOLOR

English: Chestnut-breasted Negrofinch; *French:*Bengali brun à ventre roux; *German:* Zweifarbenschwärzling; *Spanish:* Negrita pechirroja
Range: Sierra Leone to Uganda, Angola and Zaire
Identification: Length 11.5cm. **Male:** Forehead, face, and entire underparts chestnut-maroon. Crown and upperparts slaty-grey. Wings dark brown, tail blackish. **Female:** Paler and more brownish-slate on upperparts.
Young birds: Like female but duller.
Bare parts: Bill black. Legs and eyes dark brown.
Geographical variation:
Birds from Nigeria to Uganda and Angola have sooty greyish-brown crown and upperparts.
List: III Ghana
Lit.: 7, 95, 129, 172, 173, 268, 410

2. NIGRITA C. CANICAPILLA

English: Grey-crowned Negrofinch, Grey-headed Negro-finch; *French:* Bengali nègre, Sénegali nègre; *German:* Graunakkenschwärzling; *Spanish:* Negrita cana
Range: Nigeria to Zaire, Uganda, Kenya, Tanzania and Angola
Identification: Length 13.5cm. Sexes similar. Forehead, face, and entire underparts black. Hind crown, nape, and mantle silvery grey bordered whitish on crown. Rump whitish-grey. Wings and tail black, wing-coverts and three inner flight-feathers with white spots.
Young birds: Sooty-brown only with a faint spotting on wing-coverts.
Bare parts: Bill black. Legs dark grey to dark brownish. Eyes orange to red.

Subspecies:
Nigrita canicapilla emiliae
Range: Sierra Leone to Ghana
Smaller. Less white spotting on wing-coverts, and spots silvery-grey.
Nigrita canicapilla candida
Range: W. Tanzania
Crown and nape whitish-grey like rump.
Geographical variation:
Birds from S. Sudan, E. Zaire, Uganda and N. W. Tanzania have broader white band between face and upperparts. Birds from S. E. Sudan, Kenya and N. Tanzania are darker and without white spotting on greater wing-coverts. Birds from C. Kenya have mantle and rump blackish-grey.
List: III Ghana
Lit.: 7, 95, 129, 171, 172, 173, 235, 268, 321, 410

3. NIGRITA FUSCONOTA

English: White-breasted Negrofinch; *French:* Bengali brun à ventre blanc, Sénégali brun à ventre blanc; *German:* Mantelschwärzling; *Spanish:* Negrita pechiblanca
Range: Guinea to Kenya, Angola, Zaire
Identification: Length 10cm. Sexes alike. Head, nape, lower rump and upper tail-coverts glossy blue-black. Tail black. Rest of upperparts and wings brown. Chin, throat and entire underparts white.
Young birds: Top of head dull brown. Underparts greyer and usually spotted.
Bare parts: Bill black.Legs greyish, toes greenish tinge. Eyes dark brown.
Geographical variation:
Birds from Guinea to Nigeria have pale isabelline back and rump.
List: III Ghana
Lit.: 7, 95, 129, 171, 173, 268, 321, 410

4. NIGRITA LUTEIFRONS

English: Pale-fronted Negrofinch; *French:* Bengali nègre à front jaune, Sénégali nègre à front jaune; *German:* Blassstirnschwärzling; *Spanish:* Negrita frentigualda
Range: Nigeria to Zaire, Angola, Zaire
Identification: A. Length 10cm. **Male:** Forehead yellowish-white. Crown, nape and upperparts grey, rump and upper tail-coverts whitish. Wings blackish, wing-coverts dark grey. Sides of head, entire underparts and tail black. B. **Female:** Forehead buffish. Around eye black, otherwise face and entire underpart grey.
Young birds: Like female but duller grey and underparts tinged buffish.
Bare parts: Bill black. Legs flesh colour. Eyes pale grey.
Geographical variation:
Birds from Bioko are larger. Yellowish-white on forehead extended also over forecrown.
List: III
Lit.: 7, 95, 129, 173, 321, 410

5. ORTYGOSPIZA A. ATRICOLLIS

English: African Quailfinch; *French:* Astrild-caille à gorge noire, Astrild-caille; *German:* Wachtelastrild; *Spanish:* Astrilda aperdizada
Range: Senegal to Sudan and Kenya
Identification: A. Length 10cm. **Male:** Forehead, face and throat black. Crown, upperparts, and wings earth-brown. Chin and centre of belly white. Breast and flanks broadly barred white and earth-brown. Lower breast chestnut. Under tail-coverts rufous-buff, streaked white and blackish. Tail tipped white, outer tail feathers edged white.
B. **Female:** Forehead and face brown, throat grey, underparts paler.
Young birds: Paler upperparts. Breast very pale chestnut in middle, flanks indistinctly barred dirty white and brown.
Bare Parts: Bill red, upper bill often brownish. Legs pinkish brown to flesh. Eyes buffish to brown.

Subspecies:
6. Ortygospiza atricollis fuscocrissa
English: Ethiopian Quailfinch
Range: Angola to Kenya and south to Cape Province
Darker and more contrast between black and white.
Geographical variation:
Birds from Zambia have white eye-ring.
List: III Ghana
Lit.: 7, 88, 95, 129, 171, 172, 173, 268, 321, 410

PARMOPTILA RUBRIFRONS
Also known as: Parmoptila woodhousei (in part)
English: Jameson's Antpecker, Jameson's Hylia-Finch, Red-fronted Antpecker; *French:* Astrild-fourmilier à gorge rousse; *German:* Ameisenpicker
Range: Liberia, Ghana
Identification: Length 9.5cm. **Male:** Forehead and forecrown red, rest of head, nape and upperparts olive-brown with fine buff shaft streaks on head and fine buffish tips to wing-coverts. Upper throat buffish, rest of underparts chestnut. **Female:** Forehead brown, feathers with dark middle giving scaly appearance. Throat spotted and rest of underparts paler, almost white.
Young birds: Said to be like juvenile woodhousei.
Bare parts: Bill blackish. Legs flesh coloured. Eyes red-brown.
List: III Ghana

Also referenced as:
7. Parmoptila woodhousei
Range: Nigeria to Zaire and Angola
Like rubrifrons except
A. **Male** have underparts buffish-white heavily spotted blackish.
B. **Female** have chestnut throat.
Young birds: Duller brown on upperparts and dark reddish-buff underparts with dusky bars.
NO MORE LISTED
Lit.: 7, 95, 129, 172, 173, 321, 410
(Sibley and Monroe have two species, rubrifrons and woodhousei)

8. PHOLIDORNIS RUSHIAE

English: Tit Hylia; *French:* Astrild-mésange; *German:* Strichel Köpfchen
Range: Sierra Leone to Cameroon, Zaire, Uganda, Angola
Identification: Length 8cm. Sexes alike. Head and chin to breast pale buff streaked dusky. Mantle, scapulars and back dark brown with narrow buffish-white margins. Rump and upper tail-coverts yellow, Wings and tail dark brown; long flight-feathers and wing-coverts edged pale buff. Rest of underparts yellow.
Young birds: Breast without streaking.
Bare parts: Bill blackish, base of lower yellow. Legs yellow. Eyes red, female pale grey.
Geographical variation:
Birds from Sierra Leone to Ghana have wing-feathers edged pale olive not buff. Streaking below narrower.
Birds from Bioko have rump feathers with dark centres. More heavily streaked below.
List: III Ghana
Lit.: 7, 129, 173, 268, 410

PLATE 68

1. PYRENESTES OSTRINUS

Also known as: Pirenestes ostrinus; Pyrenestes frommi; Pyrenestes rothschildi

English: Black-bellied Seedcracker; *French:* Gros-bec pouceau à ventre noir; *German:* Purpurastrild; *Spanish:* Pirenestes ventrinegro

Range: Ivory Coast to Uganda, Angola
Identification:
A. Length 13cm. **Male:** Head, neck, rump, upper tail-coverts, breast, and flanks crimson. Tail dull crimson, underside of tail blackish. Rest of plumage black.
B. **Female:** Resembles male but is brown where male is black.
Young birds: Dull olive-brown with chestnut-orange rump and upper tail-coverts.
Bare parts: Huge bill bluish-black, length variable. Legs yellowish-brown. Eyes red-brown, female more brown, juvenile greyish-brown; eye-ring bluish-white to white.
List: III Ghana
Lit.: 95, 129, 173, 268, 325, 410

2. PYTILIA HYPOGRAMMICA

English: Red-faced Pytilia; *French:* Pytilia à ailes jaunes; *German:* Rotmaskenastrild

Range: Sierra Leone to Central African Republic
Identification:
A. Length 10.5cm. **Male:** Forehead, face, chin, rump, and upper tail-coverts scarlet. Wings dark brown edged dusky yellow to greenish-yellow, rarely orange-red. Rest uniformly grey, barred finely white on hind breast and belly. Under tail-coverts blackish-grey edged white.
B. **Female:** Head brownish-grey without any red; pale brown below with white barring.
Young birds: Duller than female, grey parts more brownish.
Bare parts: Bill black. Legs flesh. Eyes red.
List: III Ghana
Lit.: 95, 129, 173, 410

3. PYTILIA PHOENICOPTERA

English: Red-winged Pytilia, Crimson-winged Pytilia, Aurora Finch; *French:* Pytilie à ailes rouges, Diamant aurore; *German:* Auroraastrild; *Spanish:* Pitilia alirroja

Range: Senegal to Cameroon, Sudan, Zaire, Uganda
Identification:
A. Length 10.5cm. **Male:** Head and upperparts grey, except crimson rump and upper tail-coverts. Wings dark brown edged red, faded to orange-red in worn plumage. Underparts grey barred white, darker grey on under tail-coverts. Tail crimson.
B. **Female:** More brown where male is grey.
Young birds: Paler and browner than female and buffish barring below.
Bare parts: Bill black, base of lower often pale grey, juvenile have greyish bill. Legs pale flesh-brown. Eyes red.
List: III Ghana
Lit.: 95, 129, 171, 173, 268, 410

4. SPERMOPHAGA HAEMATINA

English: Western Bluebill; Blue-billed Waxbill, Crimson-breasted Bluebill, Blue-billed Weaver; *French:* Gros-bec sanguin; *German:* Rotbrust-Samenknacker; *Spanish:* Espermófaga hematina

Range: Senegal to Cameroon, Zaire, Angola
Identification:
A. Length 14.5cm. **Male:** Head and upperside glossy black. Wings dark brown with glossy black wing-coverts. Chin to breast and a variable amount of flank feathers glossy scarlet. Rest of underside and tail black.
B. **Female:** Forehead, side of face, and upper tail-coverts dull crimson or dull orange-red. Belly and under tail-coverts blackish spotted or barred white. Otherwise like male.
Young birds: Male: Upperparts dark grey with dull crimson-orange upper tail-coverts. Breast tipped rufous-orange. Female: Pale barring on lower underparts.
Bare parts: Bill shinning blue, tip red. Legs brownish-black to blackish-green or dark olive. Eyes dark chestnut to dark brown; eye-ring pale blue.

Geographical variation:
Birds from Togo to S. W. Nigeria have dull red upper tail-coverts in both sexes.
Birds from S. Nigeria, Cameroon to Zaire, and Angola are like last but also have some red on side of face in both sexes.
List: III Ghana
Lit.: 95, 129, 173, 268, 410

5. URAEGINTHUS BENGALUS

Also known as: Estrilda bengala

English: Red-cheeked Cordonbleu, Red-cheeked Blue Waxbill; *French:* Bengali cordon-bleu, Le Mariposa, Cordon bleu; *German:* Schmetterlings-astrild; *Spanish:* Coliazul Bengali

Range: Mauritania, Senegal to Ethiopia and south to Angola
Identification:
A. Length 12cm. **Male:** Earth brown forehead, crown, nape, upperparts and wings; rump and upper tail-coverts blue. Ear-coverts dark crimson, rest of face, chin, throat, forebreast, and flanks bright blue. Middle of lower breast, belly, and under tail-coverts pale earth brown. Tail greyish-blue.
B. **Female:** A little paler and duller than male and ear-coverts blue.
Young birds: Like female but no blue on flanks.
Bare parts: Bill pinkish to reddish, tip and cutting edges of upper bill blackish, juvenile has a black bill. Legs fleshy-brown. Eyes brown, juvenile greyish.
List: III Ghana
Lit.: 95, 129, 173, 268, 325, 410

6. LEUCOPSAR ROTHSCHILDI

Dutch: Balispreeuw; *English:* Bali Myna, Bali Mynah, Bali Starling, Rothschild's Mynah/Starling/Grackle; *French:* Martin de Rothschild; *Indonesian:* Jalak putih bali; *Italian:* Storno di Rothschild, Maina di Rothschild, Maina di Bali; *Spanish:* Estornino de Rothschild, Miná de Rothschild

Range: Bali
Identification:
Length 25cm. **Male:** Entire plumage white except for black wing- and tail tips.Crest long.
Female: Crest shorter.
Young birds: Crest shorter than female.
Bare parts: Bill dark yellowish horn, tip yellow, juvenile whole bill ivory. Legs greyish-blue. Eyes dark brown. Bare skin on face bright blue, juvenile lighter blue.
List: I
Status: Only about 200 birds on Bali 1990, but many in zoos.
Lit.: 37, 120, 129, 170, 410

PLATE 69

1. AMBLYOSPIZA ALBIFRONS

English: Grosbeak Weaver, Thick-billed Weaver, White-fronted Grosbeak; *French:* Gros-bec à front blanc; *German:* Weisstirnweber *Spanish:* Tejedor picogordo
Range: Africa south of Sierra Leone to Ethiopia
*Identification:*A. **Male:** Length 18cm. Dark sooty brown with white patches on wings. D. Often with white forehead. B. **Female:** Length 16cm. Upperparts dark brown. Underparts streaked grey-white and dark brown.
Young birds: Like female.
Bare parts: Large bill blackish horn, female paler, juvenile yellowish. Legs blackish. Eyes dark brown.
Geographical variation:
Western Africa: Birds east of Zaire and Angola have rufous-headed males.
East Africa: Birds from Sudan and Ethiopia and south to Zambia have brown-headed males. Central Africa: Birds from E. Zaire, Uganda, W. Kenya and south to W. Zambia have black-headed males.
List: III Ghana
Lit.: 7, 88, 129, 171, 172, 173, 174, 205, 268, 320, 325, 410

2. ANOMALOSPIZA IMBERBIS

English: Parasitic Weaver, Cuckoo Weaver; *French:* Tisserin-coucou; *German:* Kuckucksfink, Kuckucksweaver
Range: Sierra Leone to Ethiopia and south to South Africa
*Identification:*A. Length 13cm. **Male:** Head and above olive-yellow streaked dusky. Wings and tail dusky brown edged olive-yellow. Below plain yellow with paler tips to feathers just after moult. Lower flanks with dark shafts. Note: great variation in plumage depending on wear to feathers. B. **Female:** Buff stripe over eye. Upperparts buffy-tawny, streaked dusky. Chin, throat, and abdomen whitish. Rest of underparts buffish-brown streaked dark brown.
Young birds: Duller, more broadly streaked above. Lower flanks distinctly streaked.
Bare parts: Deep bill dark horn, juvenile lower bill yellowish. Legs brown. Eyes dark brown.
List: III Ghana
Lit.: 7, 88, 129, 171, 172, 173, 174, 320, 325, 410

3. BUBALORNIS ALBIROSTRIS

English: White-billed Buffalo Weaver, Buffalo Weaver, Black Buffalo Weaver; *French:* Alecto à bec blanc; *German:* Alektoweber, Büffelweber; *Spanish:* Tejedor Búfalo
Range: Senegal to Ethiopia, Uganda, Kenya
Identification: Length 24cm. Sexes alike. General plumage black, feathers with hidden white bases. Narrow white edges to greater flight-feathers. Many white feathers on flanks.
Young birds: Upperparts earth-brown, underparts mottled with white.
Bare parts: Bill of male white in breeding, outside breeding and female black. Legs brown. Eyes dark brown.
List: III Ghana

4. Female of: Bubalornis niger
Range: From Ethiopia and Sudan to South Africa
Male like last but bill red. Female greyish-brown above, whitish below with dusky streaks.
NO MORE LISTED
Lit.: 129, 171, 172, 173, 268, 320, 410

5. EUPLECTES A. AFER

Formerly known as: Euplectes afra
English: Yellow-crowned Bishop, Golden Bishop, Napoleon Bishop; *French:* Worabée ordinaire, Euplecte à tête jaune, Travailleur à tête rouge, Vorabé; *German:* Tahaweber, Napoleonweber; *Spanish:* Euplectes amarillo
Range: Senegal to Sudan, south to Zaire and Angola
Identification: A. Length 11.5cm. **Male:** Top of head and upperparts lemon yellow, hind neck and upper mantle mixed with black. Feathers of rump very long. Underparts black, chestnut area on mid-breast, flanks and under tail-coverts lemon. Male non breeding like female non breeding, but with more black streaks above and breast with band of heavy streaking. B. **Female:** Buff and dusky black above. Broad white or creamy stripe from bill over eye to nape. Underparts white tinged yellow, breast and flanks with some dusky streaks.
Female non breeding: Buff underparts with more streaking on breast and flanks.
Young birds: Browner than female. Buff edges to feathers of upperparts.
Bare parts: Bill black, male non breeding and female horn. Legs and eyes brown.
Subspecies:
Euplectes afer taha
Range: Sudan, Ethiopia to South Africa
Breast black with only little yellow on sides.
Geographical variation:
Birds from Sudan, Ethiopia, Uganda, Kenya and Tanzania have entire breast black.
List: III Ghana
Lit.: 88, 129, 171, 172, 174, 205, 268, 320, 325, 342, 410

6. EUPLECTES A. ARDENS

Also known as: Niobella ardens; Coliuspasser ardens
English: Red-collared Widowbird, Red-collared Whydah, Long-tailed Black Whydah; *French:* Veuve noire; *German:* Schildwida; *Spanish:* Falsa viuda pechirroja
Range: East and Southern Africa from Uganda to Cape Province
*Identification:*A. **Male:** Length 24-40cm (tail 21-29cm). Plumage black except an orange or red band across forebreast and white borders to wing feathers and upper and under tail-coverts. Tail graduated.
Male non breeding: Like female but tail longer and under tail-coverts dark with whitish margins. B. **Female:** Length 12cm. Upperparts tawny with black streaks. Eye-stripe pale buff. Underparts buff, breast darkest, chin and throat with yellow tinge. Belly whitish.
Young birds: Above broader and paler tawny streaks.
Bare parts: Bill and legs black, male non breeding and female brown. Eyes brown.

Subspecies:
Euplectes ardens concolor
English: Black Widowbird
Range: Senegal to Sudan, Uganda, Chad, south to Zambia
(Occur also at scattered localities in the range of Red-collared Widowbird)
Male in breeding without orange or red breast band.
Euplectes ardens laticauda
English: Red-naped Widowbird
Range: Sudan, Ethiopia, Kenya, Tanzania
Male in breeding-dress has red hindcrown and nape running to red breast band. Black feathers edged tawny. Female and juvenile have longer tails.
List: III Ghana
Lit.: 88, 129, 171, 172, 173, 174, 205, 268, 320, 325, 342, 410

7. EUPLECTES HORDEACEUS

Also known as: Euplectes hordeacea
English: Black-winged Bishop, Red-crowned Bishop, Fire-crowned Bishop, Black-crowned Red-Bishop; *French:* Monseigneur; *German:* Flammenweber; *Spanish:* Euplectes alinegro
Range: Senegal to Sudan, Ethiopia, Angola, Zimbabwe, Mozambique
*Identification:*A. Length 13-15cm. **Male:** Crown, neck, breast and rump scarlet, rarely orange. Mantle reddish-chestnut. Under tail-coverts white, brownish or buffish orange. Mask, ear-coverts, throat, wings, belly and tail black. Male non breeding differs from female in having black wings and tail.
B. **Female:** Above buff broadly streaked black. Stripe above eye yellowish. Below yellowish streaked buff on neck, breast and flanks. Wings and tail brown.
Young birds: Like female but with broader sandy edges to feathers of mantle.
Bare parts: Bill black, male non-breeding and female horn. Legs and eyes brown.
List: III Ghana
Lit.: 129, 171, 173, 174, 205, 268, 320, 325, 410

8. EUPLECTES M. MACROURUS

Also known as: Coliuspasser macrourus
English: Yellow-shouldered Widowbird, Yellow-mantled Whydah; *French:* Veuve à dos d'or; *German:* Gelbschulterwida; *Spanish:* Falsa viuda dorsigualda
Range: Senegal to Sudan, Kenya, Angola, Malawi
Identification: A. **Male:** Length 18-31cm. Male: Jet black except for yellow mantle and shoulders. Male non-breeding like female but with yellow shoulders.
B. **Female:** Length 14cm. Above buff streaked blackish, rump brown. Below yellowish-buff, faint dusky streaks on breast and flanks.
Young birds: Above pale buffish edges to feathers. Below tinged yellow.
Bare parts: Bill black, non-breeding and female horn. Legs black, non-breeding and female brownish. Eyes dark brown.
Subspecies:
Euplectes macrourus macrocercus
English: Yellow-shouldered Widowbird
Range: Ethiopia, Kenya, Uganda
Breeding male without yellow on mantle, only yellow shoulders. Tail longer.
List: III Ghana
Lit.: 129, 171, 173, 174, 205, 268, 320, 325, 410, 417

PLATE 70

1. EUPLECTES ORIX
Formerly known as: Euplectes franciscanus
English: Red Bishop, *French:* Ignicolore, *German:* Oryxweber

Range: Zaire, Angola, Uganda, Kenya, Tanzania to South Africa
Identification: A. Length 11cm. **Male:** Forehead, upper throat, breast and belly black. Back brownish-red. Wings and tail brown edged paler. Rest scarlet. **Female** and young birds: Like No. 2.
Bare parts: Like No. 2.
NO MORE LISTED

2. EUPLECTES FRANCISCANUS
Also known as: Euplectes orix franciscana
English: Orange Bishop, Franciscan Bishop, Northern Red Bishop; *German:* Feuerweber, Orangeweber

Range: Senegal to Somalia, Zaire, Uganda, Kenya
Identification: Length 10cm **Male:** Like No. 1 except scarlet throat (not black) and scarlet upper tail-coverts which cover whole tail in fresh plumage. (Orange-yellow males occur). Male non breeding: Like female but larger.
Female: Eye stripe buffish. Upperparts broadly streaked dusky and buff. Underparts buffish, belly whitish.
Young birds: Like female but buffish markings broader on upperparts.
Bare parts: Bill black, male non-breeding and female horn. Legs reddish-brown. Eyes dark brown.

Geographical variation:
Birds from S. E. Ethiopia and Somalia are paler orange-red. Tail-coverts shorter, not extending to tip of tail.
List: III Ghana
Lit.: 88, 129, 171, 172, 174, 195, 205, 320, 325, 410

3. MALIMBUS CASSINI
English: Black-throated Malimbe, Cassin's Malimbe; *French:* Tisserin-Malimbe de cassin; *German:* Cassinweber, Cassinprachtweber

Range: Cameroon, Gabon, Congo, Zaire
Identification: A. Length 17cm. **Male:** Forehead, crown, neck, and breast scarlet. Rest black.
B. **Female:** Entire plumage black.
Young birds: Black with orange on head and neck.
Bare parts: Bill black. Legs dark grey. Eyes dark brown.
List: III Ghana
Lit.: 72, 129, 173, 410

4. MALIMBUS SCUTATUS
English: Red vented Malimbe; *French:* Le Tisserin-Malimbe à croupion rouge, Malimbe à Queue; *German:* Schildweber; *Spanish:* Malimbo culirrojo

Range: Sierra Leone to Cameroon
Identification: A. Length 17cm. **Male** like number 3, cassini, but under tail-coverts also scarlet. **Female:** Scarlet patch on breast. Rest black.
Young birds: Pale pink where adults are scarlet.
Bare parts: Bill black, juvenile horn. Legs and eyes dark brown.

Geographical variation:
Birds from Nigeria and Cameroon have a larger scarlet breast patch.
B. Female: Red breast patch divided by a median black stripe.
List: III Ghana
Lit.: 72, 129, 173, 268, 320, 410

5. MALIMBUS MALIMBICUS
English: Crested Malimbe; *French:*Tisserin-Malimbe huppé, Malimbe Huppé; *German:* Haubenweber, Haubenprachtweber; *Spanish:*Malimbo moñudo

Range: Sierra Leone to Uganda, Zaire, Angola
Identification: A. Length 17cm. **Male:** Hind crown feathers longish, chest-like, side of head, throat and upper breast scarlet. Stripe from bill to behind eye, rest of body, wings and tail black. Belly with a brownish wash.
B. **Female:** Duller, feathers on hind crown not elongated.
Young birds: Head mixed black and scarlet.
Bare parts: Bill black, juvenile horn. Legs black. Eyes brown.

Geographical variation:
Birds from Sierra Leone to Nigeria have black belly without brownish tinge. Male has shorter crest.
List: III Ghana
Lit.: 72, 129, 173, 268, 320, 325, 410

6. MALIMBUS NITENS
English: Gray's Malimbe, Blue-billed Malimbe; *French:* Tisserin-Malimbe à bec bleu, Malimbe à Bec Bleu; *German:* Rotkehlweber; *Spanish:* Malimbo cabecinegro

Range: Senegal to Cameroon, Uganda, Zaire
Identification: Length 17cm. **Male:** Patch on neck and forebreast scarlet, rest black.
Female: Duller black.
Young birds: Red patch larger. Head black with red tinge.
Bare parts: Bill bluish. Legs dark grey. Eyes red-brown to red.
List: III Ghana
Lit.: 72, 129, 173, 268, 320, 410

7. MALIMBUS RUBRICOLLIS
English: Red-headed Malimbe; *French:* Tisserin-Malimbe à tête rouge, Malimbe à tête rouge; *German:* Kletterweber, Kletter-Prachtweber; *Spanish:* Malimbo cuellirrojo

Range: Guinea-Bissau to Sudan, Uganda, Kenya, Angola
Identification: A. Length 20cm. **Male:** Forehead, crown, hind neck, and sides of neck orange-red. Rest of plumage black.
B. **Female:** Like male but without red on forehead and fore-crown.
Young birds: Duller and paler orange on head and side of neck.
Bare parts: Bill black, juvenile horn. Legs dark brown. Eyes red to red-brown.

Geographical variation:
Birds from Sierra Leone to Ghana are purplish-crimson, not orange-red.
Birds from Benin and Nigeria have strong bills. Head crimson-red.
Birds from Bioko have longest bill: 20.5 - 24.0mm, against 19.0 - 21.5 mm in rubricollis.
List: III Ghana
Lit.: 72, 129, 171, 173, 268, 320, 410

8. ANAPLECTES R. RUBRICEPS
Formerly known as: Anaplectes melanotis; Malimbus rubriceps
English: Red-headed Weaver; *French:* Anaplecte à ailes rouges, Tisserin à ailes rouges; *German:* Scharlachweber; *Spanish:* Malimbo capirotado

Range: Angola to Mozambique, Zimbabwe, Botswana, South Africa
Identification: A. Length 15cm. **Male:** A stripe from bill to behind eye, ear-coverts, and chin black, rest of head, neck, and breast scarlet. Mantle grey-black and red. Wings and tail grey, feathers edged scarlet. Belly white. Male non breeding: Mantle grey or brownish-grey.
B. **Female:** Head brownish. Upperparts grey, edges of wing and tail feathers paler red or yellowish. Underparts white, breast with a grey tinge.
Young birds: Head olivaceous or yellow.
Bare parts: Bill red to orange-red, juvenile dusky. Legs grey-brown. Eyes, dark brown.

Subspecies:

Anaplectes rubriceps leuconotus
Range: Senegal to Somalia, Zaire, Uganda, Kenya, Tanzania
Face and chin black in males.

Geographical variation:
Male bird from Somalia and Kenya has abdomen scarlet.
Male bird from Angola to Tanzania and Mozambique has little or no black on head. Mantle black and red and wing and tail feathers edged yellow. Female has yellowish head, neck, and upper breast.
List: III Ghana
Lit.: 72, 129, 171, 174, 268, 320, 410

PLATE 71

1. PASSER GRISEUS

Also known as: Pyrgitopsis grisea
English: Grey-headed Sparrow; *French:* Moineau de l'ouest africain, Moineau gris; *German:* Graukopfsperling; *Spanish:* Gorrión pardillo

Range: Senegal to Ethiopia, Kenya, Zaire, Angola
Identification:
Length 16cm. Sexes similar. Head and neck grey. Mantle grey-brown, lower back and rump reddish-brown. Wings and tail brown edged paler. Underparts pale buffish-grey to white on chin, throat and belly.
Young birds: Dusky streaks on mantle.
Bare parts: Bill black to horn. Legs pale brown. Eyes light brown.
List: III Ghana
Lit.: 129, 173, 268, 289, 410, 417

2. PETRONIA DENTATA

Also known as: Gymnoris dentata
English: Bush Petronia, Lesser Rock Sparrow; *French:* Petit moineau soulcie; *German:* Buschsperling, Kleiner Kehlsperling; *Spanish:* Chillón de Sundewall

Range: Mauritania, Senegal to Ethiopia, Uganda, Yemen
Identification:
A. Length 12.5cm. **Male:** Forehead and crown grey. Stripe over and behind eyes and mantle red-brown. Wings and tail brown with paler edges. Below greyish-white with a small lemon-yellow patch on throat.
B. **Female:** Top of head brown, eye-stripe pale buff. Mantle brown with dusky black markings. Pale yellow spot on throat (can be difficult to detect).
Young birds: Like female, but without yellow area on throat.
Bare parts: Bill dark horn, outside breeding and female horn. Legs dark brown. Eyes brown.
List: III Ghana
Lit.: 129, 173, 268, 410, 417

3. PLOCEPASSER SUPERCILIOSUS

English: Chestnut-crowned Sparrow-Weaver, Sparrow-Weaver; *French:* Moineau-tisserin; *German:* Braunwangenweber; *Spanish:* Tejedor pardal capirotado

Range: Senegal to Ethiopia, Mali, Chad, Uganda, Kenya
Identification:
Length 17cm. Sexes alike. Head chestnut, stripe over eye white. Conspicuous black malar stripe. Upperparts earth-brown. Wings and tail brown, two bars on wing white. Underparts greyish-white, throat white.
Young birds: Chestnut on head paler.
Bare parts: Bill horn, lower paler. Legs brown. Eyes reddish-brown.
List: III Ghana
Lit.: 7, 35, 129, 171, 173, 268, 410, 417

4. PLOCEUS ALBINUCHA

Also known as: Melanoploceus albinucha
English: Maxwell's Black Weaver, White-naped Weaver, Black Weaver; *French:* Tisserin noir de Maxwell, Tisserin à nuque blanche; *German:* Trauerweber; *Spanish:* Tejedor negro albinuca

Range: Sierra Leone to Cameroon, Gabon, Zaire, Uganda
Identification:
Length 14cm. Sexes similar. Entire plumage black, only with a whitish spot on nape.
Young birds: Brown above, greyish-brown below.
Bare parts: Bill black. Legs dirty flesh. Eyes greyish-white.

Geographical variation:
Birds from Bioko have no whitish on nape. Young birds sooty-grey or yellowish-olive underparts.
Males from Nigeria to Gabon, Central African Republic and Zaire have no whitish area on nape.
List: III Ghana
Lit.: 7, 129, 173, 268, 320, 325, 410, 417

5. PLOCEUS AURANTIUS

Also known as: Textor aurantius; Xanthophilus aurantius
English: Orange Weaver; *French:* Tisserin orangé; *German:* Königsweber; *Spanish:* Tejedor naranja

Range: Sierra Leone to Cameroon, Zaire, Uganda, Angola, Tanzania
Identification:
A. Length 15cm. **Male:** Head, rump and underparts orange-yellow. Black spot between bill and eye. Mantle olive. Wings black broadly edged with yellow. Throat orange-yellow washed with chestnut. Tail blackish.
B. **Female:** Upperparts olive-green. Yellow eye-stripe. Underparts white with irregular yellow on breast, throat yellow.
Young birds: Like female.
Bare parts: Bill blackish brown, female brownish horn, paler below. Legs pale brown. Eyes red to reddish-brown.

Geographical variation:
Birds from Uganda and Tanzania are more yellow above and without black between bill and eye.
List: III Ghana
Lit.: 7, 129, 171, 173, 268, 320, 325, 410, 417

6. PLOCEUS C. CUCULLATUS

Also known as: Textor cucullatus; Ploceus nigriceps
English: Village Weaver, Black-headed Weaver, Spot-backed Weaver; *French:* Tisserin des villages, Gendarme, Tisserin gendarme; *German:* Textor, Dorfweber; *Spanish:* Tejedor cogullado

Range: Senegal to Ethiopia, Congo, Zaire, Kenya, Tanzania
Identification:
A. Length 17cm. **Male:** Head and most of neck black. Collar chestnut. Upperparts yellow with a V-shaped black band; rump yellow. Wings blackish edged yellow. Underparts yellow. Tail green. Male non breeding: Top and sides of head olivaceous-yellow. Upperparts olive-brown with dusky streaks. Underparts yellow, belly white.
B. **Female:** Like male non breeding.
Young birds: Like female, but mantle olivaceous earth-brown.
Bare parts: Bill black, non breeding and female horn to dark brown. Legs pinkish-brown. Eyes red to orange.

Subspecies:

Ploceus cucullatus collaris
English: Mottled Weaver
Range: Gabon to Angola
No chestnut collar, breast chestnut.

Ploceus cucullatus nigriceps
English: Layard's Black-headed Weaver
Range: Zaire, Zambia, Tanzania, Kenya, Somalia, Mozambique
No chestnut collar. Upperparts mottled yellow and black.

Ploceus cucullatus spilonotus
English: Spot-backed Weaver
Range: South Africa
Forehead and crown yellow.
List: III Ghana
Lit.: 129, 173, 174, 205, 268, 320, 325, 410, 417

7. PLOCEUS HEUGLINI

English: Heuglin's Masked Weaver; *French:* Tisserin masqué; *German:* Heuglingweber, *Spanish:* Tejedor de Heuglin

Range: Mauritania, Senegal to Sudan, Uganda, Kenya
Identification:
A. Length 14cm. **Male:** Sides of face and throat black. Forehead, crown and underparts yellow. No chestnut on crown and throat.
Male non breeding: Lacks black mask, throat yellow.
B. **Female:** Without black on face and throat. Above olivaceous-green with dusky streaks. Below yellow, on breast with buffish tinge.
Young birds: Like female.
Bare parts: Bill black, female horn. Legs brown. Eyes pale yellow.
List: III Ghana
Lit.: 129, 173, 268, 325, 410

PLATE 72

1. PLOCEUS LUTEOLUS

Formerly known as: Sitagra luteola

English: Little Weaver, Little Masked Weaver; *French:* Tisserin minulle, Le Tisserin à bec grêle; *German:* Zwergweber, Zwergmaskenweber; *Spanish:* Sitagra chico común

Range: Senegal to Ethiopia, Zaire, Uganda, Kenya, Tanzania

Identification: A. Length 11cm. **Male:** Forehead and forecrown, face, and throat black. Hind crown, sides of neck, and underparts yellow. Mantle olive-green. Male non-breeding: Head dull olivaceous. Upperparts ashy with dusky streaks. Underparts white tinged with buff on breast and flanks, under tail-coverts yellow.

B. **Female:** Top of head, face, chin to breast pale yellow. Rest of underparts yellow. Female non-breeding: Like male in non-breeding dress.

Young birds: Like adult in non-breeding.

Bare parts: Bill black, adult non-breeding horn. Legs bluish-grey. Eyes hazel.

List: III Ghana

Lit.: 129, 173, 268, 320, 325, 410, 417

2. PLOCEUS M. MELANOCEPHALUS

Also known as: Textor melanocephalus; Sitagra melanocephala

English: Black-headed Weaver; *French:* Tisserin à tête noire, *German:* Schwarzkopfweber, Kleiner Textor; *Spanish:* Tejedor cabecinegro

Range: Senegal to Chad and south to Zambia

Identification: A. Length 15cm. **Male:** Entire head and throat black. Collar, rump, and upper tail-coverts yellow. Back olive-yellow. Wings black edged yellow. Breast and belly yellow. Tail olivaceous-green.

Male non-breeding: Head olive-green with yellow eye-stripe. Back earth-brown with dusky stripes. Buffish below, throat and centre of belly white.

B. **Female:** Like male non-breeding, but smaller.

Young birds: Duller than female.

Bare parts: Bill black. Legs flesh. Eyes brown to pale yellow.

Subspecies:

Ploceus melanocephalus capitalis

English: Yellow-collared Weaver

Range: Nigeria to Tanzania, Kenya, Sudan, Ethiopia

Conspicuous yellow hind neck band and chestnut on breast and flanks.

List: III Ghana

Lit.: 129, 173, 268, 320, 325, 410, 417

3. PLOCEUS N. NIGERRIMUS

Also known as: Textor nigerrimus; Melanopteryx nigerrimus; Cinnamopteryx castaneofuscus

English: Vieillot's Black Weaver; *French:* Tisserin noir de Vieillot; *German:* Mohrenweber; *Spanish:* Tejedor negro de Vieillot

Range: Nigeria, Cameroon to Kenya, Zaire, Angola

Identification: A. 18cm. **Male** Uniformly black. B. **Female:** Upperparts mottled brown and blackish, underparts olive.

Young birds: Browner above than female.

Bare parts: Bill black, female horn. Legs pale brown. Eyes pale yellow to whitish.

Subspecies:

4. Ploceus nigerrimus castaneofuscus

English: Chestnut-and-Black Weaver

Range: Guinea to Nigeria

Male has chestnut mantle, rump, and belly.

List: III Ghana

7, 129, 173, 268, 320, 325, 410, 417

5. PLOCEUS N. NIGRICOLLIS

Also known as: Hyphanturgus nigricollis

English: Black-necked Weaver, Spectacled Weaver; *French:* Tisserin à lunettes, Tisserin à cou noir; *German:* Kurzflügelweber; *Spanish:* Tejedor cuellinegro

Range: Central African Republic to Somalia, Kenya, Tanzania, Zaire, Angola

Identification: A. Length 15cm. **Male:** Between bill and eye, round eye, throat, and hind neck black. Rest of head yellow to chestnut on forecrown. Upperparts olivaceous black, rump and upper tail-coverts greenish. Wings dusky edged greenish. Underparts yellow. Tail above dusky, below yellow.

B. **Female:** Like male, but lacks black throat. Top of head, neck, and stripe through eye black.

Young birds: Entire upperparts olivaceous yellow-green.

Bare parts: Bill black. Legs grey-brown. Eyes brown.

Subspecies:

6. Ploceus nigricollis brachypterus

English: Swainson's Weaver

Range: Senegal to Cameroon and Bioko

No black hind neck and entire upperparts yellowish-green. Female: Crown and upperparts olive-green.

List: III Ghana

Lit.: 7, 129, 173, 268, 320, 325, 410, 417

7. PLOCEUS P. PELZELNI

Also known as: Sitagra pelzelni

English: Slender-billed Weaver; *French:* Tisserin Nain, Tisserin à bec grêle de loango; *German:* Mönchsweber; *Spanish:* Sitagra de Pelzeln

Range: Sudan, Uganda, Kenya, Tanzania, Zaire

Identification: A. Length 12cm. Forehead, forecrown, face, and throat black. Hind crown, nape, and sides of neck yellow. Upperparts and tail green. Wings dusky edged green. Underparts yellow.

B. **Female:** Like male but whole head yellow with a fine streak above eye.

Young birds: Like female except bill colour.

Bare parts: Slender bill black, juvenile horn. Legs dark brown. Eyes brown.

Subspecies:

Ploceus pelzelni monachus

English: West African Weaver

Range: Senegal to Zaire, Angola, Zambia

Smaller, wing 54-59mm, against 58-66mm in pelzelni. Legs grey.

List: III Ghana

Lit.: 6, 129, 173, 268, 320, 325, 410, 417

8. PLOCEUS PREUSSI

Also known as: Phormoplectes preussi

English: Preuss's Weaver, Golden-backed Weaver; *French:* Tisserin de Preuss; *German:* Preussweber

Range: Sierra Leone, Cameroon to Congo, Zaire

Identification: A. Length 15cm. **Male:** Top of head golden chestnut. Face, ear-coverts, and throat black. Body yellow with a tinge of chestnut on lower throat. Wings and tail black.

B. **Female:** Like male but with a broad black forehead and only a slight tinge of chestnut on head and throat.

Young birds: Like female.

Bare parts: Bill black, juvenile horn. Legs brownish-pink. Eyes dark red-brown.

List III Ghana

Lit.: 7, 129, 173, 410, 417

9. PACHYPHANTES SUPERCILIOSUS

Also known as: Ploceus superciliosus

English: Compact Weaver; *French:* Tisserin Gros-bec; *German:* Braunbürzelweber; *Spanish:* Tejedor rollizo

Range: Sierra Leone to Ethiopia, Uganda, Kenya, Angola, Zambia

Identification: A. Length 15cm. **Male:** Forehead chestnut. Crown and behind ear-coverts yellow. Between bill and eye, ear-coverts and throat black. Nape and upperparts olive-green streaked dark brown.

Wings and tail dusky edged buffish-white. Underparts yellow. Male non-breeding: Top of head and streak through eye blackish. Stripe over eye, face and below buffish-brown. Rest brown streaked dusky.

B. **Female:** Top of head mixed blackish-green and yellow. Otherwise like male in breeding. Female non-breeding: Like male non-breeding.

Young birds: Like non-breeding dress of parents but with a stripe over eye, sides of neck and below pale yellow.

Bare parts: Short thick bill blackish-slate, non-breeding horn or slate. Legs brown. Eyes brown.

List: III Ghana

Lit.: 129, 173, 268, 410

PLATE 73

1. PLOCEUS TRICOLOR

Also known as: Melanoploceus tricolor; Melanoploceus fuscocastaneus
English: Yellow-mantled Weaver; *French:* Tisserin tricolore, *German:* Dreifarbenweber; *Spanish:* Trejedor tricolor
Range: Sierra Leone to Cameroon, Sudan, Kenya to Angola and Zaire
Identification: Length 16cm. **Male:** Entire head, throat, hind mantle, rump, wings, and tail black. Upper mantle bright yellow. Breast and belly chestnut. Under tail-coverts dark brown. Note: has no non-breeding dress.
Female: Like male except for colour on legs and eyes.
Young birds: Rufous-brown, hind mantle, rump, wings and tail dull black.
Bare parts: Bill black. Legs dark brown, female pale brown. Eyes brown, female dark brown.

Geographical variation:
2. Female birds from Uganda and Zaire are sooty-black on breast and belly. Male does not vary. Young birds more blackish.
List: III Ghana
Lit.: 129, 173, 268, 320, 325, 410, 417

PLOCEUS VITELLINUS

Also known as: Ploceus velatus; Ploceus katangae; Ploceus reichardi; Ploceus ruweti; Textor vitellinus
English: Vitelline Masked Weaver; *French:* Tisserin à tête rousse; *German:* Dotterweber; *Spanish:* Tejedor velado
Range: Mauritania, Senegal to Somalia, Uganda, Kenya, Tanzania
Identification: Length 15cm. **Male:** Forehead chestnut, crown golden chestnut, nape yellow. Black mask and upper throat. Upperparts olive-green. Wings dusky edged yellow. Underparts bright yellow. Tail olive-green. Male non-breeding: Upperparts olivaceous-green streaked blackish, underparts pale yellow, abdomen white. **Female:** Like male non breeding but breast and flanks buffish and yellow.
Young birds: Like female.
Bare parts: Bill black, male non-breeding and female horn. Legs pinkish-brown. Eyes orange to yellow.
List: III Ghana

3. Ploceus velatus

English: Southern Masked Weaver
Range: South Africa
A. **Male:** Differs from last one by having black forecrown and larger area of black on throat.
B. **Female:** Like other female Ploceine Weavers.
NO MORE LISTED
Lit.: 41, 88, 129, 172, 173, 174, 268, 320, 325, 410, 417

4. QUELEA ERYTHROPS

Also known as: Queleopsis erythrops
English: Red-headed Quelea, Red-headed Dioch; *French:* Travailleur à tête rouge; *German*: Rotkopfweber; *Spanish*: Laborioso cabecirrojo
Range: Senegal to Ethiopia and south to Cape Province
Identification: A. Length 12cm. **Male:** Entire head and throat scarlet, often with some black on chin and throat. Upperparts buff streaked with blackish-brown. Wings brown, flight-feathers edged yellowish-white. Underparts buffish-white. Male non-breeding: Lacks red on head. Top of head brownish with dusky streaks. Stripe over eye and ear-coverts yellowish-buff. Upperparts buffish streaked blackish. Underparts buffish with fine streaks on flanks, belly white. Males with red tinge on head are rarely seen.
B. **Female:** Like male in non-breeding dress.
Young birds: Upperparts pale sandy with blackish streaks.
Bare parts: Bill black, male non-breeding and female horn. Legs pale brown. Eyes brown.

Geographical variation:
Birds from Tanzania to Mozambique and Gabon to Natal have a darker crown, more vinous-red. Throat patch not so sharply contrasted and upperparts browner.
List: III Ghana
Lit.: 88, 129, 173, 174, 205, 268, 325, 410, 417

5. SPOROPIPES FRONTALIS

English: Speckle-fronted Weaver, Scaly-fronted Weaver; *French:* Tisserin à front pointillé, Moineau quadrillé; *German:* Schuppenköpfchen; *Spanish:* Tejedorcito punteado
Range: Senegal, Mali, Niger, Nigeria, Chad to Ethiopia, Uganda, Kenya, Tanzania
Identification: Length 12cm. Sexes similar. Forehead, forecrown and malar stripe black with tiny white spots. Hind crown and nape cinnamon-rufous. Face and upperparts pale brown. Underparts whitish, breast greyish. Wings and tail dusky edged buffish-white.
Young birds: Hind crown and nape paler tawny.
Bare parts: Bill pale brown. Legs pale brown. Eyes brown.
List: III Ghana
Lit.: 129, 171,173, 268, 320, 325, 410

6. AMADINA FASCIATA

English: Cut-throat, Cut-throat Finch/Weaver; *French:* Cou coupé; *German:* Bandamadine, Bandfink; *Spanish:* Amadina gorgirroja

Range: Senegal, Mali, Niger, Nigeria to Somalia, Kenya to South Africa
Identification: Length 11cm. **Male:** Forehead whitish, barred black. Crown, nape, and upperparts pale cinnamon barred black. A conspicuous crimson-red (cut-throat) band from below ear-coverts to lower throat. Rest of underparts pale buffish, middle of abdomen chestnut, flanks and upper belly barred black. Tail blackish with white tips and white edges on outermost feathers.
Female: Paler than male and lacks crimson-red band. Belly white or white with a paler chestnut spot.
Young birds: Males with paler red band and paler chestnut abdomen. Females like adult females.
Bare parts: Bill pale grey tinged pink. Legs pale brown. Eyes red-brown.

Geographical variation:
Birds from Ethiopia, Somalia, Kenya and Tanzania are darker and more coarsely barred.
Birds from Malawi, Zambia, Zimbabwe, Mozambique and Transvaal are browner above, less reddish.
List: III Ghana
Lit.: 88, 129, 172, 173, 174, 268, 320, 325, 410

7. AMANDAVA SUBFLAVA

Also known as: Estrilda subflava; Sporaeginthus subflavus; Neisna subflava
English: Zebra Waxbill, Orange-breasted/Golden-breasted Waxbill, Goldbreast; *French:* Astrild à flancs rayés, Ventre orange; *German:* Goldbrüstchen, *Spanish:* Astrilda pechigualda

Range: Senegal to Chad and Ethiopia, Kenya to South Africa
Identification: A. Length 9cm. **Male:** Top of head,upperparts and wings brown with a green wash. Stripe from bill to behind eye, upper and under tail-coverts, and a patch on forebreast crimson. Underparts yellow tinged orange, chin and throat bright yellow, flanks brownish-green barred yellow. Tail blackish.
B. **Female:** Lacks red eye-stripe, and underparts paler yellow.
Young birds: Like female but underside buff without barred flanks.
Bare parts: Bill red, black on ridge of upper bill and underside of lower bill, juvenile has blackish bill. Legs pale brownish-pink. Eyes red.

Geographical variation:
Birds from Angola to Mozambique and South Africa: Male yellow underside, less orange, but red on breast and under tail-coverts.
List: III Ghana
Lit.: 88, 129, 172, 174, 268, 320, 410, 417

8. POEPHILA C. CINCTA

English: Black-throated Finch, Parson Finch; *German:* Gürtelamadine, Gürtelgrasfink
Range: Queensland, New South Wales, (Australia)
Identification: Length 10cm. **Male:** Head blue-grey, between bill and eye, chin, and throat black. Upperparts and wings pale red-brown-vinaceous. A black conspicuous band from rump to thighs. Upper tail-coverts white characteristic of this race. Short tail black.
Female: Smaller, black patch on throat rounder.
Young birds: Duller.
Bare parts: Bill black. Legs orange-red. Eyes dark.
List: II
Lit.: 129, 278, 410

Note: the common Black-throated Finch, Poephila cincta atropygialis, with black upper tail-coverts is not listed.

PLATE 74

1. VIDUA C. CHALYBEATA

Also known as: Hypochera chalybeata; Vidua amauropteryx; Vidua centralis; Vidua neumanni; Vidua okavangoensis; Vidua ultramarina
English: Village Indigobird, Green Indigobird; *French:* Combasson du Sénégal; *German*: Rotfuss-Atlaswitwe; *Spanish*: Tejedor bruñido
Range: Senegal to Cameroon
Identification: A. Length 11cm. **Male:** plumage steel blue, often with greenish, green-blue or violet wash. Wing- and tail feathers blackish edged narrow brown on long flight-feathers. Innermost under-wing feathers white. Male non-breeding: like female. B. **Female:** Crown brown with broad pale central streak. Upperparts pale buff with dark brown centres to feathers. Underparts buffish, throat and belly white.
Young birds: No central streak on crown. Browner and not so clear streaked.
Bare parts: Bill pinkish-white. Legs varying from bright orange to salmon pink. Eyes brown.
Brood host: Red-billed Firefinch, Lagonosticta senegala

Subspecies:
Vidua chalybeata ultramarina
English: Purple Indigobird
Range: Chad to Sudan and Ethiopia
Strongly glossy pure blue. Female more sharply streaked above.

Vidua chalybeata amauropteryx
English: South African Indigobird
Range: Somalia, Uganda, Kenya, Zaire, Angola and N. E. South Africa
Wing- and tail feathers paler brown. Bill coral red. Legs usually reddish.
List: III Ghana
Lit.: 82, 129, 171, 172, 173, 174, 268, 320, 410, 422

2. VIDUA W. WILSONI

Also known as: Hypochera wilsoni; Hypochera funerea wilsoni; Vidua chalybeata wilsoni
English: Pale-winged Indigobird, Wilson's Indigobird; *French:* Combasson noir, Combasson noir de Wilson; *German:* Wilsonatlaswitwe; *Spanish:* Tejedor bruñido de Wilson
Range: Senegal to Ethiopia
Identification: A. Length 11cm. **Male:** Like last, Village Indigobird, but general colour is mostly glossy purplish and wing- and tail feathers greyish-brown. Male non-breeding resembles female. B. **Female:** Crown blackish-brown with broad central stripe. Upperparts rufous-buff with dusky feather-centres. Underparts dark buff, white in centre of belly.
Young birds: Like female.
Bare parts: Bill whitish to pale pink. Legs pinkish-white, pale purplish to dull red. Eyes dark brown.
Brood host: Bar-breasted Firefinch, Lagonosticta rufopicta.

Subspecies:
Vidua wilsoni incognita
English: Zaire Indigobird.
Range: S. E. Zaire
Larger, wing length 69mm (1 measured) against 60-65mm (16 measured) in last. **Victim host:** Brown Firefinch, Lagonosticta nitidula.
List: III Ghana
Lit.: 82, 129, 173, 220, 268, 320, 410, 417, 422

3. VIDUA RARICOLA (new species)

Formerly known as: Hypochera nigeriae; Hypochera chalybeata camerunensis; Hypochera chalybeata sharii.
English: Jambandu Indigobird
Range: Sierra Leone to Cameroon, Zaire (?), Sudan (?)
Identification: A. Length 11cm. **Male:** Like Vidua wilsoni, but wing feathers pale brown, vanes grey-brown. Head and body blackish with a glossy green (predominant) or blue tinge. Male non-breeding: Like female. B. **Female** and young birds: Resembles that of other indigobirds.
Nestling: Five black spots in mouth surrounded by two reddish spots, and blue and grey-pink colour like that of **brood host:** Black-bellied Firefinch, Lagonosticta rara, characterize this species. Song of adult birds also mimics L. rara.
Bare parts: Bill white. Legs reddish-grey. Eyes dark brown.
List: III Ghana
Lit.: 220, 410, 422

4. VIDUA LARVATICOLA (new species)

Formerly known as: Hypochera nigeriae; Hypochera chalybeata camerunensis; Hypochera chalybeata sharii
English: Baka Indigobird
Range: Probably widespread from Guinea Bissau to Sudan and Ethiopia
Identification: A. Length 11cm. **Male:** Like Vidua raricola, but with white flank patch. Flight-feathers pale brown, edged a little more grey, vanes more brown than raricola. Male non-breeding like female. B. **Female** and young birds: like other females of indigobirds.
Nestlings: Five black spots in mouth surrounded by blue, yellow, and orange, just like mouth pattern of **brood host:** Black-throated Firefinch, Lagonosticta larvata, characterize this species. Song of adult birds also mimics L. larvata.
Bare parts: Bill white. Legs light purplish with slight brown tinge. Eyes dark brown.
List: III Ghana
Lit.: 220, 410, 422

5. VIDUA MACROURA

English: Pin-tailed Whydah; *French:* Veuve dominicaine; *German:* Dominikanerwitwe; *Spanish:* Viuda colicinta
Range: Senegal to Somalia and south to South Africa
Identification: A. **Male:** 12-29cm, (four elongated central tail-feathers 15-17cm long). Top of head, nape, mantle, wings, patch on each side of breast, and tail glossy black. Face, hind neck, area on forewings and entire underparts white. Shorter tail feathers edged white. D. **Male non-breeding:** Nearly all black feathers edged broadly with tawny. Tail without long central feathers. B. **Female:** Length 11cm. Like male in non-breeding dress but streaking above not so distinct. Only very seldom with blackish-brown elongated tail feathers.
Young birds: Upperparts earth-brown with faint streaking on mantle. Underparts buff. Tail without white.
Bare parts: Bill red to orange, non-breeding and female pinkish, juvenile black to reddish. Legs grey to brownish-black. Eyes brown.
Brood host: Common Waxbill, Estrilda astrild, probably also other waxbills.
List: III Ghana
Lit.: 129, 173, 268, 410

6. Vidua paradisaea
See below.

7. VIDUA ORIENTALIS

Also known as: Vidua paradiseae orientalis; Steganura paradiseae
English: Northern Paradise-Whydah, Broad-tailed Paradise Whydah; *French:* Veuve à collier d'or; *German:* Senegal-Paradisswitwe, Spitzschwanz-Paradiswitwe, Grosse Paradiswitwe; *Spanish:* Viuda del paraíso oriental
Range: Senegal to Ethiopia
Identification: A. **Male:** Length 13-36cm, (tail 19-36cm). Head, throat, upperparts, wings and tail black, except tawny-rufous hind neck, and brown long flight-feathers. Breast chestnut, belly creamy. Shaft of elongated broad-tail feathers about in middle. Male non-breeding: Broad creamy central streak from forehead to nape bordered black. Side of head and throat creamy white. Upperparts tawny with dusky streaks. Breast and flanks deep tawny with black streaks. No elongated tail feathers. B. **Female:** Length 12cm. Sparrow-like.
Young birds: Pale earth-brown on wings and tail.
Bare parts: Bill black. Legs black to brown. Eyes brown.
Brood host: Green-winged Pytilia, Pytilia melba.
List: III Ghana
Lit.: 107, 129, 173, 207, 219, 222, 268, 320, 410, 417

VIDUA TOGOENSIS (not depicted)
Formerly known as: Vidua orientalis togoensis
English: Togo Paradise-Whydah
Range: Sierra Leone to Togo
Identification:
Like last, Vidua orientalis, but central tail feathers about 29-36cm long und hind neck dark golden chestnut.
Brood host: Red-faced Pytilia, Pytilia hypogrammica
List: III Ghana
Lit.: 107, 129, 173, 207, 219, 222, 268, 320, 410, 417

VIDUA INTERJECTA (not depicted)
Also known as: Vidua orientalis interjecta; Steganura interjecta
English: Long-tailed Paradise-Whydah; *German:* Langschwanz-Paradieswitwe
Range: Nigeria to Central African Republic and Zaire
Identification: Like Vidua orientalis, but central tail feathers longer, about 26-36cm, and broader. Hind neck golden brown.
Brood host: Red-winged Pytilia, Pytilia phoenicoptera.
List: III Ghana
Lit.: 107, 129, 173, 207, 219, 222, 268, 320, 410, 417

6. Vidua paradisaea
Also known as: Steganura paradiseae
English: Eastern Paradise Whydah, *German:* Schmalschwanz Paradieswitwe
Range: Sudan, Ethiopia to South Africa
Identification: A. Length 13-41cm. Like number 7, orientalis, but elongated tail feathers narrower. Narrower towards tip and shaft not in middle. Female and young birds sparrow-like and not distinguishable from other Whydahs.
Bare parts: Like orientalis.

NO MORE LISTED
Lit.: 107, 129, 173, 207, 325, 410, 417

PLATE 75

Family:
PARADISAEIDAE

English: Birds of Paradise; *French:* Paradisier; *German:* Paradiesvögel

Identification:
Length 15 to 107cm. In most species sexes are not alike, female duller, often barred without metallic colour of male and tail shorter. In this family are some of the most ornate and colourful assemblage of birds of the world. Bill and tail very variable. Legs crow-like. Wings medium and rounded.

Skins of this family arrived in Europe in large quantity in the commercial plume trade from 1522, until 1924 when the trade was banned. Because native-made trade skins all had legs removed people supposed the birds were never to alight, but to live on the wing and fly continually towards the sun and paradise! This gave the whole family its name. In 1758 Linnaeus gave one of the best known species, Greater Bird of Paradise, the name Paradisaea apoda, which means the footless Paradise Bird. (No. 20)

PLATE 76

LAWE'S PAROTIA 16

All Birds of Paradise are listed:

	Melampitta lugubris	Lesser Melampitta
	Melampitta gigantea	Greater Melampitta
2.	Loboparadisea sericea	Yellow-breasted Bird-of-Paradise
3.	Cnemophilus macgregorii	Crested Bird-of-Paradise
1.	Cnemophilus loriae	Loria's Bird-of-Paradise
4.	Macgregoria pulchra	MacGregor's Bird-of-Paradise
5.	Lycocorax pyrrhopterus	Paradise-Crow
	Manucodia atra	Glossy-mantled Manucode
	Manucodia chalybata	Crinkle-collared Manucode
6.	Manucodia comrii	Curl-crested Manucode
	Manucodia jobiensis	Jobi Manucode
7.	Manucodia keraudrenii	Trumpet Manucode
9.	Semioptera wallacii	Standardwing
	Paradigalla carunculata	Long-tailed Paradigalla
11.	Paradigalla brevicauda	Short-tailed Paradigalla
13.	Epimachus fastuosus	Black Sicklebill
	Epimachus meyeri	Brown Sicklebill
12.	Epimachus albertisi	Black-billed Sicklebill
	Epimachus bruijnii	Pale-billed Sicklebill

PLATE 77

15. Lophorina superba	Superb Bird-of-Paradise
Parotia sefilata	Western Parotia
Parotia carolae	Carola's Parotia
16. Parotia lawesii	Lawe's Parotia
Parotia helenae	Eastern Parotia
Parotia wahnesi	Wahnes's Parotia
8. Ptiloris magnificus	Magnificent Riflebird
Ptiloris victoriae	Victoria's Riflebird
Ptiloris paradiseus	Paradise Riflebird
19. Cicinnurus magnificus	Magnificent Bird-of-Paradise
Cicinnurus respublica	Wilson's Bird-of-Paradise
18. Cicinnurus regius	King Bird-of-Paradise
Astrapia nigra	Arfak Astrapia
Astrapia splendidissima	Splendid Astrapia
14. Astrapia mayeri	Ribbon-tailed Astrapia
Astrapia stephaniae	Stephanie's Astrapia
Astrapia rothschildi	Huon Astrapia
17. Pteridophora alberti	King of Saxony Bird-of-Paradise
10. Seleucidis melanoleuca	Twelve-wired Bird-of-Paradise
Paradisaea rubra	Red Bird-of-Paradise
Paradisaea minor	Lesser Bird-of-Paradise
20. Paradisaea apoda	Greater Bird-of-Paradise
Paradisaea raggiana	Raggiana Bird-of-Paradise
Paradisaea decora	Goldie's Bird-of-Paradise
Paradisaea guilielmi	Emperor Bird-of-Paradise
Paradisaea rudolphi	Blue Bird-of-Paradise

List: II
Lit: 3, 9, 43, 50, 129, 233, 307, 330, 410
Reading List:
Cooper, W. T. & Forshaw, J. M. 1977: *The Birds of Paradise and Bower Birds.* Collins, Sydney & London.

PLATE 78

BIRDS ADDED TO THE LISTS OF CITES AT THE SEVENTH MEETING OF THE CONFERENCE OF THE PARTIES, SWITZERLAND, OCTOBER 1989.

1. CACATUA MOLUCCENSIS

English: Salmon-crested Cockatoo; *French:* Cacatoès à huppe rouge; *German:* Molukkenkakadu; *Italian:* Cacatua delle Molucche; *Spanish:* Cacatúa de las Molucas

Range: Moluccas
Identification:
Length 52cm. Sexes similar. General coloration pale salmon-pink. Long backward-curved crest with some pale orange feathers on underside.
Young birds: Like adults.
Bare parts: Bill grey-black. Legs grey. Eyes dark brown, female reddish-brown. Skin around eye bluish-white.
List: I
Lit.: 37, 47, 81, 112, 129, 410

2. ARA MARACANA

Brazilian: Maracanã-do-buriti; *English:* Blue-winged Macaw, Illiger's Macaw; *French:* Ara d'Illiger; *German:* Marakana

Range: Brazil, Argentina
Identification:
Length 43cm. Sexes similar. Forehead red. Rest of head greenish-blue, darkest on crown. Body green with a red area on lower back and belly. Wings blue and green. Tail blue, base reddish-brown. Underside of wings and tail olive-yellow.
Young birds: Red areas smaller and paler. Upperparts duller, more greyish.
Bare parts: Bill black. Legs yellowish-brown. Eyes orange. Naked facial skin pale yellow.
List: I
Lit.: 81, 112, 129, 410, 431

3. AMAZONA TUCUMANA

English: Tucuman Amazon; *French:* Amazone de Tucuman; *German:* Tucuman Amazone

Range: Bolivia, Argentina
Identification:
Length 31cm. Sexes similar. Forehead red. Plumage otherwise green, all feathers edged black. Upper- and under tail-coverts yellowish-green. Wings green, near tip blue and a small red area in middle. Underside olive-green, thighs orange. Tail green tipped yellowish-green.
Young birds: Thighs green, otherwise like adults.
Bare parts: Bill horn yellowish. Legs pale greyish-pink. Eyes orange-yellow.
List: I
Lit.: 81, 112, 129, 410

4. SEMNORNIS RAMPHASTINUS

English: Toucan-Barbet; *French:* Cabézon toucan; *German:* Tukanbartvogel

Range: Colombia, Ecuador
Identification:
A. Length 22cm. **Male:** Upper head, elongated feathers on hindcrown and nape shining black. Broad white eyebrow. Side of head, throat, and breast blue-grey. Upperside dusky brown with yellow rump. Wings dark bluish-grey and brown. Underside yellowish-grey, stained with red on breast and belly. Tail grey-blue.
B. **Female:** Like male but lacks elongated head feathers.
Young birds: Undescribed.
Bare parts: Bill pale yellow with dark tip and brown area at base. Legs greenish. Eyes red-brown.

Geographical variation
Birds from Colombia have lower breast and belly yellow.
List: III Colombia
Lit.: 77, 112, 125, 129, 189, 403, 410

5. CEPHALOPTERUS ORNATUS

Brazilian: Anambé-preto; *English:* Amazonian Umbrellabird, Ornate Umbrellabird, Umbrellabird; *French:* Coracine ornée; *German:* Kurzlappen-Schirmvogel; *Spanish:* Pájaro paraguas

Range: Colombia, Venezuela, Ecuador, Peru, Bolivia, Brazil
Identification:
Male: Length 51cm. Entire plumage black glossed with blue. Tall, upstanding, umbrella-shaped crest, overhanging bill with white shafts basally. From lower throat hangs a flat, wide, feathered wattle up to 15cm long including feathers.
Female: Length 46cm. Like male but duller and with smaller crest and wattle.
Young birds: Undescribed.
Bare parts: Bill black, lower plumbeous, tip paler. Legs black. Eyes whitish.
List: III Colombia
Lit.: 112, 125, 129, 189, 192, 284, 410, 431

6. CEPHALOPTERUS PENDULIGER

English: Long-wattled Umbrellabird; *French:* Coracine casquée; *German:* Langlappen-Schirmvogel

Range: Colombia, Ecuador
Identification:
Male: Length 51cm. Similar to Amazonian Umbrellabird except for crest which is smaller and has black shafts, and wattle, which is much longer, up to 35cm and round.
Female: Length 46cm. Wattle shorter or almost absent.
Young birds: Like female.
Bare parts: Bill slaty-grey to bluish. Legs black. Eyes dark brown.
List: III Colombia
Lit.: 112, 129, 189, 284, 410

PLATE 79

BIRDS ADDED TO THE LISTS OF CITES AT THE SEVENTH MEETING OF THE CONFERENCE OF THE PARTIES, SWITZERLAND, OCTOBER 1989.

1. CRAX ALBERTI

English: Blue-knobbed Curassow, Blue-billed Curassow; *French:* Hocco du Prince Albert; *German:* Blaulappenhokko

Range: Colombia
Identification:
A. Length 91cm. **Male:** Curly black crest. General coloration black with a little bluish or greenish-blue gloss; lower part of underside white. Tail black with broad white tip.
Female differs very much from male and has two phases:
B. Rufous phase: Plumage glossy black with narrow, wavy white lines. Long flight-feathers, front of thighs and lower underparts chestnut. Tip of tail white.
D. Barred phase: Like rufous phase but crest longer with many more white bands and only lower underparts rufous. Rare.
Young birds: Undescribed.
Bare parts: Bill horny grey, base blue in male, knob-like wattles at base of lower bill blue. Female without wattles. Legs pink to reddish. Eyes dark brown.
List: III Colombia
Lit.: 13, 62, 125, 129, 410

2. CRAX DAUBENTONI

English: Yellow-knobbed Curassow; *French:* Hocco de daubenton; *German:* Gelblappenhokko, *Spanish:* Pauji de copete

Range: Colombia, Venezuela
Identification:
A. Length 84cm. **Male:** Curly black crest highly developed. General coloration black with a green sheen. Abdomen, thigh-tufts, and under tail-coverts white. Broad white band at tip of tail.
B. **Female:** Thighs black and like crest, breast, and flanks narrowly barred white. Wing-coverts also barred white but more fine, hairline markings.
Young birds: Browner.
Bare parts: Bill black. Yellow knob over bill (increases with age) and yellow knob-like wattles at base of lower bill. Female without yellow knob and wattles. Legs greyish, greenish or blackish. Eyes brown.
List: III Colombia
Lit.: 13, 27, 62, 129, 192, 410

3. CRAX GLOBULOSA

Formerly also known as: Crax carunculata; Crax yarrellii

Brazilian: Mutum-fava; *English:* Wattled Curassow; *French:* Hocco globuleux; *German:* Karunkelhokko

Range: Colombia, Peru, Bolivia, Brazil
Identification:
A. Length 85-95cm. **Male:** Curly black crest. Plumage black glossed dull greenish-blue. Abdomen, thigh tufts, and under tail-coverts white.
B. **Female:** Crest narrowly barred white. Lower underparts rufous, no white.
Young birds: Undescribed.
Bare parts: Bill black. Scarlet knob and wattles, female without knob and wattles but base of bill (cere) reddish. Legs dark brownish, greyish-blue, or red to pink. Eyes dark brown, female lighter brown.
List: III Colombia
Lit.: 13, 62, 129, 410, 431

4. PAUXI PAUXI

Also known as: Crax pauxi

English: Helmeted Curassow, Northern Helmeted Curassow; *French:* Hocco à pierre; *German:* Helmhokko

Range: Colombia, Venezuela
Identification:
A. Length 85-96cm. **Male:** General plumage glossy greenish-blue black. Belly, under tail-coverts, and broad band on tip of tail white.
Female has two phases: Common phase like male. B. Rufous phase: Head and neck black. Rest rufous-brown barred and vermiculated black, many feathers edged white. Lower underparts white. Tail broadly tipped buffy-white.
Young birds: Brown vermiculations on underparts.
Bare parts: Bill red. Large blue casque on forehead, juvenile has smaller casque.

Geographical variation:
Birds from Venezuela-Colombia border have smaller casque and are also somewhat different in shape.
List: III Colombia
Lit.: 13, 62, 129, 192, 410

PLATE 80

BIRDS ADDED TO THE LISTS OF CITES AT THE EIGHTH MEETING OF THE CONFERENCE OF THE PARTIES, JAPAN, MARCH 1992.

1. ANAS FORMOSA

English: Baikal Teal; *French:* Sarcelle élégante; *German:* Gluckente; *Italian:* Alzavola asiatica; *Japanese:* Tomoe-gamo; *Russian:* Kloktun; *Spanish:* Cerceta del Baikal

Range: Siberia, winters in Burma, India, China, Japan, Korea
Identification:
A. Length 39-43cm. **Male:** Unmistakable head pattern of white, creamy-buff, green, and black. Mantle grey vermiculated brown and black, rest of upperparts and tail brown. Wings brown with a green and black mirror edged white in front. Long narrow, pointed shoulder-feathers black in middle, edged chestnut and white. Breast pinkish with small black spots and white vertical line. Lower breast and abdomen white, flanks greyish and under tail-coverts black. Male non-breeding: Like female but more uniform and without white head-spots. Some pointed shoulder-feathers may remain.
B. **Female:** Like a Green-winged Teal female, Anas crecca (plate 11), but with conspicuous round white spot at base of bill surrounded by dark feathers.
Young birds: Like female but less red-brown and white head-spot less distinct.
Bare parts: Bill dark grey. Legs greyish to yellowish-grey. Eyes brown.
List: II
Lit.: 51, 60, 79, 129, 176, 191, 195, 297, 323, 410, 428

2. CACATUA GOFFINI

English: Goffin's Cockatoo, *French:* Cacatoès de Goffin; *German:* Goffinkakadu; *Italian:* Cacatua di Goffin; *Spanish:* Cacatúa de Goffin

Range: Kepulauan Tanimbar Island, Kepulauan Kai Island, (Moluccas)
Identification:
Length 32cm. Sexes alike. Whole body white, head-feathers at base salmon-pink. Ear-coverts, underside of wings, and tail with a yellow tinge.
Young birds: Like adults but feathers on head without salmon-pink wash.
Bare parts: Bill greyish-white. Legs grey. Eyes, male and juvenile dark brown, female red-brown. Naked skin around eye white.
List: I
Lit.: 37, 81, 129, 168, 195, 389, 410

3. CACATUA HAEMATUROPYGIA

English: Red-vented Cockatoo, Philippine Cockatoo; *French:* Cacatoès des Philippines; *German:* Rotsteisskakadu; *Italian:* Cacatua a sottocoda rosso; *Spanish:* Cacatúa de cola sangrante

Range: Philippine and Sulu Archipelago
Identification:
Length 31cm. Sexes similar. A white cockatoo except for pale yellow-pink ear-coverts and base of short crest-feathers which are yellow to rose-pink. Wing underside pale yellow and tail underside yellow. Most distinctive is orange-red under tail-coverts.
Young birds: Like adults.
Bare parts: Bill greyish-white. Legs grey. Eyes dark brown, female brownish-red, juvenile brown. Naked skin round eye white, rarely tinted with blue.
List: I
Lit.: 37, 66, 81, 129, 168, 389, 410

4. LICHENOSTOMUS MELANOPS CASSIDIX

Formerly known as: Meliphaga cassidix

English: Helmeted Honeyeater, *German:* Helmhonigfresser

Range: Yallock Creek, near Yellingbo, Victoria,(Australia)
Identification:
Length 17-23cm. Sexes similar but female smaller. Forehead, crown, nape, chin, throat, and ear-tufts yellow, centre of throat distinctively blackish. Feathers on forehead longer, forming a short helmet. Rest of head black. Upperparts, wings, and tail in various shades of olive-brown, more yellowish-green on long flight-feathers. Underparts green-yellow.
Young birds: Duller.
Bare parts: Bill black. Legs dark grey. Eyes red-brown.
List: I
Status: Total population estimated at about 100-150 birds!
Lit.: 36, 129, 225, 233, 278, 410
Note: **Lichenostomus m. melanops**
Range: S. E. Australia
Differs from cassidix by having no helmet crest on forehead.
NOT LISTED

5. GRACULA R. RELIGIOSA

English: Hill Myna; *German:* Beo; *Hindi:* Pahari myna; *Indonesia:* Beo, Tiong

Range: South Asia and Malay Archipelago
Identification:
Length 25cm. Sexes alike. Entire plumage jet-black, crown and mantle with purple gloss, rest with a greenish wash except blue-black throat. Broad white band on long flight-feathers.
Young birds: Duller with little gloss.
Bare parts: Bill orange, tip yellow, juvenile yellowish-orange with some dusky areas. Legs yellow. Eyes dark brown, juvenile brown. Bare skin on head and wattles bright yellow, variable according to locality.

Subspecies:

Gracula religiosa indica
English: Southern Hill Myna
Range: India and Sri Lanka
Smaller. Bare skin below eye separated from bare skin behind eye by black feathers.
List: III Thailand
Lit.: 2, 66, 129, 150, 170, 191, 197, 410, 419, 420

6. BAILLONIUS BAILLONI

English: Saffron Toucanet; *French:* Aracari de baillon; *German:* Goldtukan

Range: Brazil
Identification:
Length 35-43cm. **Male:** Forehead and forecrown bright yellow-green, hindcrown, nape, upperparts, wings and long tail dark olive-green, rump crimson. Underparts saffron-yellow.
Female: Smaller and duller in colours.
Young birds: Undescribed.
Bare parts: Bill green-yellow with red areas near base. Legs green-blue in male, female greenish. Eyes yellow. Bare facial skin red.
List: III Argentina
Lit.: 190, 334, 370, 375, 410

7. SELENIDERA MACULIROSTRIS

English: Spot-billed Toucanet; *French:* Aracari à bec tacheté; *German:* Flecken-Arassari

Range: Central South America
Identification:
A. Length c. 33cm. **Male:** Forehead, crown and nape jet-black. Ear-coverts orange and yellow. Yellow-green collar. Upperparts, wings and tail rich green, tip of tail feathers chestnut. Underparts black except orange flanks, yellowish-green belly, rufous and green thighs, and crimson under tail-coverts.
B. **Female:** Differs from male in having crown, nape, throat, and breast rufous. Ear-coverts duller and flanks pale orange.
Young birds: Undescribed.
Bare parts: Bill horn-grey, tip and top of upper bill yellowish with 3-6 black transverse bands on upper bill and one near tip of lower bill. Legs olive grey-blue to pale green. Eyes yellow to green-yellow. Bare skin round eyes greenish-blue, female said to have pale green skin.
List: III Argentina
Lit.: 190, 334, 370, 375, 410

PLATE 81

BIRDS ADDED TO THE LISTS OF CITES AT THE EIGHTH MEETING OF THE CONFERENCE OF THE PARTIES, JAPAN, MARCH 1992.

1. PTEROGLOSSUS ARACARI

Brazilian: Araçari-de-bico-branco; *English:* Black-necked Araçari; *French:* Araçari grigri; *German:* Schwarzkehl-Arassari, *Spanish:* Tilingo guellinegro; *Surinam*: Reddie-bantie-koejake

Range: Venezuela, Guyana, Surinam, French Guiana, Brazil
Identification:
Length 46-48cm. Sexes alike. Head and neck glossy black. Upperparts, wings and tail dark olive-green, rump crimson. Underparts greenish-yellow with a crimson band across upper belly and olive-green thighs.
Young birds: Undescribed.
Bare parts: Bill ivory, ridge of upper bill and whole lower bill black. Legs pale grey-green. Eyes brown. Skin around eye dark grey.
List: II
*Lit.:*114, 129, 192, 254, 334, 370, 375, 410, 431

2. PTEROGLOSSUS CASTANOTIS

Brazilian: Araçari-castanho; *English:* Chestnut-eared Araçari; *French:* Araçari oreillons roux; *German:* Braunohr-Arasari

Range: Colombia, Ecuador, Peru, Bolivia, Brazil, Argentina
Identification:
Length 46cm. **Male:** Top of head glossy black. Sides of head and nape chestnut. Upperparts, wings and tail dark slaty olive-green, rump crimson. Underparts yellow with crimson belly band. Thighs chestnut.
Female: A little smaller and chestnut on sides of head a little paler.
Young birds: Undescribed.
Bare parts: Bill black with buffish band on upper bill shading to yellow toward tip, base of bill and "teeth" yellow. Legs grey-green. Eyes whitish to pale yellow. Skin around eyes bright blue.
List: III Argentina
*Lit.:*125, 190, 254, 334, 370, 375, 410, 431

3. PTEROGLOSSUS VIRIDIS

Brazilian: Araçari-miudinho; *English:* Green Araçari; *French:* Araçari vert; *German:* Grünarassari; *Spanish:* Tilingo limón; *Surinam:* Stonkoejake

Range: Venezuela, Guianas, Brazil, Bolivia
Identification:
A. Length 30cm. **Male:** Head, neck and forebreast glossy black. Upperparts, wings and tail dark green, rump crimson. Underparts yellow, thighs green and buff.
B. **Female:** Head, neck and forebreast bright chestnut. Otherwise like male.
Young birds: Undescribed.
Bare parts: Bill black, yellow and red, base dull orange. Legs olive-green. Eyes reddish. Skin around eyes blue and behind eyes red.
List: II
*Lit.:*114, 190, 192, 254, 334, 370, 375, 410, 431

4. RAMPHASTOS DICOLORUS

Brazilian: Tucano-de-bico-verde; *English:* Red-breasted Toucan, Green-billed Toucan; *French:* Toucan à ventre rouge; *German:* Bunttukan

Range: Brazil, Paraguay, Argentina
Identification:
Length 45cm. Sexes similar. Small patch on forehead yellow, rest of upperhead, nape, upperparts, wings and tail glossy black except crimson rump. Chin and throat pale yellow becoming rich orange on lower throat and fading to pale yellow band. Belly and under tail-coverts red.
Young birds: Like adult.
Bare parts: Bill light green, base black with narrow line near cutting edges of upper bill reddish, juvenile pale green. Legs blue. Eyes greenish-yellow.
List: III Argentina
Lit.: 190, 254, 334, 370, 375, 410, 431

5. RAMPHASTOS TOCO

Brazilian: Tucanotoco, Tucanuçu; *English:*Toco Toucan, *French:* Toucan toco; *German:* Riesentukan

Range: Guianas, Brazil, Bolivia, Paraguay, Argentina
Identification:
Length 60cm. Sexes alike but female smaller. Plumage black except white side of head and throat gradually becoming pale yellow on upperbreast; upper tail-coverts also white. Under tail-coverts crimson. Young birds resemble adult.
Bare parts: Bill largest among all toucans, 17-24cm, orange-red and greenish-yellow, frequently with dusky vertical stripes near cutting edges, base and oval area near tip of upper bill black, juvenile paler. Legs greenish-blue. Eyes dark.
List: II
Lit.: 190, 254, 334, 370, 375, 410, 431

6. RAMPHASTOS TUCANUS

Brazilian: Tucano-grande-de-papo-branco; *English:* Red-billed Toucan, White-throated Toucan; *French:* Toucan de cuvier; *German:* Weiss-brusttukan; *Spanish:* Piapoco pico rojo

Range: Colombia, Venezuela, Guianas, Brazil, Bolivia
Identification:
Length 53cm. Sexes similar. Top of head, hindneck, upperparts, wings and tail black, upper tail-coverts yellow. Sides of head, throat and breast white. Narrow red band between breast and black belly. Under tail-coverts crimson.
Young birds: Undescribed.
Bare parts: Bill dark red with a variable amount of black, tip, ridge and base of upper bill yellow, base of lower bill light blue. Legs blue. Eyes dark brown. Facial skin blue.
List: II
Lit.: 125, 190, 192, 254, 334, 370, 375, 410, 431

7. RAMPHASTOS V. VITELLINUS

Brazilian: Tucano-de-bico-preto; *English:* Channel-billed Toucan; *French:* Toucan ariel; *German:* Dottertukan; *Spanish:* Diostedé pico acanalado

Range: Trinidad, Venezuela, Guianas, Colombia, Brazil
Identification:
Length 49cm. Sexes similar. Top of head, hindneck, upperparts, wings and tail black. Upper and under tail-coverts crimson. Chin and sides of neck white. Throat pale yellow gradually becoming orange on upper breast. Broad crimson breast band. Abdomen black.
Young birds: Undescribed.
Bare parts: Bill black, base blue. Legs blue. Eyes dark brown. Facial skin greenish-blue.

Subspecies:

Ramphastos vitellinus ariel
English: Ariel Toucan
Range: Brazil
Base of bill orange-yellow. Basal fourth of ridge on upper bill blue.

Geographical variation:
Some birds from eastern Venezuela and Guianas have orange bills. Also in other parts of the area there is considerable variation and all hybridize to varying degrees.
List: II
Lit.: 125, 190, 192, 254, 334, 370, 375, 410, 431

PLATE 82

BIRDS ADDED TO THE LISTS OF CITES AT THE EIGHTH MEETING OF THE CONFERENCE OF THE PARTIES, JAPAN, MARCH 1992.

1. ACEROS NIPALENSIS

English: Rufous-necked Hornbill; *French:* Calao à cou roux; *German:* Nepalhornvogel

Range: Nepal, Burma, Thailand, Laos, Vietnam, China
Identification: A. Length 114-122cm. **Male:** Head, neck and breast rufous to orange-red. Chin and throat red with blue patch at lower throat. Upperparts glossy black with green tinge, five longest flight-feathers with white tips. Underside red-brown. Tail black in middle, rest white.
B. **Female:** Like male but smaller and head, neck and underparts black.
Young birds: Both male and female like adult male.
Bare parts: Bill without casque yellowish to ivory with up to eight black, transverse ridges, base of lower bill olive. Legs black. Eyes red. Skin around eyes and at base of lower bill bright blue.
List: I
Lit.: 2, 147, 150, 191, 258, 283, 394, 410, 419, 421

2. ACEROS SUBRUFICOLLIS

Also known as: Aceros plicatus subruficollis; Rhyticeros subruficollis
English: Plain pouched Hornbill, Blyth's Hornbill; *German:* Sundajahrvogel; *Malay:* Burong jawa, Enggang

Range: Burma, Thailand, Sumatra and Borneo
Identification: A. Length 86-89cm. **Male:** Crown and crest chestnut, rest of head and neck white. Tail white, rest of plumage black with a greenish wash.
B. **Female:** Head and crest black. Smaller size. otherwise like male.
Young birds: No data available.
Bare parts: Bill creamy white without corrugations on sides with a small, brownish, wrinkled casque. Legs blackish. Eyes red, female yellow. Naked skin around eyes red and pouch on throat yellow without any black band, female have bare skin on throat bluish.
List: I
Lit: 47, 150, 283, 306, 367, 410, 417, 419

3. ACEROS CORRUGATUS

Also known as: Rhyticeros corrugatus
English: Wrinkled Hornbill; *French:* Calao à casque rouge; *German:* Runzelhornvogel; *Malay:* Enggang

Range: Malaya, Borneo, Sumatra, Batu Islands
Identification: A. Length 81cm. **Male:** Top of head, small crest and nape black. Side of head and front of neck white. Base of tail black, terminal two-thirds white, buffish or chestnut. Rest of body black glossed dark green.
B. **Female:** Smaller. Head and neck black.
Young birds: Like adult male.
Bare parts: Bill yellow, base of upper bill and casque red, base of lower bill buffish; females have yellow bill and casque, juveniles also have yellow bill but casque smaller. Legs dull black. Eyes red-brown, juvenile buffish. Skin around eyes blue, gular pouch white, female blue.
List: II
Lit.: 47, 150, 185, 258, 282, 306, 367, 377, 394, 410, 417, 419, 421

4. ACEROS UNDULATUS

Also known as: Rhyticeros undulatus
English: Wreathed Hornbill; *French:* Calao festonné; *German:* Furchenjahrvogel

Range: E. India, Bangladesh, Burma, Thailand, Indochina, Malaya, Sumatra, Java, Bali, Borneo
Identification: A. Length 100cm. **Male:** Top of head and crest chestnut, side of head and neck white often spotted yellow. Entire tail white. Rest of plumage black glossed green on upperparts.
B. **Female:** Smaller. Head and neck black. Otherwise like male.
Young birds: Like adult male but without casque or with a smaller one. Often black spots in otherwise white tail.
Bare parts: Bill ivory, base reddish, casque with 4-9 furrows, female 0-3. Legs black. Eyes red, female red to red-brown, juvenile orange. Naked skin around eyes orange-red. Gular pouch bright yellow with a broad transverse black band, female has blue gular pouch with black transverse band, juvenile has a grey band on gular pouch.
List: II
Lit.: 2, 150, 258, 282, 367, 394, 410, 417, 419, 421

5. ACEROS PLICATUS

Also known as: Rhyticeros plicatus
English: Blyth's Hornbill; *French:* Calao papou; *German:* Papua-Jahrvogel

Range: Indonesia, New Guinea region, Moluccas, Bismarck Arch. and Solomon Islands
Identification: A. Length 76-91cm. **Male:** Head, neck and upper breast golden buff to reddish-brown. Entire tail white, rest of plumage black, on upperparts glossed green.
B. **Female:** Like male but smaller and head and neck black.
Young birds: Resemble adult male but casque absent or smaller. Often black spots in otherwise white tail.
Bare parts: Bill ivory, base red-brown, up to 8 furrows on casque. Legs black. Eyes red, female brown. Skin around eyes pale blue, gular pouch blue-white.
List: II
Lit.: 9, 105, 150, 183, 258, 321, 393, 394, 402, 410, 417

6. ACEROS LEUCOCEPHALUS

Also known as: Rhyticeros leucocephalus
English: Writhed Hornbill; *French:* Calao de vieillot; *German:* Mindanao-Hornvogel

Range: Mindanao, Camiquin Island, Panay, Guimaras, Negros Island, (all Philippines)
Identification: A. Length 74cm. **Male:** Head and neck buffish, crest chestnut. Tail white, often spotted rufous, with terminal black band. Rest of plumage black with bluish-green gloss on upperparts.
B. **Female:** Smaller, otherwise like male except black head and neck.
Young birds: Like adult male but casque smaller.
Bare parts: Bill and casque red with furrows, female without furrows. Legs black. Eyes red, juvenile red-brown. Naked skin around eyes, chin and throat red.
List: II
Lit.: 61, 66, 104, 353, 358, 394, 410, 417

7. ACEROS WALDENI

Also known as: Rhyticeros leucocephalus waldeni; Aceros leucocephalus waldeni
English: Writhed-billed Hornbill

Range: Panay, Guimaras, Negros Island (C. Philippines)
Identification: A. Length 68-76cm. **Male:** Head, crest and neck dark chestnut. Tail white often spotted rufous, terminal band black. Otherwise black with green gloss on upperparts.
B. **Female:** Smaller and head and neck black.
Young birds: Like adult male.
Bare parts: Bill and casque red. Legs black. Eyes red, juvenile red-brown. Naked skin round eyes, chin and throat red.
List: II
Lit.: 66, 104, 183, 258, 353, 385, 394, 410, 417

8. ACEROS CASSIDIX

Also known as: Rhyticeros cassidix
English: Knobbed Hornbill; *French:* Calao à cimier; *German:* Helmhornvogel

Range: Sulawesi
Identification: A. Length 96cm. **Male:** Crown red-brown, rest of head, neck and breast yellowish-white. Upperparts glossy black. Belly and under tail-coverts black. Tail white.
Female: Smaller. Entire plumage black except for white tail.
Young birds: Like adult male.
Bare parts: Bill yellow, base and casque chestnut, base of bill with furrows, female have casque and tip of bill orange-yellow. Legs black. Eyes red to orange, female dark brown, juvenile dark brown with yellow outer-ring. Naked skin around eyes and throat green-blue.
List: II
Lit.: 258, 321, 394, 410, 417

PLATE 83

BIRDS ADDED TO THE LISTS OF CITES AT THE EIGHTH MEETING OF THE CONFERENCE OF THE PARTIES, JAPAN, MARCH 1992.

1. ACEROS COMATUS

Also known as: Berenicornis comatus
English: White-crowned Hornbill; *French:* Calao coiffé; *German:* Langschopf Hornvogel; *Malay:* Enggang bulu
Range: Burma, Thailand, Vietnam, Malaya, Sumatra, Borneo
Identification: A. Length 90cm. **Male:** Long fluffy crest. Head, neck, underparts and tail white except blackish flanks and under tail-coverts. Upperparts blackish-brown, wings glossy black, tips of long flight-feathers white.
B. **Female:** Smaller. Neck and underparts blackish-brown, otherwise like male.Young birds: Tail black tipped white, rest of white feathers often with some black, otherwise like adult. Bare parts: Bill blackish, base bluish, with small casque, juvenile yellow bill. Legs black. Eyes pale yellow, eyelids blue, juveniles have green-yellow eyes.
List: II; *Lit.:* 150, 258, 282, 367, 410, 419

2. ACEROS EVERETTI

Also known as: Rhyticeros everetti
English: Sumba Hornbill, Everett's Hornbill; *French:* Calao de Sumba; *German:* Sumbahornvogel
Range: Lesser Sunda Islands (Sumba)
Identification: Length 70cm. **Male:** Head and nape dark brown, foreneck buffish. Upperparts glossy black, underparts black. **Female:** Smaller. Head and neck black. Young birds: No information available. Bare parts: Bill red, tip and base yellow, no true casque. Eyes red-brown.
List: II Status: Probably the most threatened of all hornbills.
Lit.: 196, 258, 321, 410, 417

3. ANORRHINUS T. TICKELLI

Also known as: Ptilolaemus tickelli
English: Brown Hornbill, Tickell's Hornbill; *French:* Calao brun; *German:* Rostbauch-Hornvogel
Range: Thailand, Laos, Vietnam, Burma, China
Identification: Length 68-74cm. **Male:** Top of head and crest dull dark brown edged rufous and with pale shaft-streaks. Side of head, neck and underparts rufous to rufous-buff, darkest on head and paling towards belly. Upperparts and middle tail feathers brown. Flight-feathers and rest of tail feathers black tipped white. **Female:** Smaller and more dark brown than rufous. Young birds: Like male but casque and crest smaller. Bare parts: Bill and small casque dull yellowish, female blackish. Legs black. Eyes brown, juvenile grey. Naked skin round eyes blue.

Subspecies:
Anorrhinus tickelli austena
English: Assam Hornbill
Range: Assam
Differs in having cheeks, throat, sides of neck and upper breast whitish. Female has also, like male, yellow bill.
List: II; *Lit.:* 2, 150, 191, 258, 283, 394, 410, 419

4. ANORRHINUS GALERITUS

English: Bushy-crested Hornbill; *French:* Calao largup; *German:* Kurzschopf-Hornvogel; *Malay:* Burong kawan, Burong mati sa kawan, Enggang buloh
Range: Burma, Thailand, Malaya, Sumatra, Borneo
Identification: Length 89cm. Sexes similar but female smaller. Thick, drooping crest diagnostic. Plumage mainly blackish-brown, above with greenish tinge. Tail brownish-grey, terminal third black. Young birds: Like adult but underside pale grey to pale brown. Bare parts: Bill black, female sometimes dull yellow or with yellow lines on upper bill, juvenile greenish at base, tip red-brown. Legs black. Eyes red, female often red-brown, juvenile grey to bluish. Naked skin around eyes and throat whitish-blue.
List: II;
Lit.: 150, 258, 282, 367, 392, 394, 410, 419

5. ANTHRACOCEROS CORONATUS

Also known as: Anthracoceros malabaricus
English: Malabar Pied-Hornbill; *French:* Calao de Malabar; *German:* Malabarhornvogel; *Hindi:* Dhan chiri
Range: South and South-East India
Identification: Length 69-92cm. **Male:** Black plumage glossed greenish blue, except white area on side of head, tip of flight-feathers and lower underparts. Two central tail feathers black, outer tail feathers all white or black with white tips. **Female:** Smaller. See also bare parts. Young birds: Outer tail feathers often more black and casque smaller. Bare parts: Bill and casque ivory-white with a large black area on casque extending along upper casque ridge. Casque high ridge-like and ending in a single point in front. Legs grey, greenish or black. Eyes orange-red, female brownish. Naked skin round eyes blue-black, female white or fleshy. Bare throat patch flesh coloured.
List: II; *Lit.:* 2, 150, 258, 366, 390, 394, 410

6. ANTHRACOCEROS A. ALBIROSTRIS

Formerly known as: Anthracoceros malabaricus; Anthracoceros coronatus albirostris; Anthracoceros coronatus leucogaster
English: Oriental Pied Hornbill, Indian Pied Hornbill; *French:* Calao oie; *German:* Malabarhornvogel
Range: N.E. India, Burma, Thailand, Indochina, Malaya, China
Identification: Length 70-74cm. Sexes alike but female smaller and casque much smaller. Like number 5, Malabar Pied-Hornbill but smaller. Outer tail feathers black with white tips, rarely with some black spots. Young birds: Like adult except colour of bill and eyes. Bare parts: Bill and large casque ivory, casque high ridge-like and ending in front in a single point; black spot only on forepart of casque, female like male but top and tip of upper bill black and often some black on lower bill and base red-brown; juveniles have much less black on bill and casque. Legs dark grey to greenish-grey. Eyes red, female red-brown to brown, juvenile brown. Naked skin around eyes and throat blue-white with some black above and in front of eye.

Subspecies:
Anthracoceros albirostris convexus
English: Malaysian Pied Hornbill
Range: S. Thailand, Malaya, Sumatra, Borneo, Natuna Islands, Java
Outer pair of tail feathers mainly white often with some black spots, central tail feathers up to 30mm longer than rest. Little or no black above and in front of eye.
List: II;
Lit.: 2, 191, 258, 282, 364, 365, 366, 367, 390, 410

7. ANTHRACOCEROS MALAYANUS

English: Black Hornbill; *French:* Calao charbonnier, *German:* Malaienhornvogel
Range: Thailand, Malaya, Sumatra, Borneo
Identification: A. Length 76cm. **Male:** Entire plumage black except white tips to outer tail feathers. D. Some birds have a broad white or greyish stripe from behind either eyes meeting on nape. B. **Female:** Smaller. See bare parts.
Young birds: Like adult, but casque smaller. Bare parts: Bill and large casque ivory, female blackish, juvenile whitish. Legs black. Eyes red, female red-brown, juvenile dark grey to brown. Naked skin around eyes and throat bluish-black, females have pinkish skin around eyes.
List: II
Lit.: 150, 258, 282, 364, 366, 367, 410

8. ANTHRACOCEROS MARCHEI

English: Palawan Hornbill; *French:* Calao de Palawan; *German:* Palawanhornvogel
Range: Calamian Islands, Palawan, Balabac, (all Philippines)
Identification: Length 71cm. Sexes similar but female smaller. Plumage black glossed green except white tail, shaft of tail feathers creamy. Young birds: Like adults but tail feathers with some black spots. Bare parts: Bill and casque ivory, base of lower bill black. Legs black. Eyes red-brown, female brown, juvenile grey. Naked skin round eyes and throat pale blue.
List: II
Lit.: 66, 104, 258, 366, 394, 410

9. ANTHRACOCEROS MONTANI

English: Sulu Hornbill; *French:* Calao de Sulu; *German:* Suluhornvogel
Range: Jolo, Tawitawi Islands (Philippines)
Identification: Length 81cm. Sexes similar but female smaller. Head and body black glossed green on upperparts. Tail pure white.
Young birds: Undescribed.
Bare parts: Bill, casque and legs black. Eyes whitish, female dark brown. Naked skin on face black.
List: II
Status: Is believed to be extinct on Sulu Islands.
Lit.: 61, 66, 104, 258, 358, 366, 410

PLATE 84

BIRDS ADDED TO THE LISTS OF CITES AT THE EIGHTH MEETING OF THE CONFERENCE OF THE PARTIES, JAPAN, MARCH 1992.

1. PENELOPIDES MANILLAE

Also known as: Penelopides manilloe; Penelopides panini manillae

English: Luzon Hornbill; *French:* Calao de Manille; *German:* Tariktikhornvogel

Range: Luzon, Polillo Islands, Catanduanes, Marinduque, (all Philippines)
Identification: A. Length 53cm. **Male:** Top of head, crest, neck and underparts yellow-white. Side of head and feathered part of throat black. Upperparts and wings dark brown glossed dull green, most flight-feathers edged buff. Tail dark brown with a white to red-brown 10-14mm band between basal dark brown and black tip.
B. **Female:** Smaller. Like male except dark brown head and neck, and brownish grey underparts.
Young birds: Like adult.
Bare parts: Bill and casque reddish-brown with four or five yellow marks on upper and lower bill; juveniles lack casque and yellow marks. Legs black. Eyes red, juvenile brown. Naked skin round eyes and throat yellow-orange, female purple.
List: II
Lit.: 61, 66, 104, 258, 353, 364, 387, 394, 410

2. PENELOPIDES MINDORENSIS

Also known as: Penelopides panini mindorensis

English: Mindoro Hornbill; *French:* Calao de Manille; *German*: Tariktikhornvogel

Range: Mindoro Island, (Philippines)
Identification: A. Length 57cm. **Male:** Like male of number 1, Luzon Hornbill, except for black upperparts and wings glossed green and more white on tail. Some birds of both sexes have more black on foreheads.
B. **Female:** Smaller. See bare parts.
Young birds: Like adults.
Bare parts: Bill like number 1 except yellow tip and a large black area on lower bill instead of yellow marks. Legs dark brown. Eyes red-brown, female paler brown. Naked skin round eyes and throat whitish, female dark blue.
List: II
Lit.: 66, 258, 353, 394, 398, 404, 410

3. PENELOPIDES PANINI

Also known as: Penelopides p. panini

English: Tarictic Hornbill; *French:* Calao de Manille; *German:* Tariktikhornvogel

Range: Ticao, Masbate, Panay, Guimaras, Negros, Pan de Azucar, Sicogon (all Philippines)
Identification: A. Length 60cm. **Male:** Differs from number 1, Luzon Hornbill, in having a brighter blue-green gloss on black upperparts and wings. Rump, upper- and under tail-coverts and belly bright chestnut. Tail white to rufous-buff, terminal one third black glossed green.
B. **Female:** Smaller. Head and upperparts black.
Young birds: Like adult male except brown upper tail-coverts and tip of tail.
Bare parts: Bill and casque dull reddish with yellow marks like Luzon Hornbill, juveniles have brown bill. Legs dark brown. Eyes red, juvenile brown. Naked skin on head and throat flesh coloured, female bluish.
List: II
Lit.: 66, 258, 353, 355, 364, 394, 410

4. PENELOPIDES SAMARENSIS

Also known as: Penelopides panini samarensis

English: Samar Hornbill; *French:* Calao de Manille, *German:* Tariktikhornvogel

Range: Samar, Leyte, Bohol (all Philippines)
Identification: A. Length 50cm. **Male:** Like number 1, Luzon Hornbill, but upperparts and wings black glossed dark green. Upper tail-coverts and underparts buff-white instead of yellow-white. Tail white often stained rufous with a one third black terminal band and often some black at base of tail.
B. **Female:** Like female of number 1, Luzon Hornbill, but dull black with green metallic gloss on upperparts. Upper tail-coverts buffish.
Young birds: Like adult but tail with some brown on black band.
Bare parts: Bill and casque horn-coloured, top of casque and base of upper bill blackish. Only yellow marks on lower bill. Legs dark brown. Eyes red, juvenile brown. Naked skin round eyes and throat yellow-white, female bluish.
List: II
Lit.: 66, 258, 353, 394, 398, 399, 401, 410

5. PENELOPIDES AFFINIS

Also known as: Penelopides panini affinis

English: Mindanao Hornbill; *French:* Calao de Manille; *German*: Tariktikhornvogel

Range: Dinagat, Siargao, Mindanao, Basilan, (all Philippines)
Identification: A. Length 50cm. **Male:** Differs from number 4, Samar Hornbill, only in having upper tail-coverts black.
Female: Like female of number 4, Samar Hornbill.
Young birds: Like adults, but upper tail-coverts chestnut and some brown on black band at tip of tail.
Bare parts: Like number 4, Samar Hornbill.

Geographical variation:
Birds from Basilan have flesh colour on base of bill. Some individuals (from Basilan) have more black at base of tail and outer tail feathers black to middle.
List: II
Lit.: 66, 258, 353, 358, 394, 398, 410

6. PENELOPIDES EXARHATUS

English: Sulawesi Hornbill, Celebes Hornbill; *French:* Calao à cannelures; *German:* Celebeshornvogel

Range: Sulawesi
Identification: A. Length 50cm. **Male:** Side of head and throat white, rest of plumage black glossed green on upperparts, wings and tail.
B. **Female:** Smaller. Entire plumage black glossed green.
Young birds: Like adult male.
Bare parts: Bill and casque ivory with a black area over basal half of upper bill, casque with several transverse ridges, female golden brown casque. Legs black. Eyes red, juvenile dark brown. Naked skin round eyes and throat yellow, female black with a narrow yellow stripe below eyes.

Geographical variation:
Birds from C., E., S. E., and S. Sulawesi, Muna and Buton have black area on lower bill broken into black and yellow bars.
List: II
Lit.: 258, 321, 394, 410

PLATE 85

REFERENCE LIST

1 Ali, S. 1960: The Pink-headed Duck. Wildfowl Trust 11:55-60.

2 Ali, S. & Ripley, S. D. 1983: Handbook of the Birds of India and Pakistan. Compact Edition. Oxford University Press, Delhi.

3 Austin, O. L. & Singer, A. 1961: Birds of the World. Hamlyn, London.

4 Baker, E. C. S. 1914: Ithaginis cruentus tibetanus. Bull. Brit. Ornithol. Club 35:18.

5 Baker, R. H. 1951: The Avifauna of Micronesia. Its Origin, Evolution, and Distribution. University Kansas Publ. Mus. Nat. Hist. Vol. 3, No. 1.

6 Bangs, O. & Phillips, J. C. 1925: A New Race of Pelzehn's Weaver-Finch. Occ. Papers Boston Soc. Nat. Hist. 5:177.

7 Bannerman, D. A. 1930-51: The Birds of Tropical West Africa. Vols. 1-8. Crown Agents for the Colonies, London.

8 Bannerman, D. A. & Bannerman, W. M. 1963-68: Birds of the Atlantic Islands. Vol. I-IV. Oliver & Boyd, Edinburgh.

9 Beehler, B. M., Pratt, T. K. & Zimmerman, D. A. 1986: Birds of New Guinea. Princeton University Press, Princeton.

10 Behnke-Pedersen, M. 1972: Kolibrier. Skibby Bøgerne, Skibby, Denmark

11 Bennett, S. 1985: The Distribution and Status of the Black-breasted Button-quail Turnix melanogaster (Gould 1837). Emu 85:157-162

12 Biricik, M. et al.: 1989: Fortpflanzungsverhalten der Palmtaube (Streptopelia senegalensis): Paarbildung bis Eiablage. J. Ornithol. 130:217-228.

13 Blake, E. R. 1977: Manual of Neotropical Birds. Vol. 1. University of Chicago Press, Chicago & London.

14 Bond, J. & Meyer de Schauensee, R. 1939: Paroaria capitata fuscipes. Notulae Naturae Acad. Nat. Sci.. Philadelphia 12:2

15 Bond J. 1979: Birds of West Indies. Collins, London.

16 Britton, D. & Hayashida, T. 1981: The Japanese Crane. Kodansha International, Tokyo, New York & San Fransisco.

17 Brooke, R. K. 1984: South African Red Data Book - Birds. South African National Sci. Programmes Report No. 97, Pretoria.

18 Brown, L. 1979: Die Greifvögel. Ihre Biologie und Ökologie. Verlag Paul Parey, Hamburg & Berlin.

19 Brown, L. & Amadon, D. 1968: Eagles, Hawks and Falcons of the World. Vol. 1-2. Hamlyn, Feltham, UK.

20 Brown, L. H., Urban, E. K. & Newman, K. 1982: The Birds of Africa. Vol. 1. Academic Press, London.

21 Brown, P. B. & Wilson, R. I. 1982: The Orange-bellied Parrot. in: Groves, R. H. & Ride, W. D. L. (eds.) Species at Risk. Research in Australia. Springer Verlag, Berlin, Heidelberg, New York.

22 Brudenell-Bruce, P. G. C. 1975: The Birds of New Providence and the Bahama Islands. Collins, London.

23 Bruggemann, 1876: Pitta Kochi. Abh. Nat. Ver., 5:65.

24 Bruning, D. 1983: Breeding Condors in Captivity for release into the Wild. Zoo Biology 2:245-252.

25 Bruning, D. 1989: Saved! But where to now? Birds International 1 (1):73-78.

26 Bruun, B. & Singer, A. 1979: Fuglene i Europa. Gyldendalske Boghandel, Copenhagen.

27 Buchholz, R. 1991: Older males have bigger knobs: Correlates of ornamentation in two species of Curassow. Auk 108:153-160.

28 Bucknill, J. A. 1924: Disappearance of the Pink-headed Duck (Rhodonessa caryophyllacea Lath.). Ibis 66:146-151.

29 Butler, P. J. 1981: The St. Lucia Amazon (Amazona versicolor): its changing Status and Conservation. in: Pasquier, R. F. (ed.) Conservation of New World Parrots. Smithsonian Institution Press for ICBP. Techn. Publ. 1.:171-180. Washington D. C..

30 Burton, J. A. (ed.) 1984: Owls of the World. Collins, Glasgow.

31 Burton, P. & Boyer, T. 1989: Birds of Prey of the World. Dragon World, Limpsfield.

32 Cabanis, 1869: Pharomacrus costaricensis. J. Ornithol. 17:313

33 Cade, T. J. 1982: The Falcons of the World. Collins, London.

34 Campbell, B. & Lack, E. 1985: A Dictionary of Birds. Poyser, Calton.

35 Cave, F. O. & Macdonald, J. D. 1955: The Birds of Sudan. Oliver & Boyd, London.

36 Cayley, N. W. 1964: What Bird is that? Angus & Robertson, Sydney.

37 CITES, Dollinger, P. (ed.) 1982 - Identification Manual. Vol. 2 & 2a. Secretariat of the Convention, Lausanne.

38 Clancey, P. A. 1957: The South African races of the Common Waxbill, Estrilda astrild (Linnaeus). Durban Mus. Nov. 5: 14-15.

39 Clancey, P. A. 1964: The Birds of Natal and Zululand. Oliver & Boyd, Edinburgh & London.

40 Clancey, P. A. 1966: Serinus gularis mendosus. Durban Mus. Nov. 7:612.

41 Clancey, P. A. 1974: Miscellaneous taxonomic notes on African birds. 38: Further on the races of the Masked Weaver, Ploceus velatus, Vieillot, with special reference to the status of Hyphantornis shelleyi, Sharpe, 1890, and Ploceus finschi, Reichenov, 1903. Durban Mus. Nov. 10 (5): 67-79.

42 Clancey, P. A. 1986: Breeding season and subspecific variation in the Red-headed Quelea. Ostrich 57:207-210.

43 Coates, B. J. 1985-1990: The Birds of Papua New Guinea. Vol. 1-2. Dove Publication, Alderley.

44 Coats, S. & Phelps, W. H. 1985: The Venezuelan Red Siskin: case history of an endangered species. in: Buckley, P. A., Forster, M. S., Morton, E. S., Ridgely, R. S. & Buckley, F. G. (eds.) Neotropical Ornithology. Pp. 977-985. AOU. Washington D. C.

45 Collar, N. J. & Stuart, S. N. 1985: Threatened Birds of Africa and Related Islands. ICBP & IUCN. Cambridge & Gland.

46 Collar, N. J., Round, P. D. & Wells, D. R. 1986: The past and future of Gurney's Pitta, Pitta gurneyi. Forktail 1:29-51.

47 Collar, N. J. & Andrew, P. 1988: Birds to watch. ICBP. Cambridge. Technical Publication No. 8.

48 Collar, N. J. 1991: Red Data Bird: Great Bustard. World Birdwatch 13 (1):13.

49 Conway, W. G. 1961: In Quest of the rarest Flamingo. Nat. Geo. Vol. 120, (1):91-102.

50 Cooper, W. T. & Forshaw, J. M. 1977: The Birds of Paradise and Bower Birds. Collins, Sydney & London.

51 Cramp, S. (ed.) 1977-1988: Handbook of the Birds of Europe, the Middle East and North Africa.5 vols. Oxford University Press, Oxford.

52 Crivelli, A. 1981: The Dalmatian Pelican, Pelecanus crispus Bruch 1832. A recently world-endangered species. Biol. Conservation 20:297-310.

53 Danks, A. 1990: Red Data Bird: Noisy Scrub-bird. World Birdwatch 12, No. 4:13.

54 Davies, S. J. J. F., Smith, G. T. & Robinson, F. N. 1982: The Noisy Scrub-bird in Western Australia. in: Groves, R. H. & Ride, W. D. L. (eds.) Species at Risk. Research in Australia. Springer Verlag, Berlin, Heidelberg, New York.

55 Davison, G. W. H. 1974: Geographical Variation in Lophophorus sclateri. Bull. Brit. Ornithol. Club 94:163-164.

56 Dee, T. J. 1986: The Endemic Birds of Madagascar. ICBP Cambridge.

57 Deignan, H. G. 1946: A new Pitta from the Malay Peninsula. Proc. Biol. Soc., Washington D. C. 59:55.

58 Delacour, J. 1927: Tropicoperdix charltoni tonkinensis, subsp. nov.. Bull. Brit. Ornithol. Club 47:152

59 Delacour, J. 1965: The Pheasants of the World. Country Life, London.

60 Delacour, J. 1973: The Waterfowl of the World. Vol. I-IV. Country Life, London & Arco Publ. Co. Inc., New York.

61 Delacour, J. & Mayr, E. 1946: Birds of the Philippines. MacMillan, New York.

62 Delacour, J. & Amadon, D. 1973: Curassows and Related Birds. American Mus. Nat. Hist., New York.

63 Dementiew, G. P. 1960: Der Gerfalke. A. Ziemsen Verlag, Wittenberg Lutherstadt.

64 Dementiev, G. P. & Gladkov, N. A. (eds.) 1966-1968: Birds of the Soviet Union. Vol. I-VI. Israel Program for Scientific Translations, Jerusalem.

65 Dennis, R. 1989: Soaring above Mountains, Moors and Glens. Birds International 1 No. 3:71-81.

66 duPont, J. E. 1971: Philippine Birds. Delaware Museum of Nat. Hist., Greenville.

67 Eley, J. W. 1982: Systematic relationships and zoogeography of the White-winged Guan (Penelope albipennis) and related forms. Wilson Bull. 94:241-259.

68 Ellis, D. H. 1985: The Austral Peregrine Falcon: color variation, productivity, and pesticides. Natl. Geogr. Res. 1:388-394.

69 Falla, R. A., Sibson, R. B. & Turbott, E. G. 1981: The new Guide to the Birds of New Zealand and outlying Islands. Collins, Auckland & London.

70 Farrand, J. 1983: The Audubon Society Master Guide to Birding. Vol. I-III. Alfred A. Knopf, New York.

71 ffrench, R. 1973: A Guide to the Birds of Trinidad and Tobago. Livingston Publ. Co., Wynnewood, Penn..

72 Field, G. D. 1979: A new species of Malimbus sighted in Sierra Leone and a review of the genus. Malimbus 1:2-13.

73 Fischer, W. 1959: Die Seeadler. Ziemsen Verlag, Wittenberg.

74 Fisher, C. T. 1986: A type specimen of the Paradise Parrot Psephotus pulcherrimus (Gould 1845). Australian Zoologist Vol. 22 (3):10-12.

75 Fisher, J., Simon, N. & Vincent, J. 1969: The Red Book. Wildlife in Danger. Collins, London.

76 Fittgen, H. 1985: Haben juvenile Wanderfalken (Falco peregrinus) blaugraue Füsse? Orn. Mitt. 37:267-268.

77 Fjeldså, J. & Krabbe, N. 1990: Birds of the High Andes. Zoological Museum & Apollo Books, Copenhagen & Svendborg.

78 Fleming, R. L. Sr. + Jr. & Bangdel, L. S. 1979: Birds of Nepal with reference to Kashmir and Sikkim. Avalok Publ., Kathmandu, Nepal.

79 Flint, V. E., Boehme, R. L., Kostin, Y. V. & Kuznetsov, A. A. 1984: A Field Guide to Birds of the USSR. Princeton University Press, Princeton.

80 Forshaw, J. M. & Cooper, W. T. 1981: Australian Parrots. Landsdowne, Melbourne.

81 Forshaw, J. M. & Cooper, W. T. 1989: (rev. ed.): Parrots of the World. Blandford Press, London.

82 Friedmann, H. 1960: The Parasitic Weaverbirds. Smithsonian Institution, Washington D. C..

83 Fry, C. H., Keith, S. & Urban, E. K. 1988: The Birds of Africa. Vol. III. Academic Press, London.

84 Garson, P. J. 1983: The Cheer Pheasant, Catreus wallichii in Himachal Pradesh, Western Himalayas: an update. World Pheasant Assoc. J. 8:29-39.

85 Gaston, A. J. & Singh, J. 1980: The status of the Cheer Pheasant, Catreus wallichii in the Chail Wildlife Santuary, Himachal Pradesh. World Pheasant Assoc. J. 5:68-73.

86 Gensbøl, B. 1984: Rovfuglene i Europa, Nordafrika og Mellemøsten. Gad, Copenhagen.

87 Gewalt, W. 1959: Die Grosstrappe. A. Ziemsen Verlag, Wittenberg Lutherstadt.

88 Ginn, P. J., McIlleron, W. G. & le S. Milstein, P. 1989: The complete book of Southern African birds. Struik, Cape Town.

89 Glutz von Blotzheim, U. N. et al. 1966- 1991: Handbuch der Vögel Mitteleuropas. Vol. 1-12. Aula Verlag, Wiesbaden.

90 Goriup, P. D. & Karpowicz, Z. J. 1985: A review of the past and recent status of the Lesser Florican. Bustard Studies 3:163-182.

91 Goriup, P. D. & Vardhan, H. (eds.) 1983: Bustards in decline. Jaipur, Tourism and Wildlife Society of India.

92 Gollop, J. B., Barry, T. W. & Iversen, E. H. 1986: Eskimo Curlew. A vanishing species? Special Publ. No. 17. Saskatchewan Nat. Hist. Soc., Saskatchewan.

93 Gonzalez, L. M., Hiraldo, F., Delibes, M. & Calderon, J. 1989: Zoogeographic support for the Spanish Imperial Eagle as a distinct species. Bull. Brit. Ornithol. Club 109:86-93.

94 Goodwin, D. 1970: Pigeons and Doves of the World. British Mus. Nat. Hist. & Comstock Publ.Ass., Ithaca & London.

95 Goodwin, D. 1982: Estrildid Finches of the World. British Mus. Nat. Hist. & Oxford University Press. Oxford.

96 Gould, J. 1836: Zosterops albogularis. Proc. Zool. Soc. London, 4:75-76.

97 Gray, G. R. 1867: List of the specimens of birds in the collection of the British Museum, p. 102. Brit. Mus., London.

98 Greenway, J. C. 1967: Extinct and Vanishing Birds of the World. Dover, New York.

99 Gretton, A. 1987: Gurney's Pitta: the struggle to survive. World Birdwatch 9 (4):6-7.

100 Griscom, 1929: Podilymbus gigas. American Mus. Novit. No. 379:5.

101 Grossman, M. L. & Hamlet, J. 1965: Birds of Prey of the World. Cassell, London.

102 Grote, 1928: Falco newtoni aldabranus. Orn. Monatsber. 36:78.

103 Gunning, J. W. B. & Roberts, A. 1911: Vinago orientalis. Ann. Transv. Mus. Pretoria 3:109.

104 Hachisuka, M. 1931-35: The Birds of the Philippine Islands. Vol. I-II. Witherby, London.

105 Hadden, D. 1981: Birds of the North Salomons. Wau Ecology Institute. Papua New Guinea. (Handbook No. 8).

106 Hald-Mortensen, P. 1970: A new subspecies of the Senegal Firefinch (Lagonosticta senegalae (L.)) from West Africa. Dansk Orn. Forenings Tidskrift 64:113-117.

107 Hall, B. P. & Moreau, R. E. 1970: An Atlas of Speciation in African Passerine Birds. Trustees of Brit. Mus. Nat. Hist., London.

108 Hancock, J. & Elliott, H. 1978: The Herons of the World.London Editions, London.

109 Hancock, J. & Kushlan, J. 1984: The Heron Handbook. Croom Helm, London.

110 Harris, M. 1982: A Field Guide to the Birds of Galapagos. (Rev. ed.) Collins, London.

111 Harrison, C. J. O. 1964: The taxonomic status of the African Silverbill Lonchura cantans and Indian Silverbill L. malabarica. Ibis 106:462-468.

112 Harrison, C. J. O. 1974: Plumage variation of the Banded Pitta. Avicultural Mag. 80 (2):41.

113 Harrison, P. 1989: Seabirds, an identification guide.Christopher Helm, London.

114 Haverschmidt, F. 1968: Birds of Surinam. Oliver & Boyd, Edinburgh & London.

115 Hayman, P., Marchant, J. & Prater, T. 1988: Shorebirds. An identification guide to the waders of the World. Christopher Helm, London.

116 Hays, C. 1984: The Humboldt Penguin in Peru. Oryx 18:92-95.

117 Heinroth, O. & Heinroth, M. 1966-67: Die Vögel Mitteleuropas. Band I-IV. Verlag für Kunst und Wissenschaft, Leipzig.

118 Heinzel, H., Fitter, R. & Parslow, J. 1977: Pareys Vogelbuch. Alle Vögel Europas, Nordafrikas und des Mittleren Ostens. Paul Parey, Hamburg & Berlin.

119 Helander, B., Ekman, B., Hägerroth, J-E., Hägerroth, P-Å. & Israelsson, J. 1989: Dräktkaraktärer hos havsörnar med känd ålder. Vår Fågelvärld 48:319-334.

120 Helvoort, B. van 1987: Status and conservation needs of the Bali Starling. Bull. Oriental Bird Club 5:9-12.

121 Henry, G. M. 1955: A Guide to the Birds of Ceylon. Geoffrey Cumberlese & Oxford University Press, London.

122 Hickey, J. J. (ed.) 1965: Peregrine Falcon Populations. Their Biology and Decline. University of Wisconsin Press, Madison, Milwaukee & London.

123 Hillgarth, N., Stewart-Cox, B. & Thouless, C. 1986: The decline of the Green Peafowl Pavo muticus. in: Ridley, M. (ed.) Pheasants in Asia 1986. World Pheasant Association. Basildon, UK.

124 Hicks, J., Olsen, P. & Greenwood, D. 1990: Saving the last survivor. Boobook Owls brought back to the forests of Norfolk Island. Birds International 2 (1):66-73.

125 Hilty, S. L. & Brown, W. L. 1986: A Guide to the Birds of Colombia. Princeton University Press, Princeton.

126 Hollom, P. A. D., Porter, R. F., Christensen, S. & Willis, I. 1988: Birds of the Middle East and North Africa. Poyser, Calton.

127 Hoogerwert, A. 1963: The differences between Pitta g. guajana and P. g. affinis. Bull. Brit. Ornithol. Club 83:96-98.

128 Holyoak, D. 1970: Comments on the classification of the Old World Ibises. Bull. Brit. Ornithol. Club 90:67-73.

129 Howard, R. & Moore, A. 1984: A complete Checklist of the Birds of the World. Macmillan, London.
130 Howes, J. 1988: Nordmann's Greenshank Tringa gutifer: status and threats. Asian Wetland News 1:12.
131 Hutton, 1874: Ocydromus hectori (Gallirallus australis hectori). Trans. New Zealand Inst. 6:110.
132 ICBP, 1990: An extraordinary bird from Caledonia. Birds International 2 (4):50-51.
133 Ilicev, V. D. & Flint, V. E. 1989: Handbuch der Vögel der Sowjetunion. Band 4. Galliformes, Gruiformes, & 1990: Band 6. Charadriiformes (Lari). Aula-Verlag, Wiesbaden.
134 Ingels, J., Parkes, K. C. & Farrand, J. 1981: The status of the macaw, generally but incorrectly called Ara caninde (Wagler). Gerfaut 71:283-294.
135 Inskipp, C. & Collar, N. J. 1984: The Bengal Florican: its conservation in Nepal. Oryx 18:3o-35.
136 Inskipp, C. & Inskipp, T. 1983: Results of a preliminary survey of Bengal Floricans Houbaropsis bengalensis in Nepal and India, 1982. ICBP, Cambridge.
137 Inskipp, C. & Inskipp, T. 1985: A guide to the birds of Nepal. Croom Helm, London.
138 Inskipp,C. & Inskipp, T. 1985: A survey of Bengal Floricans in Nepal and India.1982. Bustards Studies 3:141-160.
139 Johnsgard, P. A. 1981: The Plovers, Sandpipers, and Snipes of the World. University of Nebraska Press, Lincoln & London.
140 Johnsgard, P. A. 1983: The Grouse of the World. Croom Helm, London & Canberra.
141 Johnsgard, P. A. 1983: Cranes of the World. Croon Helm. London.
142 Johnsgard, P. A. 1986: The Pheasants of the World. Oxford University Press, Oxford.
143 Johnsgard, P. A. 1988: The Quails, Partridges, and Francolins of the World. Oxford University Press, Oxford.
144 Johnson, A. W. & Goodall, J. D. 1965: The birds of Chile and adjacent regions of Argentina, Bolivia and Peru. Vol. I. Platt Estab. Graf., Buenos Aires.
145 Kahl, M. P. 1971: Social behaviour and taxonomic relationships of the Storks. The Living Bird:151-170. Cornell Laboratory of Ornithology, Ithaca.
146 Kear, J. & Williams, G. 1978: Waterfowl at risk. Wildfowl 29:5-21.
147 Kemp, A. C. 1978: A review of the Hornbills: biology and radiation. The Living Bird: 105-136. Cornell Laboratory of Ornithology, Ithaca.
148 Kennedy, R. S. 1981: Saving the Philippine Eagle. Nat. Geo. 159 No. 6:847-856.
149 Khan, M. A. R. 1986: The threatened White-winged Wood Duck Cairina scutulata in Bangladesh. Forktail 2:97-101.
150 King, B. F. & Dickinson, E. C. 1975: A Field Guide to the Birds of South-East Asia. Houghton Mifflin Co., Boston.
151 King, W. B. 1978-1979: Red data book. 2 Aves. IUCN, Morges.
152 King, W. B. 1981: Endangered Birds of the World. Smithsonian Institution Press & ICBP, Washington D. C..
153 Kitson, A. R.. 1980: Larus relictus - a review. Bull. Brit. Ornithol. Club 100 (3):178-185.
154 Kloss, C. B. 1926: Two Neglected Bird Names: Eucichla guajana (P. L. S. Mull.) and Chloropsis cochinchinensis (Gm.). Journ. of the Malayan Branch Royal Asiatic Soc. Vol. IV Part 1:161-163.
155 Knoder, C. E. 1983: Elliot's Pheasant conservation. World Pheasant Assoc. J. 8:11-28.
156 Kothe, K. 1907: Rhynchotus pallescens. J. Ornithol. 55:164.
157 Kothe, K. 1907: Rhynchotus rufescens pallescens. J. Ornithol. 55:164.
158 Krabbe, T. N. 1934-1935: De grønlandske Jagtfalke. Videnskabelige Meddelser 98:49-109.
159 Kumerloeve, H. 1984: The Waldrapp, Geronticus eremita (Linnaeus, 1758): historical review, taxonomic history, and present status. Biol. Conserv. 30:363-373.
160 Lambert, F. 1983: Report of an expedition to survey the status of the St. Vincent Parrot Amazona guildingii. ICBP, Cambridge. (Study Report 3).
161 Lambert, F. 1985: The St. Vicent Parrot, an endangered Caribbean bird. Oryx 19:34-37.
162 Langrand, O. 1990: Guide to the Birds of Madagascar. Yale University Press, New Haven & London.
163 Larsen, T. 1986: Juvenile vandrefalker - ikke nödvendigvis brune? Vår Fuglefauna 9:36-37.
164 Lawson, P. W. & Lanning, D. V. 1981: Nestling and status of the Maroon-fronted Parrot (Rhynchopsitta terrisi). Pp. 385-392. in: Pasquier, R. F. (ed.) Conservation of New World parrots. Smithsonian Institution Press for ICBP. Washington D. C.
165 Lever, C. 1987: Naturalized Birds of the World. Longman Scientific and Technical, Harlow, UK.
166 Lister, J. J. 1888/89: Ninox natalis, sp. nov. Proc. Zool. Soc. London:525-528.
167 Lönneberg, E. 1931: Larus melanocephalus relictus. Ark. Zool. 23B, No. 2:2 + fig. 1 p. 5. Stockholm.
168 Low, R. 1984: Endangered Parrots. Blandford Press, Poole.
169 Lyster, S. 1985: International Wildlife Law. Grotius, Cambridge.
170 MacKinnon, J. 1990: Field Guide to the Birds of Java and Bali. Gadjah Mada University Press, Yogyakarta, Indonesia.
171 Mackworth-Praed, C. W. & Grant, C. H. B. 1960: Birds of Eastern and North-Eastern Africa. Vol. I-II. Longmans, London.
172 Mackworth-Praed, C. W. & Grant, C. H. B. 1963: Birds of the Southern Third of Africa. Vol. I-II. Longmans, London.
173 Mackworth-Praed, C. W. & Grant, C. H. B. 1973: Birds of West Central and Western Africa. Vol. I-II. Longmans, London.
174 McLachlan, G. R. & Liversidge, R. 1982: Roberts Birds of South Africa. The Trustees of the John Voelcker Bird Book Fund, Cape Town.
175 McNulty, F. 1966: The Whooping Crane. Longmans, London.
176 Madge, S. & Burn, H. 1988: Wildfowl: an identification guide to the ducks, geese and swans of the world. Helm, London.
177 Manry, D. E. 1985: Distribution, abundance and conservation of the Bald Ibis Geronticus calvus in southern Africa. Biol. Conservation 33:351-362.
178 Marchant, S. & Higgins, P. J. (eds.) 1990: Handbook of Australian, New Zealand and Antarctic Birds. Oxford University Press, Melbourne.
179 Marshall, J. T. 1949: The Endemic Avifauna of Saipan, Tinian, Guam, and Palau. Condor 51:200-221.
180 Massey-Stewart, J. 1987: The "lily of birds": the success story of the Siberian white crane. Oryx 21:6-10.
181 Mathews, G. M. 1912: Ninox boobook royana, subsp. n. Austr. Av. Rec. 1:120.
182 Mathews, G. M. 1933: On the name of the Boobook Owl... Ibis 75:353.
183 Mayr, E. 1978: Birds of the South-west Pacific. Charles E. Tuttle Co., Rutland, Vermont & Tokyo.
184 Mayr, E. & Vuilleumier, F. 1983: New species of birds described from 1966 to 1975. J. Ornithol. 124:217-232.
185 Medway, L. & Wells, D. R. 1976: The Birds of the Malay Peninsula. Vol. V. Conclusion and Survey of every Species. Witherby, London.
186 Mees, G. F. 1969: A systematic review of the Indo-Australian Zosteropidae, Part III. Zool. Verh., Leiden.
187 Meyer, A. B. & Wiglesworth, L. W. 1898: The Birds of Celebes and Neighbouring Islands. Friedländer, Berlin.
188 Meyer de Schauensee, R. 1958: Pitta guayana bangkae de Schauensee. Proc. Acad. Nat. Sci., Philadelphia 110:286.
189 Meyer de Schauensee, R. 1964: The Birds of Colombia and Adjacent Areas of South and Central America. Livingston Publ. Co., Narberth, Penn.
190 Meyer de Schauensee, R. 1970: A Guide to the Birds of South America. Oliver & Boyd, Edinburgh.
191 Meyer de Schauensee, R. 1984: The Birds of China. Oxford University Press, Oxford.
192 Meyer de Schauensee, R. & Phelps, W. H. 1978: A Guide to the Birds of Venezuela. Princeton University Press, Princeton.
193 Mikkola, H. 1983: Owls of Europe. Poyser, Calton.
194 Milligan, A. W. 1910: Description of a New Bristle Bird (Sphenura). Emu 10:67-69
195 Morony, J. J., Bock, W. J. & Farrand, J. 1975: Reference List of the Birds of the World. with Corrections and Additions 1978. American Mus. Nat. Hist., New York.
196 Mountford, G. 1988: Rare Birds of the World. Collins & ICBP, Cambridge.
197 National Geographic Soc. 1983: Field Guide to the Birds of North America. Nat. Geogr. Soc., Washington D. C.

198 Nature Conservancy Council, 1990: World Checklist of Threatened Birds. NCC, Peterborough, UK.

199 Nelson, 1900: Cyrtonyx montezumä mearnsi, subsp. nov.. Auk 17:255-256.

200 Nelson, J. B. 1971: The biology of Abbott's Booby. Ibis 113: 429-467.

201 Nelson, J. B. 1978: The Sulidae - Gannets and Boobies. Oxford University Press, Oxford.

202 Neumann, O. 1905: Lagonosticta senegala abayensis. J. Ornithol. 53:349-350.

203 Neumann, O. 1905: Estrilda astrild nyansae, nov. subsp.. J. Ornithol. 53:596-597.

204 Newman, A. 1865: Drymoeca rodericana, sp. nov.. Proceedings Zoological Soc. of London:47-48.

205 Newman, K. 1990: Birds of Southern Africa. Southern Book Publ., Johannesburg.

206 Nichols, H. A. & Nichols, T. D. 1973: St. Vincent Parrot: plumage polymorphism, juvenile plumage and nidification. Bull. Brit. Ornithol. Club 93:120-123.

207 Nicolai, J. 1969: Beobachtungen an Paradieswitwen (Steganura paradisaea L., Steganura obtusa Chapin) und der Strohwitwe (Tetraenura fischeri Reichenow) in Ostafrika. J. Ornithol. 110:421-447.

208 Ogilvie-Grant, M. 1893: Caloperdix sumatrana. Bull. Brit. Ornithol. Club 1:5.

209 Ogilvie-Grant, M. 1893: Caloperdix borneensis. Bull. Brit. Ornithol. Club 1:5.

210 Ogilvie-Grant, M. 1907: Estrilda astrild macmillani sp. n.. Bull. Brit. Ornithol. Club 19:108

211 Ogilvie, M. & Ogilvie, C. 1986: Flamingos. Alan Sutton, Gloucester.

212 Olsen, P. & Olsen, J. 1989: Living with the World's Most Studied Raptor. Birds International 1 (3):23-31.

213 O'Neill, J. P., del Solar, R. G., Ortiz, T. E., Eley, J. W. & Williams, M. D. 1981: The White-winged Guan, Penelope albipennis, its rediscovery, status, nestling, systematics, and recommendations for its continued survival. Pp. 203-215 in: Primer Simposio Internacional de la Familia Cracidae: Memorias. Universidad Nacional Autonoma de Mexico, Mexico City.

214 Oren, D. C. & Novaes, F. C. 1986: Observations on the Golden Parakeet, Aratinga guarouba in Northern Brazil. Biological Conservation 36:329-337.

215 Osborne, P., Collar, N. & Goriup, P. 1984: Bustards. Dubai Wildlife Research Centre, Dubai.

216 Oustalet, E. 1881: Cyornis ruecki. Bull. Soc. Philomathique de Paris, 7 Ser. 5:78.

217 Palmer, R. S. (ed.) 1962: Handbook of North American Birds. Vol. I. Yale University Press, New Haven & London.

218 Pasquier, R. F. (ed.) 1980: Conservation of New World Parrots. Proceedings of the ICBP Parrot Working Group Meeting St. Lucia, 1980. Smithsonian Institution Press.

219 Payne, R. B. 1971: Paradise Whydahs, Vidua paradisaea and V. obtusa of southern and eastern Africa, with notes on differentiation of the females. Bull. Brit. Ornithol. Club 91:66-76.

220 Payne, R. B. 1982: Species Limits in the Indigobirds (Ploceidae, Vidua) of West Africa: Mouth Mimicry, Song Mimicry, and Description of new species. University Michigan Mus. Zool. Misc. Publ. 162:1-96.

221 Penny, M. 1982: The Birds of Seychelles and the Outlying Islands. Collins, London.

222 Peters, J. L. et al. 1931-1986: Checklist of Birds of the World. Vol. I-XV. Harvard University Press, Cambridge, Mass..

223 Peterson, R. T. & Chalit, E. L. 1973: A Field Guide to Mexican Birds. Houghton Mifflin Co., Boston.

224 Philip, Prince, Duke of Edinburgh, & Fisher, J. 1970: Wildlife Crisis. Hamilton, London.

225 Pizzey, G. 1981: A Field Guide to the Birds of Australia. Collins, Sydney.

226 Pool, K. G. 1989: Determining age and sex of nestling Gyrfalcons. J. Raptor Res. 23:45-47.

227 Pratt, H. D., Bruner, P. L. & Berrett, D. G. 1987: A Field Guide to the Birds of Hawaii and the Tropical Pacific. Princeton University Press, Princeton.

228 Pyle, P, Howell, S. N. G., Yunick, R. P. & De Sante, D. F. 1987: Identification guide to North American Passerines. Slate Creek Press, Bolinas, California.

229 Raethel, H.-S. 1988: Hühnervögel der Welt. Neumann-Neudamm. Melsungen.

230 Raffles, 1822: Tantalus (Mycteria) cinereus. Trans. Linn. Soc. London 13:327.

231 Rahmani, A. R. & Manaskadam, R. 1985: Present status of the Great Indian Bustard. Bustard Studies 3:123-132.

232 Reader's Digest, 1985: Complete book of New Zealand birds. Reader's Digest, Sydney.

233 Reader's Digest, 1986: Complete book of Australian birds. Reader's Digest, Sydney.

234 Reichenow, A. 1892: Spermestes stigmatophorus. J. Ornithol. 40:46.

235 Reichenow, A. 1892: Nigrita sparsimguttata. J. Ornithol. 40:132.

236 Restall, R. L. 1975: Finches and other seed-eating birds. Faber & Faber, London.

237 Reville, R., Tranter, J. & Yorkston, H. 1987: Monitoring the endangered Abbot's Booby on Christmas Island 1983-1986. Australian National Parks and Wildlife Service (Occ. Paper No. 11), Canberra.

238 Ridgely, R. S. 1981: A guide to the birds of Panama. Princeton University Press, Princeton.

239 Ridgely, R. S. 1981: The current distribution and status of mainland Neotropical parrots.in: Pasquier, R. F. (ed.) Conservation of New World parrots, pp. 233-284. ICBP, Technical Publ. No. 1. Smithsonian Institution Press.

240 Ridgely, R. S. & Tudor, G. 1989: The Birds of South America. Vol. I, The Oscine Passerines. Oxford University Press, Oxford, Tokyo.

241 Ripley, S. D. 1977: Rails of the World. David R. Godine, Boston.

242 Ripley, S. D. 1961: A synopsis of the birds of India and Pakistan. Natural History Society, Bombay.

243 Ripley, S. D. & Beehler, B. M. 1985: Rails of the World, a Compilation of New Information, 1975-1983 (Aves:Rallidae). Smithsonian Institution Press, Washington D. C.

244 Roberts, A. 1932: Estrilda astrild ngamiensis. Ann. Transvaal Museum, Pretoria 15:33.

245 Roberts, A. 1932/35: Vinago schalowi chobiensis. Ann. Transvaal Museum, Pretoria 15:25

246 Robinson, H. C. & Chasen, F. N. 1939: The Birds of the Malay Peninsula. Vol. IV. Witherby, London.

247 Robinson & Kinnear, 1928: Cyornis ruecki. Novitates Zool. 34:256-257.

248 Robinson & Kloss, 1919: Cyornis vanheysti (Niltava ruecki). Journal of the Straits Branch of the Royal Asiatic Society. Singapore 80:104.

249 Rooke, I. 1986: Survey of the White-breasted White-eye and the Norfolk Island Boobook Owl on Norfolk Island, October-November 1985. Royal Australasian Ornithologist's Union. Victoria.

250 Rothschild, L. W. 1892: Anas Laysanensis, sp. n.. Bull. Brit. Ornithol. Club 1:XVII.

251 Rothschild, L. W. 1917: Melanoperdix nigra borneensis, subsp. nov.. Bull. Brit. Orniithol. Club 38:3.

252 Round, P. D. 1988: Resident forest birds in Thailand. Their status and conservation. ICBP Monograph 2. Cambridge.

253 Rowan, M. K. 1983: The Doves, Parrots, Lories and Cuckoos of Southern Africa. Croom Helm, London & Canberra.

254 Rutgers, A. 1972: John Gould. Birds of South America. Eyre Methuen, London.

255 Sahin, R. 1980: Erfolgreiche Volierenbrut der Waldrappen in der Türkei. Orn. Mitt. 32:72-74.

256 Salomonsen, F. 1967: Fuglene på Grønland. Rhodos, Copenhagen.

257 Salomonsen, F. & Gitz-Johansen, Å. 1950: The Birds of Greenland. Ejnar Munksgaard, Copenhagen.

258 Sanft, K. 1960: Bucerotidae. Das Tierreich. Lieferung 76. Walter de Gruyter & Co., Berlin.

259 Sankaran, R. 1987: The Lesser Florican. Sanctuary 7:26-37.

260 Schiøler, E. L. 1925-1931: Danmarks Fugle. Vol. 1-3. Gyldendalske Boghandel, Copenhagen.

261 Scholten, C. J. 1990: The timing of moult in relation to age, sex, and breeding status in a group of captive Humboldt Penguins (Spheniscus humboldti), at Emmen Zoo, The Netherlands. Netherlands J. Zool. 39:113-125.

262 Scott, D. A. & Brooke, M. de L. 1985: The endangered avifauna of southeastern Brazil: a report on the BOU/WWF expedition of 1980/81 and 1981/82. Pp. 115-139 in: Diamond, A. W. &

Lovejoy, T. E. (eds.) Conservation of tropical forest birds. ICBP, (Technical Publication 4.) Cambridge.
263 Schuchmann, K-L. 1978: Allopatrische Artbildung bei der Kolibrigattung Trochilus. Ardea 66:156-172.
264 Schuchmann, K-L. 1978: Notes on the Rufous-capped Thornbill Chalcostigma ruuficeps, a new hummingbird species for Colombia. Bull. Brit. Ornithol. Club 98:115-116.
265 Schuchmann, K-L. 1980: Ökologie und evolution der Trochilidenfauna auf den ozeanischen Inseln der Karibischen See. Bonner Zool. Beitr. 31:289-309.
266 Schuchmann, K-L. 1984: Two hummingbird species, one a new subspecies, new to Bolivia. Bull. Brit. Ornithol. Club 104:6.
267 Schuchmann, K-L. 1990: Biologie, Haltung und Pflege wenig bekannter Kolibriarten (Trochilidae) - Teil I. Trochilus 11 (2):45-51.
268 Serle, W., Morel, G. J. & Hartwig, W. 1988: A field guide to the birds of West Africa. Collins, London.
269 Severinghaus, S. R. 1978: Recommendations for the conservation of the Swinhoe's and Mikado Pheasants in Taiwan. World Pheasant Assoc. J. 3:79-89.
270 Severinghaus, S. 1986: The adaptability of Mikado and Swinhoe's Pheasants to disturbed habitats in Taiwan. Unpaginated. in: Ridley, M. (ed.) Pheasants in Asia 1986. World Pheasant Ass., Basildon.
271 Severinghaus, S. R. & Blachshaw, K. T. 1976: A new guide to the birds of Taiwan. Me Ya Publications, Taipei.
272 Sharpe, R. B. & Chubb, C. 1906: Arboricola graydoni, sp. n. Ornis 13:133.
273 Sherrod, S. K., White, C. M. & Williamson, F. S. L. 1976: Biology of the Bald Eagle on Amchitka Island, Alaska. Living Bird 15:143-182.
274 Short, L. L. 1982: Woodpeckers of the World. Delaware Mus. Nat. Hist., Greenville.
275 Sick, H. & Teixeira, D. M. 1983: The discovery of the home of the Indigo Macaw Anadorhynchus leari Bonaparte, 1856. Hornero, No. extraord.:109-112.
276 Siegfried, N. R. 1978: Social behaviour of the African Comb Duck. Living Bird 12:85-104.
277 Silvius, M. J. & Verheugt, W. J. M. 1989: The status of Storks, Ibises and Spoonbills in Indonesia. Kukila 4 (3-4): 119-132.
278 Simpson, K. & Day, N. 1984: The Birds of Australia. Lloyd O'Neil, South Yarra, Victoria.
279 Slater, P. 1980: Rare and vanishing Australian birds. Rigby, Adelaide.
280 Smithe, F. B. 1975: Naturalist's Color Guide. American Mus. Nat. Hist., New York.
281 Smythies, B. E. 1953: The Birds of Burma. Oliver & Boyd, Edinburgh.
282 Smythies, B. E. 1968: The Birds of Borneo. Oliver & Boyd, Edinburgh & London.
283 Smythies, B. E. 1986: The Birds of Burma. Nimrod Press, Liss, Hants, UK and Silvio Mattacchione & Co., Pickering, Ontario.
284 Snow, D. 1982: The Cotingas: bellbirds, umbrellabirds, and their allies. Oxford University Press, and Brit. Mus. Nat. Hist., Oxford.
285 Snyder, H. A. & Snyder, N. F. R. 1990: The comeback of the California Condor. Birds International 2 (2):10-23.
286 Sophasan, S. & Dobias, R. 1984: The fate of the "princess bird", or White-eyed River Martin (Pseudochelidon sirintarae). Nat. Hist. Bull. Siam Soc. 32 (1):1-10.
287 Staub, F. 1976: Birds of the Mascarenes and Saint Brandon. Organisation Normale des Enterprises Ltee, Mauritius.
288 Stoddart, D. R. 1981: Abbott's Booby on Assumption. Atoll Res. Bull. 255:27-32.
289 Summers-Smith, J. D. 1988: The Sparrows. Poyser, Calton.
290 Svensson, L. 1984: Identification guide to European Passerines. Lars Svensson, Stockholm.
291 Tacha, T. C., Vohs, P. A. & Warde, W. D. 1985: Morphometric variation of Sandhill Cranes from mid-continental North America. J. Wildl. Manage 49 (1):246-250.
292 Tarboton, W. 1984: The status and conservation of the Wattled Crane in the Transvaal. Proc. V. Pan Afr. Ornithol. Congr.:665-678.
293 Taynton, K. & Jeggo, D. 1988: Factors affecting breeding success of Rothschild's Mynah, Leucopsar rothschildi at the Jersey Wildlife Preservation Trust. Dodo (J. Jersey Wildl. Preserv. Trust 25:66-76.
294 Temple, S. A. 1986: Recovery of the endangered Mauritius Kestrel from an extreme population bottleneck. Auk 103:632-633.
295 Terres, J. K. 1980: The Audubon Society Encyclopedia of North American Birds. Alfred A. Knopf, New York.
296 Thomsen, J. B. & Munn, C. A. 1988: Cyanopsitta spixii: a non-recovery report. Parrotletter 1 (1):6-7.
297 Todd, F. S. 1979: Waterfowl. Ducks, Geese and Swans of the World. Sea World Press Publ., New York & London.
298 Trans. Linn. Soc. London 1821: Myiothera (Pitta) affinis. Java. 13:154.
299 Tristram, F. R. S. 1879: Description of a New Species of Woodpecker from the Island of Tzus Sima, near Japan. in: Proc. Zool. Soc. London 386-387.
300 Truslow, F. K. 1961: Eye to Eye with Eagles. Nat. Geo. 119 (1):123-148.
301 Tuck, G. & Heinzel, H. 1980: Die Meeresvögel der Welt. Paul Parey, Hamburg & Berlin.
302 Turbott, E. G. (ed.) 1967: Buller's Birds of New Zealand. East-West Center Press, Honolulu.
303 Turner, A. & Rose, C. 1989: A Handbook to the Swallows and Martins of the World. Christopher Helm, London.
304 Tye, H. 1986: The erectile crest and other head feathering in the genus Picathartes. Bull. Brit. Ornithol. Club 106: 90-93.
305 Urban, E. K., Fry, C. H. & Keith, S. 1986: The Birds of Africa. Vol. II. Academic Press, London and Florida.
306 Van Marle, J. G. & Voous, K. H. 1988: The Birds of Sumatra. BOU Checklist No. 10, BOU, Tring.
307 Van Tyne, J. & Berger, A. J. 1976: Fundamentals of Ornithology. John Wiley & Sons, New York.
308 Vardhan, H. 1985: A report on the status of Bustards in India. Bustards Studies 3:113-118.
309 Vaurie, C. 1953: A Generic Revision of Flycatchers of the Tribe Muscicapini. Bull. Amer. Mus. Nat. Hist. Vol. 100: Article 4, New York.
310 Vaurie, C. 1959-65: The Birds of the Palaearctic Fauna. Vol. I, Passeriformes. Vol. II, Non-Passeriformes. Witherby, London.
311 Vaurie, C. 1962: The status of Larus relictus and of other hooded gulls from Central Asia. Auk 79:303-309.
312 Vaurie, C. 1968: Taxonomy of the Cracidae (Aves). Bull. Amer. Mus. Nat. Hist., Vol. 138:Article 4, New York.
313 Wick, F. 1980: Birds of Prey of the World. Paul Parey, Hamburg & Berlin.
314 Verheugt, W. J. M. 1987: Conservation status and action program for the Milky Stork (Mycteria cinerea). Colonial Waterbirds 10 (2):211-220.
315 Vincent, J. 1964-71: Red Data Book, 2 Aves. IUCNN, Morges, Switzerland.
316 Voous, K. H. & Cameron, A. 1989: Owls of the Northern Hemisphere. The MIT Press, Cambridge, Mass..
317 Wahlstedt, J. 1988: Farväl till atitlandoppingen. Vår Fågelvärld 47 (5):266.
318 Warner, R. E. 1963: Recent History and Ecology of the Laysan Duck. Condor 65:2-23.
319 Watling, D. 1982: Birds of Fiji, Tonga and Samoa. Millwood Press, Wellington, New Zealand.
320 White, C. M. N. 1963: A revised check list of African Flycatchers, Tits, Tree Creepers, Sunbirds, White-Eyes, Honey Eaters, Buntings, Finches, Weavers and Waxbills. The Government Printer, Lusaka.
321 White, C. M. N. & Bruce, M. D. 1986: The Birds of Wallacea. (Sulawesi, the Moluccas and Lesser Sunda Island, Indonesia): an annotated check-list. BOU, London.
322 Whitlock, R. 1981: Birds at Risk. a comprehensive world-survey of threatened species. Moonraker Press, Wiltshire.
323 Wild Bird Society of Japan. 1983: A field guide to the birds of Japan. Wild Bird Soc. of Japan, Tokyo.
324 Williams, G. R. & Given, D. R. 1981: The red data book of New Zealand. Nature Conservation Council, Wellington, New Zealand.
325 Williams, J. G. & Arlott, N. 1992: A field guide to the Birds of East Africa. Collins, London.
326 Wiseley, A. N. & Pinel, H. 1987: Occurrence, distribution, and plumage of Gyrfalcons in the Calgary area. Alberta Nat. 17:37-45.
327 Witherby, H. F., Jourdain, F. C. R., Ticehurst, N. F. & Tucker, B.

W. 1938-1941: The Handbook of British Birds. Vol. 1-5. Witherby, London.
328 Yamashita, C. 1987: Field observations and comments on the Indigo Macaw (Anodorhynchus leari), a highly endangered species from north-eastern Brazil. Wilson Bull. 99:280-282.
329 Yamashina, Y. 1948: Notes on the Marianas Mallard. Pacific Science 2:121-124.

ADDENDA

330 Anon. 1992: Great Philippine Eagle bred in Captivity. Oriental Bird Club 15:14.
331 Amos, E. J. R. 1991: A guide to the birds of Bermuda. Corncrake, Bermuda.
332 Argeloo, M. 1992: The Maleo more than a symbol. World Birdwatch 14 (1):8--9.
333 Blakers, M., Davies, S. J. J. F. & Reily, P. V. (eds.) 1984: The atlas of Australian birds. Melbourne Univ. Press, Melbourne.
334 Boetticher, H. von 1959: Die Pfefferfresser. Die Neue Brehm Bücherei. A. Ziemsen Verlag, Wittenberg Lutherstadt.
335 Bradley, P. 1985: Birds of the Cayman Islands. P. E. Bradley, George Town, Grand Cayman.
336 British Ornithologist's Union Records Committee. 1988: Suggested change to the English names of some Western Palaearctic birds. Ibis 130 (Supplement) Brit. Birds 81:355-377.
337 Brooke, M. & Birkhead, T. (eds.) 1991: The Cambridge Encyclopedia of Ornithology. Cambridge University Press, Cambridge.
338 Browning, M. R. 1992: Comments on the Nomenclature and dates of Publication of some Taxa in Bucerotidae. Bull. Brit. Ornithol. Club 112 (1):22-27.
339 Bruning, D. 1989: Saved! But where to now? Birds International 1 (1):73-78.
340 Collar, N. J. 1992: Red Data Bird: Red Siskin. World Birdwatch 14 (1):13
341 Cooper, J. E., Needham, J. R., Applebee, K. & Jones, C. G. 1988: Clinal and pathological studies on the Mauritian Pink Pigeon, Columba mayeri. Ibis 130:57-64.
342 Craig, A. J. F. K. 1992: The identification of Euplectes species in non-breeding plumage. Bull. Brit. Ornithol. Club 112 (2):102-108.
343 Delacour, J. 1966: Guide des oiseaux de la Nouvelle-Calédonie et de ses dépendances. Delachaux et Niestlé, Neuchâtel.
344 Devillers, P. 1976: Projet de Nomenclature Française des Oiseaux du Monde.1.Struthionides aux Phoenicopterides. Gerfaut 66:153-168.
345 Devillers, P. 1976: Projet de Nomenclature Française des Oiseaux du Monde.2. Anhimides aux Otidides. Gerfaut 66:391-421.
346 Devillers, P. 1977: Projet de Nomenclature Française des Oiseaux du Monde.3. Jacanides aux Psittacidés. Gerfaut 67:171-200.
347 Devillers, P. 1977: Projet de Nomenclature Française des Oiseaux du Monde.4. Musophagides aux Coliides. Gerfaut 67: 337-365.
348 Devillers, P. 1977: Projet de Nomenclature Française des Oiseaux du Monde.5. Trogonidés aux Picidés. Gerfaut 67:469-489.
349 Devillers, P. 1978: Projet de Nomenclature Française des Oiseaux du Monde. 6. Eurylaimidés aux Furnariidés. Gerfaut 68: 129-136.
350 Devillers, P. 1978: Projet de Nomenclature Française des Oiseaux du Monde. 7. Formicariides et Rhinocryptides. Gerfaut 68:233-240.
351 Devillers, P. 1978: Projet de Nomenclature Française des Oiseaux du Monde. 8. Cotingidae aux Xenicidae. Gerfaut 68:703-720.
352 Devillers, P. 1980: Projet de Nomenclature Française des Oiseaux du Monde. 9. Alaudidae aux Prunellidae. Gerfaut 70:121-146.
353 Dickinson, E. C., Kennedy, R. S. & Parkes, K. C. 1991: The Birds of the Philippines. BOU Check-list No. 12. BOU, Tring.
354 Duff, D. G., Bakewell, D. N. & Williams, M. D. 1991: The Relict Gull Larus relictus in China and elsewhere. Forktail 6:43-65.
355 duPont, J. E. 1972: Notes on Philippine Birds (no. 2). Birds of Ticao. Nemouria 6:1-13. Occ. Papers Delaware Mus. Nat. Hist..
356 duPont, J. E. 1976: South Pacific Birds. Delaware Mus. Nat. Hist..
357 duPont, J. E. & Rabor, D. S. 1973: South Sulu Archipelago Birds. An Expedition Report. Nemouria 9:1-63. Delaware Mus. Nat. Hist..
358 duPont, J. E. & Rabor, D. S. 1973: Birds of Dinagot and Siargao, Philippines. Nemouria 10:1-111. Delaware Mus. Nat. Hist..
359 Eck, S. & Busse, H. 1973: Eulen. Die neue Brehm Bücherei. A. Ziemsen Verlag, Wittenberg Lutherstadt.
360 Etchécopar, R. D. & Hüe, F. 1978: Les Oiseaux de Chine. Non passereaux. Ed. du Pacifique, Paris.
361 Evans, P. G. H. 1991: Status and conservation of Imperial and Red-necked Parrots Amazona imperialis and A. arausiaca on Dominica. Bird Conservation International 1:11-32.
362 Faanes, C. A. & Senner, S. E. 1991: Status and Conservation of the Eskimo Curlew. Am. Birds 45:237-239.
363 Favre, D. S. 1989: International Trade in Endangered Species. A guide to CITES. Martinus Nijhoff Publishers, Dordrecht.
364 Frith, C. B. & Dauglas, V. E. 1978: Notes on ten Asian Hornbill species (Aves:Bucerotidae); with particular reference to growth and behaviour. Nat. Hist. Bull. Siam Soc. 27:35-82.
365 Frith, C. B. & Frith, D. W. 1978: Bill growth and development in the Northern Pied Hornbill. Avicultural Mag. 84:20-31.
366 Frith, C. B. & Frith, D. W. 1983: A systematic Review of the Hornbill Genus Anthracoceros (Aves, Bucerotidae). Zoological J. Linn. Soc. 78:29-71.
367 Glenister, A. G. 1971: The Birds of the Malay Peninsula, Singapore and Penang. Oxford University Press, London.
368 Gnam, R. S. 1990: Conservation of the Bahama Parrot. Am. Birds 44:32-36.
369 Gore, M. E. & Gepp, A. R. M. 1978: Las Aves del Uruguay. Mosca Hnos. Edit., Montevideo.
370 Gould, J. 1988: A Monograph of the Ramphastidae or Family of Toucans. Hell House Publ., Melbourne and Brit. Mus. Nat. Hist., London.
371 Gould, J. 1990: John Gould's Hummingbirds. Wordsworth, Ware. With a partial Synonymy to Gould's Hummingbirds by Paul R. Clyne, 420 Smith Avenue/Chapel, HM NC 27516, USA.
372 Greenewalt, C. H. 1960: Hummingbirds. American Mus. Nat. Hist., New York.
373 Greenewalt, C. H. 1966: The Marvellous Hummingbird rediscovered. Nat. Geo. 139 No. 1:98-101.
374 Gretton, A. 1990: Slender-billed Curlew. World Birdwatch 12 (3):11
375 Haffer, J. 1974: Avian Speciation in Tropical South America with a Systematic Survey of the Toucans (Ramphastidae) and Jacamars. Mass. Publ. Nuttall Ornithol. Club no. 14. Cambridge, Mass..
376 Herremans, M., Louette, M. & Stevens, J. 1991: Conservation status and vocal and morphological description of the Grand Comoro Scops Owl O. p. Benson 1960. Bird Conservation Int. 1:123-133.
377 Holmes, D. & Nash, S. 1990: The Birds of Sumatra and Kalimantan. Oxford University Press, Oxford.
378 Hüe, F. & Etchécopar, R. D. 1970: Les Oiseaux du Proche et du Moyen Orient. Boubée, Paris.
379 Hume, R. & Boyer, T. 1991: Owls of the World. Dragonsworld, Limpsfield.
380 Hutton, I. 1991: Birds of Lord Howe Island. Past and Present. Ian Hutton, Coffs Harbour Plaza.
381 Immelmann, K. 1965: Australian Finches in Bush and Aviary. Angus and Robertson, Sydney.
382 Johnsgard, P. A. 1991: Bustards, Hemipodes, and Sandgrouse. Birds of dry places. Oxford University Press. Oxford.
383 Jørgensen, H. I. 1958: Nomina Europaearum. Ejnar Munksgaard, Copenhagen.
384 Juniper, A. T. & Yamashita, C. 1991: The Habitat and Status of Spix's Macaw. Bird Conservation Int. 1:1-9.
385 Kemp, A. C. & Growe, T. M. 1985: The Systematics and Zoogeography of Afrotropical Hornbills (Aves:Bucerotidae). Pp.279-324. Proc. Intern. Symp. African Vertebr., Zool. Forsch. Mus. A. Koenig, Bonn.
386 Konig, C. & Straneck, R. 1989: A new Owl (Aves: Strigidae) from northern Argentina. Stuttgarter Beitr. Naturkd. Ser. A. (Biol.) 428, 20 pp.
387 Lieras, M. 1983: A Bird's Eye View. Zoonooz 56 (3):4-10.
388 Low, R. 1972: The Parrots of South America. John Gilford Ltd., London.
389 Low, R. 1980: Parrots, their Care and Breeding. Blandford, Poole.
390 Lowther, E. H. N. 1942: Notes on some Indian Birds. VII. Hornbills. J. Bombay Nat. Hist. Soc. 43:389-401.

391 MacKinnon, J. 1979: A Glimmer of Hope for Sulawesi. Oryx 15 (1):55-59.
392 Madge, S. G. 1969: Notes on the Breeding of the Bushy-crested Hornbill Anorrhinus galeritus. Malay Nat. J. 23:1-6
393 Mayr, E. & Gilliard, E. T. 1954: Birds of Central Guinea. Bull. American Mus. Nat. Hist., New York 103 (4):313-374.
394 Ogilvie-Grant, W. R. 1892: Suborder Bucerotes:347-428, in: Sharpe, R. B. & Ogilvie-Grant, W. R.: Catalogue of the Birds in the British Museum. Vol. XVII, London.
395 Olrog, C. C. 1984: Las Aves Argentinas. "Una Nueva Guia de Campo". Admin. de Parques Nacionales, Buenos Aires.
396 Olsen, S. L. & Warheit, K. I. 1988: A new genus for Sula abbotti. Bull. Brit. Ornithol. Club 108:9-12.
397 O'Neill, J. P., Munn, C. A. & Franke, I. J. 1991: Nannopsittaca dachilleae, a new species of Parrotlet from eastern Peru. Auk 108:225-229.
398 Parkes, K. C. 1971: Taxonomic and distributional notes on Philippine Birds. Nemouria 4:1-67. Delaware Mus. Nat. Hist..
399 Parkes, K. C. 1973: Annotated List of the Birds of Leyte Island, Philippines. Nemouria 11:1-73. Delaware Mus. Nat. Hist..
400 Phelps, W. H. & Meyer de Schauensee, R. 1978: Una Guia de las Aves de Venezuela. Princeton University Press, Princeton, New Jersey.
401 Rand, A. L. & Rabor, D. S. 1960: Birds of the Philippine Islands: Siquijor, Mount Malindang, Bohol, and Samar. Fieldiana: Zoology Vol. 35 Number 7:225-441. Chicago Nat. Hist. Mus., Chicago.
402 Reichenow, A. 1899: Die Vögel der Bismarckinseln. Mitt. Zool. Sammlung Mus. für Naturkunde, Berlin.
403 Restrepo, C. & Mondrgon, M. L. 1988: Der Tukanbartvogel Semnornis ramphastinus - eine gefährdete Art. Trochilus 9:116-117.
404 Ripley, S. D. & Rabor, D. S. 1958: Notes on a Collection of Birds from Mindoro Island, Philippines. Peabody Mus. Nat. Hist., Yale University Bull. 13, New Haven.
405 Scheithauer, W. 1966: Kolibris, fliegende Edelsteine. Bayerischer Landwirtschaftsverlag, München.
406 Schouteden, H. 1954-60: Fauna du Congo Belge et du Ruande-Urundi: Oiseaux. Vol. 1-3, Tervuren.
407 Schröder, W. 1988: Zur Biologie und zum Status der Kubaamazone (Amazona leucocephala) auf Great Abaco (Bahamas). Trochilus 9:3-7.
408 Schulz, H. 1986: Agonistisches Verhalten, Territorialverhalten und Balz der Zwergtrappe (Tetrax tetrax). J. Ornithol. 127: 125-204.
409 Sibley, C. G. & Ahlquist, J. E. 1990: Phylogeny and Classification of Birds. A Study in Molecular Evolution. Yale University Press, New Haven.
410 Sibley, C. G. & Monroe, B. L. 1990: Distribution and Taxonomy of Birds of the World. Yale University Press, New Haven.
411 Sick, H. 1985: Ornitologia brasileira. Brasilia.
412 Skutch, A. F. 1984: Aves de Costa Rica. San Jose.
413 Snyder, N. F. R., King, W. B. & Kepler, C. B. 1981: Biology and Conservation of the Bahama Parrot. The Living Bird 19:91-114. Cornell University, Ithaca.
414 Van den Berg, A. B. 1988: Identification of Slender-billed Curlew and its occurrence in Morocco in winter of 1987/88. Dutch Bird 10: 45-53.
415 Watson, J. 1992: Nesting Ecology of the Seychelles Kestrel Falco araea on Mahé, Seychelles. Ibis 134:259-267.
416 Willey, J. W. 1991: Status and Conservation of Parrots and Parakeets in the Greater Antilles, Bahama Islands, and Cayman Islands. Bird Conservation Int. 1:187-214.
417 Wolters, H. E. 1975-82: Die Vogelarten der Erde. Eine Systematische Liste mit Verbreitungsangaben sowie Deutschen und Englischen Namen. Paul Parey, Hamburg.
418 Yun-Sun, Z., Chang-Jiang, L., Lü, T. & He, B. 1991: Recent Records of the Relict Gull Larus relictus in western Nei Mongol autonomous Region, China. Forktail 6:66-67.

Last Addenda
(not alphabetic)

419 Lekagul, B. & Round, P. D. 1991: A Guide to the Birds of Thailand. Saha Karn Bhaet Co., Bangkok.
420 Long, J. L. 1981: Introduced Birds of the World. David & Charles, Newton.
421 Riley, J. H. 1938: Birds from Siam and the Malay Peninsula in the United States National Museum collected by DRS. Hugh M. Smith and William L. Abbott. Smithsonian Inst. Bull. 172, Washington D. C.
422 Nicolai, J. 1972: Zwei neue Hypochera-Arten aus West-Afrika (Ploceidae, Viduinae). J. Ornithol. 113: 229-240.
423 Gonzalez, L. M., Hiraldo, F., Delibes, M. & Calderon, J. 1989: Zoogeographic support for the Spanish Imperial Eagle as a distinct species. Bull. Brit. Ornithol. Club 109 (2):86-93.
424 Roberts, T. J. 1991: The Birds of Pakistan. Vol. 1. Oxford University Press, Oxford.
425 Beebe, W. 1990: A Monograph of the Pheasants. Vol. I-IV. Dover Publications, New York.
426 McGregor, R. C. 1909: A Manual of Philippine Birds. Vol. I-II. Bureau of Science, Manila.
427 Gretton, A. 1991: Conservation of the Slender-billed Curlew. ICBP Monograph No. 6, Cambridge.
428 del Hoyo, J., Elliott, A. & Sargatel, J. (eds.) 1992: Handbook of the Birds of the World. Vol. 1. Lynx Editions, Barcelona.
429 Ash, J. S. 1991: The Grey-necked Picathartes, Picathartes oreas and Ibadan Malimbe, Malimbus ibadanensis in Nigeria. Bird Conservation Int. 1 (2): 93-106.
430 Anon. 1992: Bolivian find. World Birdwatch 14 (4):3.
431 Sick, H. 1993: Birds in Brazil. Princeton University Press, Princeton.
432 Howard, R & Moore, A. 1991: A Complete Checklist of the Birds of the World. Academic Press, London.

APPENDIX
TEXT OF CITES

THE CONVENTION ON INTERNATIONAL TRADE IN ENDANGERED SPECIES OF WILD FAUNA AND FLORA 1973 (ALSO KNOWN AS THE WASHINGTON CONVENTION)

The Contracting States, recognizing that wild fauna and flora in their many beautiful and varied forms are an irreplaceable part of the natural systems of the earth which must be protected for this and the generations to come; conscious of the ever growing value of wild fauna and flora from aesthetic, scientific, cultural, recreational and economic points of view; recognizing that peoples and States are and should be the best protectors of their own wild fauna and flora; recognizing in addition, that international co-operation is essential for the protection of certain species of wild fauna and flora against over exploitation through international trade; convinced of the urgency of taking appropriate measures to this end; have agreed as follows:

ARTICLE I

Definitions for the purpose of the present Convention, unless the context otherwise requires:

a) "Species" means any species, subspecies, or geographically separate population thereof;

b) "Specimen" means:

(i) any animal or plant, whether alive or dead;

(ii) in the case of an animal: for species included in Appendices I and II, any readily recognizable part or derivative thereof; and for species included in Appendix III, any readily recognizable part or derivative thereof specified in App. III in relation to the species; and

(iii) in the case of a plant; for species included in App. I, any readily recognizable part or derivative thereof; and for species includes in App. II and III, any readily recognizable part or derivative thereof specified in App. II and III in relation to the species;

c) "Trade" means export, re-export, import and introduction from the sea;

d) "Re-export" means export of any specimen that has previously been imported;

e) "Introduction from the sea" means transportation into a State of specimens of any species which were taken in the marine environment not under the jurisdiction of any State;

f)''Scientific Authority" means a national scientific authority designated in accordance with Article IX;

g) "Management Authority" means a national management authority designated in accordance with Article IX;

h)''Party" means a State for which the present Convention has entered into force.

ARTICLE II

Fundamental Principles:

1. Appendix I shall include all species threatened with extinction which are or may be affected by trade. Trade in species of these species must be subject to particularly strict regulation in order not to endanger further their survival and must only be authorized in exceptional circumstances.

2. App. II shall include:

a) all species which although not necessarily now threatened with extinction may become so unless trade in specimens of such species is subject to strict regulation in order to avoid utilization incompatible with their survival; and

b) other species which must be subject to regulation in order that trade in specimens of certain species referred to in sub-paragraph (a) of this paragraph may be brought under effective control.

3. App. III shall include all species which any Party identifies as being subject to regulation within its jurisdiction for the purpose of preventing or restricting exploitation, and as needing the co-operation of other Parties in the control of trade.

4. The Parties shall not allow trade in specimens of species included in App. I, II and III except in accordance with the provisions of the present Convention.

ARTICLE III

Regulation of Trade in Specimens of Species Included in Appendix I:

1. All trade in specimens of species included in App. I shall be in accordance with the provisions of this Article.

2. The export of any specimen of a species included in App. I shall require the prior grant and presentation of an export permit. An export permit shall only be granted when the following conditions have been met:

a) a Scientific Authority of the State of export has advised that such export will not be detrimental to the survival of that species;

b) a Management Authority of the State of export is satisfied that the specimen was not obtained in contravention of the laws of that State for the protection of fauna and flora;

c) a Management Authority of the State of export is satisfied that any living specimen will be so prepared and shipped as to minimize the risk of injury, damage to health or cruel treatment; and

d) a Management Authority of the State of export is satisfied that an import permit has been granted for the specimen.

3. The import of any specimen of a species included in App. I shall require the prior grant and presentation of an import permit and either an export permit or a re-export certificate. An import permit shall only be granted when the following conditions have been met:

a) a Scientific Authority of the State of import has advised that the import will be for purposes which are not detrimental to the survival of the species involved:

b) a Scientific Authority of the State of import is satisfied that the proposed recipient of a living specimen is suitably equipped to house and care for it; and

c) a Management Authority of the State of import is satisfied that the specimen is not to be used for primarily commercial purposes.

4. The re-export of any specimen of a species included in App. I shall require the prior grant and presentation of a re-export certificate. A re-export certificate shall only be granted when the following conditions have been met:

a) a Management Authority of the State of re-export is satisfied that the specimen was imported into that State in accordance with the provisions of the present Convention;

b) a Management Authority of the State of re-export is satisfied that any living specimen will be so prepared and shipped as to minimize the risk of injury, damage to health or cruel treatment; and

c) a Management Authority of the State of re-export is satisfied that an import permit has been granted for any living specimen.

5. The introduction from the sea of any specimen of a species included in App. I shall require the prior grant of a certificate from a Management

Authority of the State of introduction. A certificate shall only be granted when the following conditions have been met:

a) a Scientific Authority of the State of introduction advises that the introduction will not be detrimental to the survival of the species involved;

b) a Management Authority of the State of introduction is satisfied that the proposed recipient of a living specimen is suitably equipped to house and care for it; and

c) a Management Authority of the State of introduction is satisfied that the specimen is not to be used for primarily commercial purposes.

ARTICLE IV

Regulation of Trade in Specimens of Species Included in Appendix II:

1. All trade in specimens of species included in App. II shall be in accordance with the provision of this Article.

2. The export of any specimen of a species included in App. II shall require the prior grant and presentation of an export permit. An export permit shall only be granted when the following conditions have been met:

a) a Scientific Authority of the State of export has advised that such export will not be detrimental to the survival of that species;

b) a Management Authority of the State of export is satisfied that the specimen was not obtained in contravention of the laws of the State for the protection of fauna and flora; and

c) a Management Authority of the State of export is satisfied that any living specimen will be so prepared and shipped as to minimize the risk of injury, damage to health or cruel treatment.

3. A Scientific Authority in each Party shall monitor both the export permits granted by that State for specimens of species included in App. II and the actual exports of such specimens. Whenever a Scientific Authority determines that the export of specimens of any such species should be limited in order to maintain that species throughout its range at a level consistent with its role in the ecosystems in which it occurs and well above the level at which that species might become eligible for inclusion in App. I, the Scientific Authority shall advise the appropriate Management Authority of suitable measures to be taken to limit the grant of export permits for specimens of that species.

4. The import of any specimen of a species included in App. II shall require the prior presentation of either an export permit or a re-export certificate.

5. The re-export of any specimen of a species included in App. II shall require the prior grant and presentation of a re-export certificate. A re-export certificate shall only be granted when the following conditions have been met:

a) a Management Authority of the State of re-export is satisfied that the specimen was imported into that State in accordance with the provisions of the present Convention; and

b) a Management Authority of the State of re-export is satisfied that any living specimen will be so prepared and shipped as to minimize the risk of injury, damage to health or cruel treatment.

6. The introduction from the sea of any specimen of a species included in App. II shall require the prior grant of a certificate from a Management Authority of the State of introduction. A certificate shall only be granted when the following conditions have been met:

a) a Scientific Authority of the State of introduction advises that the introduction will not be detrimental to the survival of the species involved; and

b) a Management Authority of the State of introduction is satisfied that any living specimen will be so handled as to minimize the risk of injury, damage to health or cruel treatment.

7. Certificates referred to in paragraph 6 in this Article may be granted on the advice of a Scientific Authority, in consultation with other national scientific authorities or, when appropriate, international scientific authorities, in respect of periods not exceeding one year for total numbers of specimens to be introduced in such periods.

ARTICLE V

Regulation of Trade in Specimens of Species Included in Appendix III:

1. All trade in specimens of species included in App. III shall be in accordance with the provisions of this Article.

2. The export of any specimen of a species included in App. III from any State which has included that species in App. III shall require the prior grant and presentation of an export permit. An export permit shall only be granted when the following conditions have been met:

a) a Management Authority of the State of export is satisfied that the specimen was not obtained in contravention of the laws of that State for the protection of fauna and flora; and

b) a Management Authority of the State of export is satisfied that any living specimen will be so prepared and shipped as to minimize the risk of injury, damage to health or cruel treatment.

3. The import of any specimen of a species included in App. III shall require, except in circumstances to which paragraph 4 of this Article applies, the prior presentation of a certificate of origin and, where the import is from a State which has included that species in App. III, an export permit.

4. In the case of re-export, a certificate granted by the Management Authority of the State of re-export that the specimen was processed in that State or is being re-exported shall be accepted by the State of import as evidence that the provision of the present Convention have been complied with in respect of the specimen concerned.

ARTICLE VI

Permits and Certificates:

1. Permits and certificates granted under the provisions of Article III, IV, and V shall be in accordance with the provisions of this Article.

2. An export permit shall contain the information specified in the model set forth in App. IV, and may only be used for export within a period of six months from the date on which it was granted.

3. Each permit or certificate shall contain the title of the present Convention, the name and any identifying stamp of the Management Authority granting it and a control number assigned by the Management Authority.

4. Any copies of a permit or certificate issued by a Management Authority shall be clearly marked as copies only and no such copy may be used in place of the original, except to the extend endorsed thereof.

5. A separate permit or certificate shall be required for each consignment of specimens.

6. A Management Authority of the State of import of any specimen shall cancel and retain the export permit or re-export certificate and any corresponding import permit presented in respect of the import of that specimen.

7. Where appropriate and feasible a Management Authority may affix a mark upon any specimen to assist in identifying the specimen. For these purposes "mark" means any indelible imprint, lead seal or other suitable means of identifying a specimen, designed in such a way as to render its imitation by unauthorized persons as difficult as possible.

ARTICLE VII

Exemptions and Other Special Provisions Relating to Trade:

1. The provisions of Articles III, IV and V shall not apply to the transit or transshipment of specimens through or in the territory of a Party while the specimens remain in Customs control.

2. Where a Management Authority of the State of export or re-export is

satisfied that a specimen was acquired before the provisions of the present Convention applied to that specimen, the provisions of Articles III, IV and V shall not apply to that specimen where the Management Authority issues a certificate to that effect.

3. The provisions of Articles III, IV and V shall not apply to specimens that are personal or household effects. This exemption shall not apply where:

a) in the case of specimens of a species included in App. I, they were acquired by the owner outside his State of usual residence, and are being imported into that State; or

b) in the case of specimens of species included in App. II:

(i) they were acquired by the owner outside his State of usual residence and in a State where removal from the wild occurred;

(ii) they are being imported into the owner's State of usual residence; and

(iii) the State where removal from the wild occurred requires the prior grant of export permits before any export of such specimens; unless a Management Authority is satisfied that the specimens were acquired before the provisions of the present Convention applied to such specimens.

4. Specimens of an animal species included in App. I bred in captivity for commercial purposes, or of a plant species included in App. I artificially propagated for commercial purposes, shall be deemed to be specimens of species included in App. II.

5. Where a Management Authority of the State of export is satisfied that any specimen of an animal species was bred in captivity or any specimen of a plant species was artificially propagated, or is a part of such an animal or plant or was derived therefrom, a certificate by that Management Authority to that effect shall be accepted in lieu of any of the permits or certificates required under the provisions of Articles III, IV or V.

6.The provisions of Articles III, IV and V shall not apply to the non-commercial loan, donation or exchange between scientists or scientific institutions registrated by a Management Authority of their State, or herbarium specimens, other preserved, dried or embedded museum specimens, and live plant material which carry a label issued or approved by a Management Authority.

7. A Management Authority of any State may waive the requirements of Articles III, IV and V and allow the movement without permits or certificates of specimens which form part of a travelling zoo, circus, menagerie, plant exhibition or other travelling exhibition provided that:

a) the exporter or importer registers full details of such specimens with that Management Authority;

b) the specimens are in either of the categories specified in paragraphs 2 or 5 of this Article; and

c) the Management Authority is satisfied that any living specimen will be so transported and cared for as to minimize the risk of injury, damage to health or cruel treatment.

ARTICLE VIII

Measures to be Taken by the Parties:

1. The Parties shall take appropriate measures to enforce the provisions of the present Convention and to prohibit trade in specimens in violation thereof. These shall include measures:

a) to penalize trade in, or possession of, such specimens, or both; and

b) to provide for the confiscation or return to the State of export of such specimens.

2. In addition to the measures taken under paragraph 1 of this Article, a Party may, when it deems it necessary, provide for any methods of internal reimbursement for expenses incurred as a result of the confiscation of a specimen traded in violation of the measures taken in the application of the provisions of the present Convention.

3. As far as possible, the Parties shall ensure that specimens shall pass through any formalities required for trade with a minimum of delay. To facilitate such passage, a Party may designate ports of exit and ports of entry at which specimens must be presented for clearance. The Parties shall ensure further that all living specimens, during any period of transit, holding or shipment, are properly cared for so as to minimize the risk of injury, damage to health or cruel treatment.

4. Where a living specimen is confiscated as a result of measures referred to in paragraph 1 of this Article:

a) the specimen shall be entrusted to a Management Authority of the State of confiscation;

b) the Management Authority shall, after consultation with the State of export, return the specimen to that State at the expense of the State, or to a rescue centre or such other place as the Management Authority deems appropriate and consistent with the purposes of the present Convention; and

c) the Management Authority may obtain the advice of a Scientific Authority, or may, whenever it considers it desirable, consult the Secretariat in order to facilitate the decision under sub-paragraph (b) of this paragraph, including the choice of a rescue centre or other place.

5. A rescue centre as referred to in paragraph 4 of this Article means an institution designated by a Management Authority to look after the welfare of living specimens, particularly those that have been confiscated.

6. Each Party shall maintain records of trade in specimens of species included in App. I, II and III which shall cover:

a) the names and addresses of exporters and importers; and

b) the number and type of permits and certificates granted; the States with which such trade occurred; the numbers or quantities and types of specimens, names of species as included in App. I, II and III and, where applicable, the size and sex of the specimens in question.

7. Each Party shall prepare periodic reports on its implementation of the present Convention and shall transmit to the Secretariat:

a) an annual report containing a summary of the information specified in sub-paragraph (b) of paragraph 6 of this Article; and

b) a biennial report on legislative, regulatory and administrative measures taken to enforce the provisions of the present Convention.

8. The information referred to in the paragraph 7 of this Article shall be available to the public where this is not inconsistent with the law of the Party concerned.

ARTICLE IX

Management and Scientific Authorities:

1. Each Party shall designated for the purpose of the present Convention:

a) one or more Management Authorities competent to grant permits or certificates on behalf of that Party; and

b) one or more Scientific Authorities.

2. A State depositing an instrument of ratification, acceptance, approval or accession shall at that time inform the Depositary Government of the name and address of the Management Authority authorized to communicate with other Parties and with the Secretariat.

3. Any changes in the designations or authorizations under the provisions of this Article shall be communicated by the Party concerned to the Secretariat for transmission to all other Parties.

4. Any Management Authority referred to in paragraph 2 of this Article shall if so requested by the Secretariat or the Management Authority of another Party, communicate to it impression of stamps, seals or other devices used to authenticate permits or certificates.

ARTICLE X

Trade with States not Party to the Convention:

Where export or re-export is to, or import is from, a State not a Party to the present Convention, comparable documentation issued by the competent authorities in that State which substantially conforms with the requirements of the present Convention for permits and certificates may be accepted in lieu thereof by any Party.

ARTICLE XI

Conference of the Parties:

1. The Secretariat shall call a meeting of the Conference of the Parties not later than two years after the entry into force of the present Convention.

2. Thereafter the Secretariat shall convene regular meetings at least once every two years, unless the Conference decides otherwise, and extraordinary meetings at any time on the written request of at least one-third of the Parties.

3. At meetings, whether regular or extraordinary, the Parties shall review the implementation of the present Convention and may:

a) make such provision as may be necessary to enable the Secretariat to carry out its duties, and adopt financial provisions; [this last phrase is the Financial Amendment which became effective April 13, 1987.]

b) consider and adopt amendments to App. I and II in accordance with Article XV;

c) review the progress made towards the restoration and conservation of the species included in App. I, II and III;

d) receive and consider any reports presented by the Secretariat or by any Party; and

e) where appropriate, make recommendations for improving the effectiveness of the present Convention.

4. At each regular meeting, the Parties may determine the time and venue of the next regular meeting to be held in accordance with the provisions of paragraph 2 of this Article.

5. At any meeting, the Parties may determine and adopt rules of procedure for the meeting.

6. The United Nations, its Specialized Agencies and the International Atomic Energy Agency, as well as any State not a Party to the present Convention, may be represented at meetings of the Conference by observers, who shall have the right to participate but not to vote.

7. Any body or agency technically qualified in protection, conservation or management of wild fauna and flora, in the following categories, which has informed the Secretariat of its desire to be represented at meetings of the Conference by observers, shall be admitted unless at least one-third of the Parties present object:

a) international agencies or bodies, either governmental or non-governmental, and national governmental agencies and bodies; and

b) national non-governmental agencies or bodies which have been approved for this purpose by the State in which they are located.

Once admitted, these observers shall have the right to participate but not to vote.

ARTICLE XII

The Secretariat:

1. Upon entry into force of the present Convention, a Secretariat shall be provided by the Executive Director of the United Nations Environment Programme. To the extend and in the manner he considers appropriate, he may be assisted by suitable inter- governmental or non-governmental, international or national agencies and bodies technically qualified in protection, conservation and management of wild fauna and flora.

2. The function of the Secretariat shall be:

a) to arrange for and service meetings of the Parties;

b) to perform the functions entrusted to it under the provision of Articles XV and XVI of the present Convention;

c) to undertake scientific and technical studies in accordance with programmes authorized by the conference of the Parties as will contribute to the implementation of the present Convention, including studies concerning standards for appropriate preparation and shipment of living specimens and the means of identifying specimens;

d) to study the reports of Parties and to request from Parties such further information with respect thereto as it deems necessary to ensure implementation of the present Convention;

e) to invite the attention of the Parties to any matter pertaining to the aims of the present Convention;

f) to publish periodically and distribute to the Parties current editions of App. I, II and III together with any information which will facilitate identification of specimens of species included in those Appendices;

g) to prepare annual reports to the Parties on its work and on the implementation of the present Convention and such other reports as meetings of the Parties may request;

h) to make recommendations for the implementation of the aims and provisions of the present Convention, including the exchange of information of a scientific or technical nature; and

i) to perform any other function as may be entrusted to it by the Parties.

ARTICLE XIII

International Measures:

1. When the Secretariat in the light of information received is satisfied that any species included in App. I or II is being affected adversely by trade in specimens of that species or that the provisions of the present Convention are not being effectively implemented, it shall communicate such information to the authorized Management Authority of the Party or Parties concerned.

2. When any Party receives a communication as indicated in paragraph 1 of this Article, it shall, as soon as possible, inform the Secretariat of any relevant facts insofar as its laws permit and, where appropriate, propose remedial action. Where the Party considers that an inquiry is desirable, such inquiry may be carried out by one or more persons expressly authorized by the Party.

3. The information provided by the Party or resulting from any inquiry as specified in paragraph 2 of this Article shall be reviewed by the next Conference of the Parties which may make whatever recommendations it deems appropriate.

ARTICLE XIV

Effect on Domestic Legislation and International Conventions:

1. The provisions of the present Convention shall in no way effect the right of Parties to adopt:

a) stricter domestic measures regarding the conditions for trade, taking, possession or transport of specimens of species included in App. I, II and III, or the complete prohibition thereof; or

b) domestic measures restricting or prohibiting trade, taking, possession, or transport of species not included in App. I, II or III.

2. The provisions of the present Convention shall in no way effect the provisions of any domestic measures or the obligations of Parties deriving from any treaty, convention, or international agreement relating to other aspects of trade, taking, possession, or transport of specimens which is in force or subsequently may enter into force for any Party including any measure pertaining to the Customs, public health, veterinary or plant quarantine fields.

3. The provisions of the present Convention shall in no way effect the provisions of, or the obligations deriving from, any treaty, convention or international agreement concluded or which may be concluded between States creating a union or regional trade agreement establishing or maintaining a common external Customs control and removing Customs control between the parties thereto insofar as they relate to trade among the States members of that union agreement.

4. A State party to the present Convention, which is also a party to any other treaty, convention or international agreement which is in force at the time of the coming into force of the present Convention and under the

provision of which protection is afforded to marine species included in App. II, shall be relieved of the obligation imposed on it under the provision of the present Convention with respect to trade in specimens of species included in App. II that are taken by ships registrated in that State and in accordance with the provisions of such other treaty, convention or international agreement.

5. Notwithstanding the provisions of Articles III, IV and V, any export of a specimen taken in accordance with paragraph 4 of this Article shall only require a certificate from a Management Authority of the State of introduction to the effect that the specimen was taken in accordance with the provisions of the other treaty, convention or international agreement in question.

6. Nothing in the present Convention shall prejudice the codification and development of the law of the sea by the United Nations Conference on the Law of the Sea convened pursuant to Resolution 2750 C (XXV) of the General Assembly of the United Nations nor the present or future claims and legal views of any State concerning the law of the sea and the nature and extent of coastal and flag State jurisdiction.

ARTICLE XV

Amendments to Appendices I and II:

1.The following provisions shall apply in relation to amendments to App. I and II at meetings of the Conference of the Parties:

a) Any Party may propose an amendment to App. I or II for consideration at the next meeting. The text of the proposed amendment shall be communicated to the Secretariat at least 150 days before the meeting. The Secretariat shall consult the other parties and interested bodies on the amendment in accordance with the provisions of sub-paragraphs (b) and (c) of paragraph 2 of this Article and shall communicate the response to all Parties not later than 30 days before the meeting.

b) Amendments shall be adopted by a two-thirds majority of Parties present and voting. For these purposes "Parties present and voting" means Parties present and casting an affirmative or negative vote. Parties abstaining from voting shall not be counted among the two-thirds required for adopting an amendment.

c) Amendments adopted at a meeting shall enter into force 90 days after that meeting for all Parties except those which make a reservation in accordance with paragraph 3 of this Article.

2. The following provisions shall apply in relation to amendments to App. I and II between meetings of the Conference of the Parties:

a) Any Party may propose an amendment to App. I or II for consideration between meetings by the postal procedures set forth in this paragraph.

b) For marine species, the Secretariat shall, upon receiving the text of the proposed amendment, immediately communicate it to the Parties. It shall also consult inter-governmental bodies having a function in relation to those species especially with a view to obtaining scientific data these bodies may be able to provide and to ensuring co-ordination with any conservation measures enforced by such bodies. The Secretariat shall communicate the views expressed and data provided by these bodies and its own findings and recommendations to the Parties as soon as possible.

c) For species other than marine species, the Secretariat shall, upon receiving the text of the proposed amendment, immediately, communicate it to the Parties, and, as soon as possible thereafter, its own recommendations.

d) Any Party may, within 60 days of the date on which the Secretariat communicated its recommendations to the Parties under sub-paragraph (b) or (c) of this paragraph, transmit to the Secretariat any comments on the proposed amendment together with any relevant scientific data and information.

e) The Secretariat shall communicate the replies received together with its own recommendations to the Parties as soon as possible.

f) If no objection to the proposed amendment is received by the Secretariat within 30 days of the date the replies and recommendations were communicated under the provisions of sub-paragraph (e) of this paragraph, the amendment shall enter into force 90 days later for all Parties except those which make a reservation in accordance with paragraph 3 of this Article.

g) If an objection by any Party is received by the Secretariat, the proposed amendment shall be submitted to a postal vote in accordance with the provisions of sub-paragraph (h), (i) and (j) of this paragraph.

h) The Secretariat shall notify the Parties that notification of objection has been received.

i) Unless the Secretariat receives the votes for, against or in abstention from at least one-half of the Parties within 60 days of the date of notification under sub-paragraph (h) of this paragraph, the proposed amendment shall be referred to the next meeting of the Conference for further discussion.

j) Provided that votes are received from one-half of the Parties, the amendment shall be adopted by a two-thirds majority of Parties casting an affirmative or negative vote.

k) The Secretariat shall notify all Parties of the result of the vote.

l) If the proposed amendment is adopted it shall enter into force 90 days after the date of the notification by the Secretariat of its acceptance for all Parties except those which make a reservation in accordance with paragraph 3 of this Article.

3. During the period of 90 days provided for by sub-paragraph (c) of paragraph 1 or sub-paragraph (l) of paragraph 2 of this Article any Party may by notification in writing to the Depositary Government make a reservation with respect to the amendment. Until such reservation is withdrawn the Party shall be treated as a State not a Party to the present Convention with respect to trade in the species concerned.

ARTICLE XVI

Appendix III and Amendments thereto:

1. Any Party may at any time submit to the Secretariat a list of species which it identifies as being subject to regulation within its jurisdiction for the purpose mentioned in paragraph 3 of Article II. App. III shall include the names of the Parties submitting the species for inclusion therein, the scientific names of the species so submitted, and any parts or derivatives of the animals or plants concerned that are specified in relation to the species for the purpose of sub-paragraph (b) of Article I.

2. Each list submitted under the provisions of paragraph I of this Article shall be communicated to the Parties by the Secretariat as soon as possible after receiving it. The list shall take effect as part of App. III 90 days after the date of such communication. At any time after the communication of such list, any Party may by notification in writing to the Depositary Government enter a reservation with respect to any species or any parts or derivatives, and until such reservation is withdrawn, the State shall be treated as a State not a Party to the present Convention with respect to trade in the species or part or derivative concerned.

3. A Party which has submitted a species for inclusion in App. III may withdraw it at any time by notification to the Secretariat which shall communicate the withdrawal to all Parties. The withdrawal shall take effect 30 days after the date of such communication.

4.Any Party submitting a list under the provisions of paragraph I of this Article shall submit to the Secretariat a copy of all domestic laws and regulations applicable to the protection of such species, together with any interpretations which the Party may deem appropriate or the Secretariat may request. The Party shall, for as long as the species in question is included in App. III, submit any amendment of such laws and regulations or any new interpretations as they are adopted.

ARTICLE XVII

Amendment of the Convention:

1. An extraordinary meeting of the Conference of the Parties shall be convened by the Secretariat on the writing request of at least one-third of the Parties to consider and adopt amendments to the present Convention. Such amendments shall be adopted by a two-thirds majority of the Parties present and voting. For these purposes "Parties present and voting" means Parties present and casting an affirmative or negative vote. Parties abstaining from voting shall not be counted among the two-thirds required for adopting an amendment.

2. The text of any proposed amendment shall be communicated by the Secretariat to all Parties at least 90 days before the meeting.

3. An amendment shall enter into force for the Parties which have accepted it 60 days after two-thirds of the Parties have deposited an instrument of acceptance of the amendment with the Depositary Government. Thereafter, the amendment shall enter into force for any other Party 60 days after that Party deposits its instrument of acceptance of the amendment.

ARTICLE XVIII

Resolution of Disputes:

1. Any dispute which may arise between two or more Parties with respect to the interpretation or application of the provisions of the present Convention shall be subject to negotiation between the Parties involved in the dispute.

2. If the dispute cannot be resolved in accordance with paragraph 1 of this Article, the Parties may, by mutual consent, submit the dispute to arbitration, in particular that of the Permanent Court of Arbitration at The Hague and the Parties submitting the dispute shall be bound by the arbitral decision.

ARTICLE XIX

Signature:

The present Convention shall be open for signature at Washington until 30th April 1973 and thereafter at Berne until 31st December 1974.

ARTICLE XX

Ratification, Acceptance, Approval:

The present Convention shall be subject to ratification, acceptance or approval. Instruments of ratification, acceptance or approval shall be deposited with the Government of the Swiss Confederation which shall be the Depositary Government.

ARTICLE XXI

Accession:

1. The present Convention shall be open indefinitely for accession. Instruments of accession shall be deposited with the Depositary Government.

2. This Convention shall be open for accession by regional economic integration organizations constituted by sovereign States which have competence in respect of the negotiation, conclusion and implementation of international agreements in matters transferred to them by their Member States and covered by this Convention.

3. In their instruments of accession, such organizations shall declare the extent of their competence with respect to the matters governed by the Convention. These organisations shall also inform the Depositary Government of any substantial modification in the extent of their competence. Notifications by regional economic integration organizations concerning their competence with respect to matters governed by this Convention and modifications thereto shall be distributed to the Parties by the Depositary Government.

4. In matters within their competence, such regional integration organizations shall exercise the rights and fulfil the obligations which this Convention attributes to their Member States, which are Parties to the Convention. In such cases the Member States of the organizations shall not be exercise such rights individually.

5. In the fields of their competence, regional economic integration organizations shall exercise their right to vote with a number of votes equal to the number of their Member States which are Parties to the Convention. Such organizations shall not exercise their right to vote if their Member States exercise theirs, and vice versa.

6. Any reference to "Party" in the sense used in Article 1 (h) of this Convention to "State"/"States" or to "State Party"/"States Parties" to the Convention shall be construed as including a reference to any regional economic integration organization having competence in respect of the negotiation, conclusion and application of international agreements in matters covered by this Convention.]**

** The paragraphs in square brackets are an amendment to the Convention which was adopted at an extraordinary meeting of the Conference of the Parties in Gaborone (Botswana) on April 30, 1983. It will enter into force when it has been formally accepted by 54 of the 80 States which were Parties to the Convention on that date. By December 31, 1986, it had been accepted by 12 of those States.

ARTICLE XXII

Entry into Force:

1. The present Convention shall enter into force 90 days after the date of deposit of the tenth instrument of ratification, acceptance, approval or accession, with the Depositary Government.

2. For each State which ratifies, accepts or approves the present Convention or accedes thereto after the deposit of the tenth instrument of ratification, acceptance, approval or accession, the present Convention shall enter into force 90 days after the deposit by such State of its instrument of ratification, acceptance, approval or accession.

ARTICLE XXIII

Reservations:

1. The provisions of the present Convention shall not be subject to general reservations. Specific reservations may be entered in accordance with the provisions of this Article and Articles XV and XVI.

2. Any State may, on depositing its instrument of ratification, acceptance, approval or accession, enter a specific reservation with regard to:

a) any species included in App. I, II or III; or

b) any parts or derivatives specified in relation to a species included in App. III.

3. Until a Party withdraws its reservation entered under the provisions of this Article, it shall be treated as a State not a Party to the present Convention with respect to trade in the particular species or parts or derivatives specified in such reservation.

ARTICLE XXIV

Denunciation:

Any Party may denounce the present Convention by written notification to the Depositary Government at any time. The denunciation shall take effect twelve months after the Depositary Government has received the notification.

ARTICLE XXV

Depositary:

1. The original of the present Convention, in the Chinese, English, French, Russian and Spanish languages, each version being equally authentic, shall be deposited with the Depositary Government, which shall transmit certified copies thereof to all states that have signed it or deposited instruments of accession to it.

2. The Depositary Government shall inform all signatory and acceding States and the Secretariat of signatures, deposit of instruments of ratification, acceptance, approval or accession, entry into force of the present Convention, amendments thereto, entry and withdrawal of reservations and notifications of denunciation.

3. As soon as the present Convention enters into force, a certified copy thereof shall be transmitted by the Depositary Government to the Secretariat of the United Nations for registration and publication in accordance with Article 102 of the Charter of the United Nations.

Note: since 1973 many Resolutions have been agreed on the biannual meetings of the Conference of the Parties. Please ask your Management Authority for information.

DICTIONARY

Translation of English words from the main text that cannot be found in the glossary or on the colour plates.

English	French	German	Spanish
above	en haut, dessus	oben	encima de
about	autour de	zirka	alrededor de
absent	absent	abwesend	ausente
according	selon	zufolge	según
adapt(ed)	adapter(-é)	anpassen(-gepasst)	adaptar (-ado)
add(ed)	ajouter(-é)	hinzufügen	añadir (-ido)
adult	adulte	erwachsen	adulto
after	après	nach	después
against	contre	gegen	contra
age	âge	Alter	edad
ago	depuis que	seit	desde
air	air	Luft	aire
alight	se poser, atterrir	herabsteigen	aterrizar
alike	pareil	gleich	idéntico
almost	presque	fast	casi
along	le long de	entlang	a lo largo de
also	aussi	auch	también
altogether	tout	ganz	todo
always	toujours	immer	siempre
amethyst	améthyste	Amethyst	amatista
among	entre	unter	entre
amount	montant	Anzahl	suma
ancestor	ancêtre	Vorfahren	antepasado
and	et	und	y
angle	angle	Gesichtswinkel	ángulo
any	quelque	einige	algún
appearance	aspect	Erscheinen	aspecto
appendages	appendice	Anhänge	apéndice
appendix(-ices)	annexe(s)	Anhang	anexo(s)
apply	appliquer	verwenden	aplicar
are	sont	sind	son, están
area	zone	Gebiet	zona
around	tout autour	rund um	alrededor de
arrow	flèche	Pfeil	flecha
as	comme	wie	que, tan, como
assemblage	assemblement	Sammlung	colección
attach(ed)	attacher(-é)	festmachen(-ge macht)	pegar(-ado)
attain	atteindre	erreichen	lograr
available	disponsible	zugänglich	disponible
average	moyen	Durchschnitt	medio
avoid(ed)	éviter(-é)	entgehen(-gangen)	evitar(-ado)
backwards	en arrière	zurück	atrás
balloon	ballon	Ballon	globo
band	bande	Band	banda, franja
bar(red)	barre(-é)	Streifen	barra(do)
bare	nu	nackt	desnudo
basal	basal	grundlegend	basal
base	base	Basis	base
because	parce que	weil	porque
become	devenir	werden	ponerse
before	avant	vor	antes de
behind	derrière	hinter	dorso
believe(d)	croire (cru)	glauben (geglaubt)	creer (-ído)
belong(ing)	appartenir à	gehören	pertenecer
below	dessous	unter	bajo
bend	pliage	Krümmung	ángulo, curvo
beneath	dessous	unter	bajo
besides	de plus	ausserdem	además
best	le meilleur	best(e)	el mejor
between	entre	zwischen	entre
big	grande	gross	grande
bifurcate(d)	bifurqué	gespalten	bifurcado
bird	oiseau	Vogel	ave
blood	sang	Blut	sangre
blotch(ed)	pâlir	erbleichen	ponerse pálido

English	French	German	Spanish
body	corps	Körper	cuerpo
bold(ly)	visible	deutlich	claro
border(ed)	bord(é)	saum/gesäumt	borde(ado)
both	les deux	beide	ambos
bottle	bouteille	Flasche	botella
breeding	nuptial	fortpflanzen	nupcial
bright	clair	klar	claro
brilliant	brillant	funkelnd	brillante
bristle(s)(-ly)	hérissé	bonstig	cepilloso
broad(er)	plus large	breit	(más) ancho
broken	(inter-)rompu	abgebrochen	interrumpido
business	commerce	Geschäft	comercio
but	mais	aber	pero
by	par	von	por
call(ed)	nommer(-é)	benennen(-nannt)	llamar(-ado)
cap	capuchon	Kapuze	capucha
captivity	captivité	Gefangenschaft	cautividad
case	cas	Fall	caso
cavity(-ties)	cavité(s)	Höhle	cavidad(es)
central	central	zentral	central
central	central	zentral	central
centre(d)	centre	Mittelpunkt	centro
century	siècle	Jahrhundert	siglo
certain(ly)	certain(-ement)	sicher	seguro(amente)
change	changement	Veränderung	cambiar(-io)
characteristics	caractéristiques	charakteristisch	características
chief(ly)	salut	hauptsächlich	principalmente
chocolate	chocolat	Schokolade	chocolate
circle	cercle	Kreis	círculo
clear	clair, net	klar	claro
cliff	falaise	Felsabhang	acantilado, risco
close(ly)	près de, étroite(ment)	dicht	cerca, denso
closing	fin	Abschluss	fin
coarse(ly)	rude	grob/rauh	grueso
collar	collier	Kragen	collar
colour(-ation)	couleur	Farbe	color(-ación
commercial	commercial	kommerziell	comercial
common	commun	gewöhnlich	común
complete(ly)	complet	komplett	completo
compress(ed)	comprimer(-é)	zusammenpressen (-gepresst)	comprimir(-ido)
conference	conférence	Konferenz	congreso
confuse(d)	confondre(-u)	verwechseln	confundir(-ido)
consider(ed)	considérer(-é)	erwägen	considerar(-ado)
considerable	considérable	bedeutend	considerable
conspicuous	visible	auffällig	conspicuo
continual(ly)	continuel(lement)	ständig	constante(-amente)
contrast(ing)	contraste	Kontrast	contraste
copper	cuivre	Kupfer	cobre
coral	corail	Koralle	coral
corrugation	ride	Runzel	arruga
courtship	la cour	Balz	cortejo
cover(ed)	couvrir(-ert)	decken (gedeckt)	cubrir (cubierto)
curl(y) (ed)	bouclé	gekräuselt	rizado
crow	corneille	Krähe	corneja
curve(d)	courbe(-é)	gekrümmt	curva(-o)
cut(-ting)	couper(-é)	schneiden (geschnitten)	cortar(-ado)
dark(-ening)	sombre (assombrir)	dunkel	scuro (oscurecéndose)
data	données	Daten	datos
decurved	recourbé	nach unten gebogen	decurvado
deep(er)	profond, foncé	tief(er)	profundo, oscuro
degree(s)	degré	Grad	grado
define(d)	définir(-i)	definieren(-t)	definir(-ido)

dense	épais	dicht	denso
depend(ing)	dépendre	abhängen	sujeto
derive(d)	dériver(-é)	ableiten (abgeleitet)	derivar(-ado)
describe	décrire	beschreiben	describir
detect	découvrir	entdecken	descubrir
develop(ed)	développer(-é)	entwickeln(-t)	desarrollar(-ado)
dewlap	lobe	Fleischlappen	papada de buey
differ	se différer	abweichen	diferenciarse
difference	différence	Unterschied	diferencia
diffuse	diffus	unklar	difuso
dim	trouble	unklar	poco claro
direct(ed)	direct	richten (gerichtet)	recto
dirty	sale	schmutzig	sucio
disappear	disparaître	verschwinden	desaparecer
disc	disque	Diskette	disco
discover(ed)	découvrir(-ert)	entdecken(-t)	descubrir(-ió)
disintegrate(d)	décomposer(-é)	auflösen(-gelöst)	disolver
display	parade	Ausstellung	parada
distal(ly)	distal(-ement)	fern	distal(mente)
distinct(ly)	distinct(-ement)	deutlich	claramente
distinguish	distinguer	unterscheiden	distinguir
distribute(d)	étendu	ausbreiten	extender
divide(d)	diviser(-é)	teilen (geteilt)	dividir
domesticate(d)	domestiquer(-é)	zähmen (gezähmet)	domesticar(-ado)
doom(ed)	juger(-é)	verurteilen(-t)	juzgar(-ado)
domicile	domicile	Heimatort	domicilio
dorsal	dorsal	Rücken	dorsal
dot	point	Flecken	punto
doubtful	douteux	zweifelhaft	dudoso
down	bas	nieder	abajo
down	duvet	Flaumfeder	plumón
drab	morne, terne	eintönig	monótono
dress	costume	Kleidung	traje
drop(s)	goutte(s)	Tropfen	gota(s)
dull(er)	(plus) terne	dunkel(er)	(más) apagado
during	pendant	während	durante
dusky	sombre	dunkel	oscuro
each	chaque	jeder	cada
early (-lier)	(plus) tôt	früh(er)	(más) temprano
earth	sol	Erde	tierra
earthquake	tremblement de terre	Erdbeben	terremoto terre
east(ern)	est (oriental)	Osten	este (oriental)
edge(d)	bord(é)	Rand	margen (bordeado)
edition	édition	Ausgabe	edición
egg	oeuf	Ei	huevo
eight	huit	acht	ocho
either	soit	beide	o, cualquier
elevate(d)	lever	heben (gehoben)	alzar
elongate	allonger	länglich	alargar
emerald	d'un vert d'é	smaragdgrün	color esmeralda
end(ing)	fin	Ende	final
English	anglais	englisch	inglés
enlargement	élargissement	Erweiterung	extención
enormous(ly)	énorme(-ément)	enorm	enorme(mente)
entire	entier	ganz	entero
erect(ile)	dressé, erectile	aufgerichtet, aufrichtbar	levantardo eréctil
especial(ly)	spéciale(ment)	besonders	especial(mente)
estimate(d)	évaluer(-é)	schätzen (geschätzt)	estimación
etc.	etc.	usw.	etc.
exact	exact	genau	exacto
examine(d)	examiner(-é)	untersuchen(-t)	examinar
except	excepté	ausgenommen	excepto
exist	exister	existieren	existir
exclude(-ding)	exclure (sans compter)	ausschliessen(d)	excluir(-yendo)
expand(able)	élargir, étendre (expansible)	erweitern	extender expandible
expansion	expansion	Expansion	expansión
explanation	explication	Erklärung	explicación
expose(d)	exposer(-é)	aussetzen(-gesetzt)	exponer (expuesto)
extant	existant	bewahren	existente
extend	élargir	ausstrecken	extender
extent	mesure	Ausdehnung	extensión
extensive	extensif	weit	extenso
extinction	extinction	Aussterben	extinción
extreme(ly)	extrême(ment)	äusserst	extremo(adamente)
eyebrow	sourcil	Augenbraue	cejas

eyelid	paupière	Augenlid	párpado
face	visage	Gesicht	cara
facial	facial	Gesichts-	facial
facing	face à	sehend	afrontar
fade(d)	passer, décoloré	verbleichen	descolorarse
faint(ly)	pâle, léger	schwach	débil, pálido
family	famille	Familie	familia
far	lointain	weit	lejos
farthest	le plus loin	weitest	lo más lejos
feet, see: foot			
female	femelle	Weibchen	hembra
few(er)	peu (moins)	wenige(r)	pocos (menos)
filamentous	fil pareil	Faden ähnlich	filamentoso
fine(ly)	fin	fein	fino
firm	ferme	sicher	firme
first	premier	erst	primero
five	cinq	fünf	cinco
flame	flammé	geflammt	arder
flap(s)	battement	Läppchen	aletazo(s)
flat(tened)	plat (applati)	flach	plano
fleck(ed)	tache	flecken (gefleckt)	pinta
flight	voler (de vol), primaires	Flug	volar (de) vuelo, primarias
flightless	voler incapable	fliegen unfähig	volar inhábil
fluff(y)	duveteux	flaumweich	velloso
fly	voler	fliegen	volar
fold(ed)	plier(-é)	falten (gefaltet)	plegar(-ado)
follow(ing)	suivre(-ant)	folgen	seguir (siguiente)
foot (feet)	pied	Fuss	pie
fore	antérieur	vordere	primero
former(ly)	ancien, autrefois	früher	anterior (antiguamente)
form(ing)	forme former(-ant)	Form	forma formar(-ando)
forward	en avant	vorwärts	adelante
four(th)	quatre(ième)	vier(te)	cuatro (cuarto)
free	libre	frei	libre
French	francais	französisch	francés
frequent(ly)	fréquent(ment)	häufig	frecuente(mente)
fresh	nouveau	frisch	nuevo
fringe(s)	bords, frange	Rand	margen(es)
from	de	von	de
front	devant	Vorderseite	frente
full	rempli	voll	completo, lleno
furrow(s)	sillon	Furche	arruga
gave see: give			
general	général	allgemein	general
genus	genre	Gattung	género
geographical	géographique	geographisch	geográfico
German	allemand	deutsch	alemán
give(-ing)	donner(ant)	geben(d)	dar(dando)
glitter(ing)	briller(ant)	glitzern(d)	relucir(-iendo)
gloss(y)	lustre(-é), luisant	Glanz	lustre(-oso)
golden	doré	golden	dorado
grade(-ding)	degré	stufe	grado
gradual(ly)	graduel(lement)	stufenweise	gradual(mente)
graduate(d)	graduer	graduieren(-t)	graduado
grass	herbe	Gras	hierba
group	groupe	Gruppe	grupo
hair	poil	Haar	pelo
half	moitié	halb	mitad
hanging	pendant	hängend	colgante
have(-ving)	avoir (ont)	haben(d)	haber, han
head	tête	Kopf	cabeza
heart	coeur	Herz	corazón
heavy(-vier)(ly)	lourd (plus) lourd fortement	schwer	pesado (más) fuerte(mente)
height	taille	Höhe	talla
held, see: hold			
helmet	casque	Helm	casco
hide(-dden)	cacher(-é)	verbergen(-orgen)	ocultar(-to)
hind	postérieur	hintere	posterior
hindmost	postérieur	ganz hinten	posterior
hold (held)	tenir (tenu)	(ge)halten	tener (tenido)
hole	trou	Loch/Höhle	hueco
hood(ed)	capuchon(né)	Kapuze	capucha
hook(ed)	recourbé crochet (crochu)	krumm	curvo gancho
horizontal	horizontal	horizontal	horizontal
horn	corne	Horn	cuerno

host	hôte	Wirt	patrón
hovering	voltiger, volant sur place	rütteln, schweben	cerniéndose
how	comment	wie	como
huge	énorme	riesig gross	enorme
hybrid(ize)	hybride	Hybride	híbrido
identification	identification	Identifizierung	identificación
ill	mal défini	unbetont	mal definido
illustration	illustration	Illlustration	ilustración
impossible	impossible	unmöglich	imposible
include(-ding)	inclure comprendre(-ant)	einschliessen(d)	incluir(-yendo)
inconspicuous	insignifiant	unscheinbar	insignificante
increase(-singly)	augmenter	zunehmen(d)	aumentar
incomplete	incomplete	unvollständig	incompleto
independent	indépendant	selbständig	independiente
indication	indication	Indikation	indicación
individual	individuel	individuell	individual
inflate(d)	gonfler(-é)	aufblasen	hinchar, inflar
inflatable	gonflable	aufblasbar	inflable
inhabit(ing)	habiter(ant)	bewohnen	habitar(-ando)
inner	intérieur	innere	interior
innermost	le plus près de	innerste	(más) interior
instead	au lieu de	statt	en lugar de
intense	intense	intensiv	intenso
interrupt(ed)	interrompre	abbrechen (abgebrochen)	romper
into	dans	herein	dentro
introduce(d)	introduire(-uit)	einführen(-geführt)	introducir(-ido)
invert(ed)	inverse	umgekehrt	inverso
iridescent	irisé	irisieren	tornasolado
irregular(ly)	irrégulier(-èrement)	unregelmässig	irregular(mente)
is	est	ist	es, está
island(s)	île(s)	Insel(n)	isla(s)
joint(s)	joint(s)	Gelenk(e)	coyuntura(s)
just	juste	eben	justamente
keel	os de la poitrine	Brustbeinmitte	esterón medio centre
knob	tuberculeux	Knoten	crisma
knowledge	connaissance	Kenntnis	conocimiento
know(n)	savoir (su) connaître (connu)	wissen (gewisst) kennen (gekannt)	saber (sabido) conocer(-ido)
lack(ing)	manque(r)	fehlen	falta(r)
lanceolate	lancéolé	lanzettförmig	lanceolado
large	grand	gross	grande
last	dernier	letzte	último
later	plus tard	später	más tarde
lateral	latéral	Seiten	lateral
latin	latin	Latein	latin
layman	profane	Laie	lego
lemon	citron	Zitrone	limón
length	longueur	Länge	longitud
less	moins	weniger	más pequeño
less	sans	ohne	sin
lie	coucher	liegen	estar, extenderse
light(est)	lumière (le plus) clair, leger	hell	luz (más) claro, ligero
light	écllairage	Beleuchtung	iluminación
like	pareil	wie	parecido a
likely	plausible	wahrscheinlich	verosimil
limb	membre	Glied	miembro
line	ligne	Linie	línea
list	liste	Liste	lista
live(-ving)	vivre (vivant)	leben(d)	vivir (vivo)
locality	localité	Lokalität	localidad
locally	local(ement)	örtlich	localmente
locate(d)	localiser, situer(-é)	lokalisieren(-t)	localizar,situar(ado)
long(est)	(le plus) long	lang	(lo más) largo
longitudinal(ly)	longitudinal	Längen-	longitud
look	ressembler à	sehen	parecer
loose(ly)	large(ment)	lose	flojo(-amente)
lower	inférieur	unter	inferior
lustrous	lustré	schimmernd	lustroso
macaques (ape)	macaque (singe)	Makat (Affe)	macaca (mono)
made, see: make			
mainly	principalement	hauptsächlich	principalmente
make (made)	faire (fait)	machen (gemacht)	hacer (hecho)
male	mâle	männlich	macho
mahogany	acajou	Mahagoni	caoba

marble(d)	marbrure	marmoriert	jaspeado
marginal	marginal	Rand	marginal
mark(ed)	marquer(-é)	Zeichnung, gezeichnet	marcar(-ado)
marking	coloration	Abzeichnung	mancha (coloración)
mask	masque	Maske	máscara
maybe	peut-être	vielleicht	quiza
mealy	farineux	mehlig	harinoso
measurement	mesure	Mass	medida
medium	moyen	mittelgross	mediano
meeting	rencontre	Begegnung	encuentro
member	membre	Mitglied	miembro
membrane	membrane	dünne Haut	membrana
mention(ed)	mentionner(-é)	erwähnen(-t)	mencionar(-ado)
merging	fondre ensemble	zusammen-schmelzen	fundir
metallic	métallique	metallisch	metálico
middle	centre, millieu	mittel-	centro, medio
mimic(s)	imiter (mimique)	nachahmen	imitar, imitador
mingle(d)	melánger(-é)	mischen (gemischt)	mezclar
misapply	mal comprendre	missverstehen	malentender
misleading	égarer	irreführen	desviar
mixed	mêlé	gemischt	mezclado
more	plus de	mehr	más
most	le plus	meist	(el) más
mottle(d)	tacheté	fleckig	moteado
mottling	marbrure	Fleckung	manchas
moult(ing)	muer	mausern	mudar
much	beaucoup	viel	mucho
mud	boue	Schlamm	lodo
museum	musée	Museum	museo
naked	nu	nackt	desnudo
name	nom	Name	nombre
narrow	étroit	schmal	estrecho
nasal	nasal	ander Nase	nasal
native	indigène	eingeboren	autóctono
nature	nature	Natur	naturaleza
near(ly)	(de) près	fast	casi
nearby	voisin	naheliegend	cercano
nest	nid	Nest	nido
nestling	oisillon, nichant	nisten	pollo
never	jamais	nie	nunca
new	nouveau	neu	nueva
niche	niche	Nische	nicho
nine	neuf	neun	nueve
non	non	nicht	no
none	aucun	kein	ningún
not	pas	nicht	no
note	remarquer	bemerken	notar
notch(ed)	entaille	Kerbe	muesca
noticeable	notable	bemerkenswert	notable
number	numéro	Nummer	número
numerous	nombreux	zahlreich	numeros
occur	exister	vorkommen	encontrarse
ocellus(-i)	ocelle(s)	Pfauenauge	ojo(s) de pavon
of	de	von	de
often	souvent	oft	a menudo
old	vieux	alt	viejo
on	sur	auf	en, sobre
one	un	ein	uno
only	seul(ement)	nur	solo, único
open(ing)	ouvert(ure)	offen/öffnen	abierto (abertura)
or	ou	oder	o, o bien
order	ordre	Order/Befehl	orden, arreglar
origin	origine	Ursprung	origen
ornamental	ornementation	Ausschmückung	decorado
ornate	orné	ausschmücken	ornado
other	autre	andere	otro
otherwise	autrement	sonst	si no
outer	extérieur	äussere	exterior
outermost	extrême	äusserst	(más) exterior
outlining	contour	Umriss	contorno
outlying	éloigné	entlegen	remoto
outside	exterieur en dehors de	Aussenseite	externo fuera
outstretch(ed)	tendu	ausgestreckt	tendido
oval	ovale	Oval	oval
over	dessus	über	por
overall	en gènèral (total)	generell	en conjunto, total

overhang	surplomber	überhängen	colgar
pair	couple	Paar	pareja
pale	pâle	blass	pálido
paradise	paradis	Paradies	paraíso
parents	parents	Eltern	padres
part	partie	Teil	parte
partial	partiel	teilweise	parcial
party(-ies)	groupe(s)	Partei	parte(s)
particular(ly)	particulier(-èment)	besonders	particular(mente)
passage	migration	(Vogel) Zug	pasaje
passing	disparaître	verschwindend	desaparecer
patch(es)	tache(s)	Fleck	mancha(s)
pattern	dessin	Muster	diseño
pearly	perle	perlartig	perla
people	gens	Leute	gente
perhaps	peut-être	vielleicht	quiza
pertain(ing)	relatif à	gehören	relacionado
place	lieu	Stelle/Platz	lugar
place(d)	mettre (mis)	anbringen (angebracht)	poner (puesto)
plain	unicolor, ordinaire	einfarbig	sencillo
plumage	plumage	Gefieder	plumaje
plume	plume	Feder	pluma
point(ed)	pointe(-u)	spitz	punta (puntiagudo)
poor(ly)	pauvre (peu)	arm	pobre (mal)
population	population	Bestand	población
portion	partie	Teil	parte
posterior	postérieur	ganz hinten	posterior
powder(ed)	poudre	Puder	polvos
preceding	précédent	vorherige	anterior
predominant	prédominant	vorherrschend	predominante
preening	polir plume	Feder putzen	pulir pluma
present	présent	anwesend	presente
presume(d)	présumer(-é)	vermuten(-t)	suponer (supuesto)
previous	antérieur	vorhergehend	anterior
prey	proie	Beute	presa
primary(-rily)	principale(ment) primaire	zuerst primär	principal(mente) primaria
probably	probablement	wahrscheinlich	probablemente
prohibit(ed)	défendu	verbieten	prohibir(-ido)
prominent	proéminent frappant	auffällig	prominente
pronounced	distinct	deutlich	distinto
proximate	le plus proche	nächst, unmittelbar	(el más) próximo
pupil	pupille	Pupille	pupila
pure	pur	rein	puro
quantity	quantité	Menge	cantidad
quarter	quart	Viertel	cuarto
quite	tout	ganz	totalmente
race	race	Rasse	raza
range	habitat, répartition	Gebiet	(zona de) distribución
rare(ly)	rare(ment)	selten	raro(-amente)
rather	assez	ziemlich	bastante
raw	brut	roh	basto
reach(ed)	atteindre, gagner(é)	erreichen(-t)	alcanzar(-ado)
reading list	liste de lecture	Lektüre	lista de lectura
rear	côte de derrière	Rückseite	cara posterior
recent(ly)	récemment	vor kurzem	recientemente
record(s)	note	Aufzeichnung	nota
rediscover(ed)	redécouvrir	wiederentdecken(-t)	redescubrir
reduce(d)	réduire(-uit)	vermindern(-t)	reducir(-ido)
reference	référence	Hinweis	referencia
regard(ed)	considérer(-é)	ansehen (angesehen)	conciderar(-ado)
regular(ly)	régulier(-èrement)	regelmässig	regular(mente)
relate(d)	concerner (apparenté)	verwandt	relacionarse (afín)
relatively	relativement	verhältnismässig	relativamente
remain	rester	bleiben	quedarse
remainder	reste	Rest	resto
remove(d)	en lever(-é)	beseitigen(-t)	quitar(-ado)
replace(d)	remplacer(-é)	ersetzen(-t)	reemplazar(-ado)
require	demander	fordern	demandar
resemble	resembler à	ähneln	parecerse a
resident	installé	ansässig	sedentario
respect(s)	égard	Beziehung	aspecto
rest	reste, repos	Rest	resto, reposo
ridge	crète	Kamm	cresta

ring	anneau, autour	Ring	anillo
round(ed)	rond	rund	redondo
row	rang	Reihe	hilera
ruff	rugueux, collerette	Halskragen, Halskrause	garguera
run(ning)	courir(-ant) s'étendre(-ant)	laufen(d)	correr(-iendo) ir, extenderse
rusty	rouillé	rostig	oxidado
sack	sac	Sack	saco
saddle	selle	Sattel	sadler
said see: say			
same	même	derselbe	mismo
sandy	de sable	sandig	arenoso
save	sauver	retten	salvar
say	dire	angeblich	decir
scale (skin)	écaille	Schuppe	escala
scale (measure)	echelle	Massstab	escama
scatter(ed)	disperser(-é)	zerstreuen(-t)	disperso
season(al)	saison(nier)	Saison	temporado
second	second	zweite	segundo
sedentary	sédentaire	standorttreu	sedentario
see	voir	sehen	ver
seem(s)	sembler	scheinen	parecer
segment	segment	Abschnitt	segmento
seldom	rarement	selten	raramente
separate(d)	séparé	separat	separado
series	série	Reihe	serie
serrate	en dent de scie	gesägt	endentado
seven	sept	sieben	siete
several	plusieurs	mehrere	varios
sex(es)	sexe(s)	Geschlecht	sexo(s)
shade(-ding)	ombre	Schatten	sombra
shape(d)	(en) forme (de)	Form	(en) forma (de)
sheen	éclat	Glanz	lustre
shield	bouclier	Schild	escudo
shine (-ning)	splendeur	Glanz	brillo
shoe	soulier	Schuh	zapato
short(er)	court	kurz	corto
side	côte	Seite	lado
silk(y)	soie (soyeux)	seidenartig	sedoso
silver(y)	argent(é)	Silber (silbrig)	plateado
similar	similaire	ähnlich	parecido(s)
since	depuis	seit dem	después
single	simple, seul	einzeln	sencillo
six	six	sechs	seis
skin	peau	Haut	piel
sky	ciel	Himmel	cielo
slender	grêle	schlank	delgado, fino
slight(ly)	petit (légèrement)	geringfügig	ligero
small	petit	klein	pequeño
smok(y)	fumée(eux)	rauchgrau	ahumado
snow	neige	Schnee	nieve
sole	seul	allein	solo
solid	solide	solid	sólide
some	quelque	einige	alguno
sometimes	parfois	ab und zu	a vezes
somewhat	assez	etwas	algo
soon	bientôt	bald	pronto
sort	espèce	Art/Sorte	especie
space	espace	Zeitraum	espacio
Spanish	espagnol	spanisch	español
sparse(ly)	clairsemé	spärlich	poco denso
spatulate	spatulé	Spartel	espatulado
species	espèce	Art	especie
speckle(d)	tache(té)	gefleckt	pinta (moteado)
spike(d)	pointe	spitz	agudo
spine	épine	Dorn	espino
spoon	cuiller	Löffel	cuchara
spot	tache	Fleck	pinta
spread	étendre déplier	ausbreiten	extender desplegar
spring	printemps	Frühling	primavera
spur	ergot	Sporen	espolón
square	carré	Quadrat	cuadrado
stain(ed)	tacher(-é)	spritzen (gespritzt)	salpicar
statement	rapport	Bericht	informe
status	statut	Zustand/status	estatus
steel	acier	Stahl	acero
stiff	rigide	steif	tieso, rígido
still	immobil	bewegungslos	inmóvil

straight	droit	gerade	recto
straw	paille	Halm	paja
streak(ed)	raie (rayé)	gestreift	lista(do)
streamer(s)	banderole	Wimpel	gallardete
striated	ridé	furchen	surco
stripe	raie	Streifen	raya
strong(ly)	fort	stark	fuerte
structure	structure	Gefüge	estructura
subspecies	sous-espèce	Unterart	raza
subterminal	subterminal	nah dem Ende	subterminál
suffused	diffus	übergezogen	difuso
sun	soleil	Sonne	sol
suppose(d)	supposer(-é)	vermuten(-t)	suponer (supuesto)
surface	surface	Oberfläche	superficie
surround(ed)	entourer(-é)	umgeben(-ge geben)	rodear(-ado)
swollen	grossissement	geschwollen	hinchazón
synonymous	synonymique	synonym	sinónimo
tall	haut	hoch	alto
teeth, see: tooth			
tell	raconter	erzählen	contar
ten	dix	zehn	diez
tend(ency)	avoir tendance à (tendance)	Tendenz	tendencia
term	terme	Ausdruck	término
terminal	terminal	äusserst	terminal
text	texte	Text	texto
texture	texture	Textur	textura
than	que	als	que
them	les, eux	sie	ellos, los, les
thick(er)	(plus) épais	dick(er)	(más) grueso
third	troisième	dritte	tercera
this	ce, ceci	dieser	éste
three	trois	drei	tres
through	par	durch	por
tinge	teinte	Schimmer	tinte
tint(ed)	teinte	Anstrich	tinte
tiny	tout petit	winzig klein	disminuto
tip	pointe	Spitze	punta
to	à	nach, zu	a, hasta
tomato	tomate	Tomate	tomatera
tooth (teeth)	dent	Zahn	diente
total	total	total/gesamt	total
toward	vers	gegen	hacia
trace	trace	Spur	pista
trade	commerce	Handel	comercio
train	s'entraîner	trainieren	entrenar
transfer(red)	transférer(-é)	übertragen(-ge tragen)	transferir(-ido)
transverse	transversal	Querlinie	transversal
travel	voyager	reisen	viajar
tree	arbre	Baum	árbol
triangular	triangulaire	dreieckig	triangular
true	vrai	wahr	verdadero
tubercule	noueux	knorrig	nudoso
tuft	touffe	Quaste	copete
two	deux	zwei	dos
type	type	Typ	tipo
umbrella	parapluie	Schirm	paraguas
under	sous	unter	bajo
underparts	dessous	Unterseite	partes inferiores
understand	comprendre	verstehen	comprender
undescribed	non dècrit, inconnu	unbeschrieben	desconocido
uniform(ly)	uniforme	gleichartig	uniforme
unknown	inconnu	unbekannt	desconocido
unlike	dissemblable	unähnlich	a diferencia de
unmistakable	ne peut être confondu	unverkennbar	inconfundible
untidy	désordonné	unordentlich	desordenado
until	jusque	bis	hasta
up(per)	en haut (supérieur)	auf (obere)	arriba (superior)
upperparts	dessus	obere Seite	patres superiores
upstanding	droit	aufrecht	erguido
upturned	en haut	nach oben kehren	vuelto arriba
upward	en haut	nach oben kehren	vuelto arriba
use(d)	employer(-é)	brauchen (gebraucht)	usar(-ado)

usually	habituellement	gewöhnlich	habitualmente
variable	variable	variabel	variable
variant	variante	Variante	variante
variation	variation	Variation	variación
various(ly)	différent	verschieden	diversos
vary(ing)	varier	variieren	variar
velvety	velouté	samtweich	aterciopelado
vertical	vertical	senkrecht	vertical
very	très	viel	muy
victim	victime	Opfer	víctima
visible	visible	sichtbar	visible
warty	verrue	warzenähnlich	verruga
washed	teinte	angehaucht	tinte
wattle(s)	gorge lobe	Halslappen	cuello pliegue
wavy	ondulé	wellenförmig	ondulante
wax	cire	Wachs	cera
way	façon	Gewohnheit	hábito
weak	faible	schwach	débil
wear	usure	Abnutzung	desgaste
wedge	coin	Keil	cuña
weight	poids	Gewicht	peso
well-defined	bien défini	deutlich definiert	bien definido, claro
were, see: is			
west	ouest	west	oeste
which	qui	welches	que, cual
whole (-ly)	entier	ganz	todo
wide	large	breit	ancho
widespread	répandu	weit verbreitet	extendido
with	avec	mit	con
within	dans, en dedans de	innerhalb	dentro
without	sans	ohne	sin
work	travaille	Arbeit	trabajo
world	monde	Welt	mundo
worn	usé	abgenutzt	gastado
wound	blessure	Wunde	herida
wrinkle(d)	ride	runzeln(d)	arruga
year(s)	année	Jahr	año
young	jeune	jung	joven

INDEX OF SCIENTIFIC NAMES

1 Alaska
99 Aldabra Island
2 Aleutian Islands
50 Algeria
32 Amazonia
120 Andaman Islands
36 Andes
84 Angola
102 Arabia
37 Argentina
106 Asia Minor
115 Assam
43 Atlantic
165 Australia
16 Bahamas
148 Bali
109 Baluchistan
121 Bangladesh
146 Batu Islands
70 Benin
114 Bhutan
92 Bioko
151 Bismarck Archipelago
38 Bolivia
25 Bonaire Island
159 Bonin Islands
143 Borneo
85 Botswana
31 Brazil
67 Burkina Faso
122 Burma
79 Burundi
5 California
168 Campbell Island
126 Cambodia (Kampuchea)
72 Cameroon
3 Canada
94 Canary Islands
89 Cape Province
93 Cap Verde Islands
149 Celebes (Sulawesi)
73 Central African Republic
57 Chad
35 Chile
129 China
154 Christmas Island
26 Colombia
98 Comoros
75 Congo
14 Costa Rica
17 Cuba
20 Dominica
33 Ecuador
53 Egypt
55 Ethiopia
41 Falkland Islands
30 French Guiana
74 Gabon
62 Gambia
68 Ghana
152 Greater Sunda Islands
42 Greenland
11 Guatemala
63 Guinea-Bissau
28 Guyana
134 Hainan Island
10 Hawaii
111 Himalayas
18 Hispaniola
12 Honduras
110 India
118 Indian Ocean
126 Indochina (Kampuchea)
142 Indonesia
104 Iran
66 Ivory Coast
160 Izu Islands
138 Japan
147 Java
7 Kansas
130 Kansu
78 Kenya
137 Korea
124 Laos
10 Laysan Island (Hawaii)
153 Lesser Sunda Islands
65 Liberia
52 Libya
8 Louisiana
162 Loyalty Islands
140 Luzon

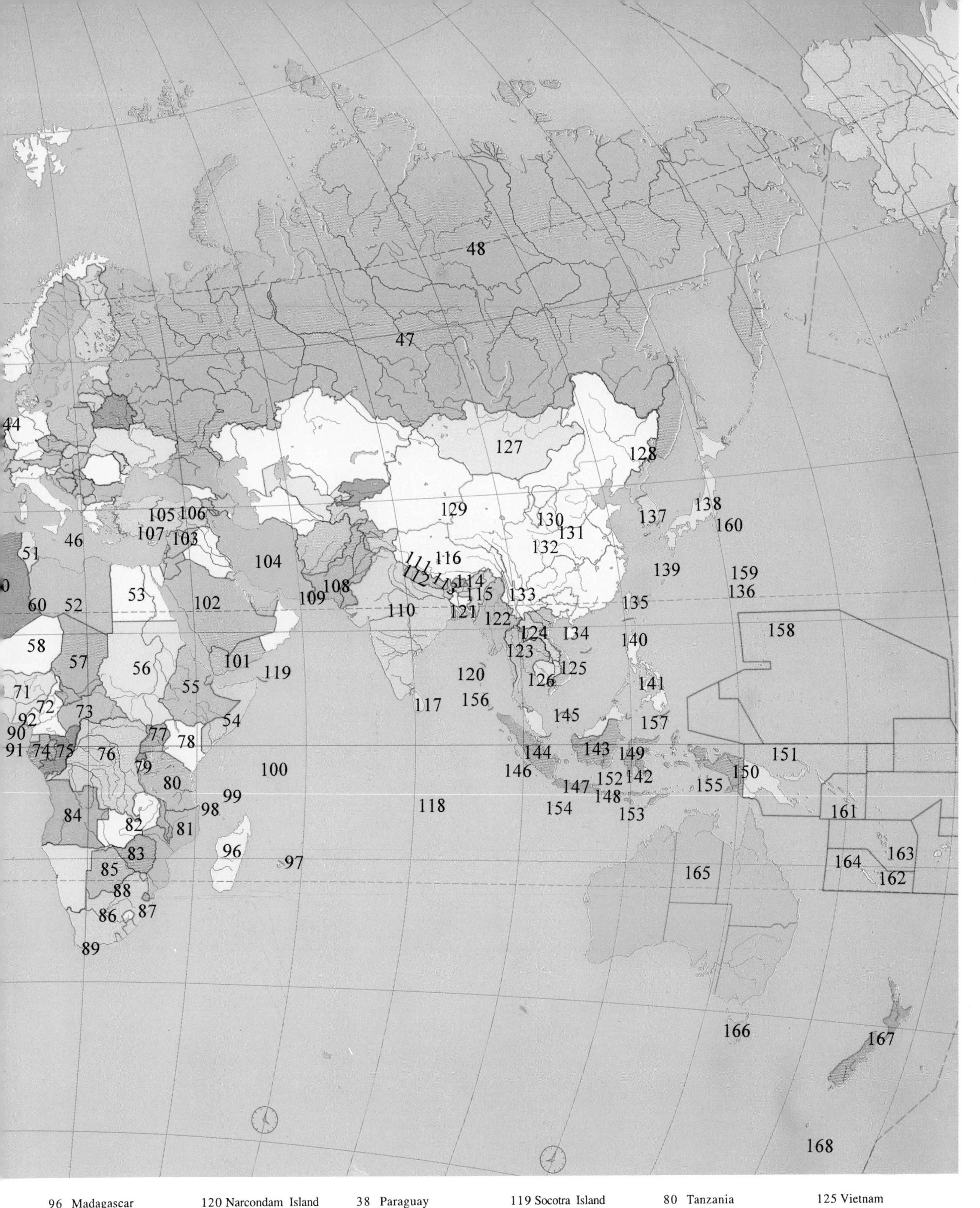

96 Madagascar
95 Madeira
145 Malaysia
59 Mali
128 Manchuria
158 Mariana Islands
97 Mascarene Islands
61 Mauritania
97 Mauritius
46 Mediterranean
9 Mexico
155 Moluccas
127 Mongolia
49 Morocco (Marokko)
81 Mozambique
120 Narcondam Island
87 Natal
112 Nepal
44 Netherlands
24 Netherlands Antilles
164 New Caledonia
150 New Guinea
163 New Hebrides
6 New Mexico
167 New Zealand
13 Nicaragua
156 Nicobar Island
58 Niger
71 Nigeria
108 Pakistan
15 Panama
38 Paraguay
34 Peru
141 Philippines
91 Principe Island
19 Puerto Rico
97 Reunion
47 Russia
139 Ryukya Islands
60 Sahara
90 Sao Thomé Island
62 Senegal
100 Seychelles
131 Shensi
48 Siberia
64 Sierra Leone
113 Sikkim
119 Socotra Island
54 Somalia
161 Solomon Islands
86 South Africa
45 Spain
117 Sri Lanka
93 St. Luzia Island
21 St. Vincent
56 Sudan
149 Sulawesi (Celebes)
157 Sulu Archipelago
144 Sumatra
29 Surinam
103 Syria
132 Szechwan
135 Taiwan
80 Tanzania
166 Tasmania
107 Taurus Mt.
7 Texas
123 Thailand
116 Tibet
40 Tierra del Fuego
69 Togo
88 Transval
22 Trinidad
51 Tunisia
105 Turkey
77 Uganda
4 United States (USA)
39 Uruguay
27 Venezuela
125 Vietnam
136 Volcano Island
149 + 153 + 155 Wallacea
23 West Indies
101 Yemen
133 Yunnan
76 Zaire
82 Zambia
83 Zimbabwe